David V. Keyson · Olivia Guerra-Santin
Dan Lockton
Editors

Living Labs

Design and Assessment of Sustainable Living

Springer

Editors
David V. Keyson
Faculty of Industrial Design Engineering
Delft University of Technology
Delft
The Netherlands

Dan Lockton
School of Design
Royal College of Art
London
UK

Olivia Guerra-Santin
Faculty of Industrial Design Engineering
Delft University of Technology
Delft
The Netherlands

ISBN 978-3-319-33526-1 ISBN 978-3-319-33527-8 (eBook)
DOI 10.1007/978-3-319-33527-8

Library of Congress Control Number: 2016943822

Printed on acid-free paper

This Springer imprint is published by Springer Nature
The registered company is Springer International Publishing AG Switzerland

Contents

Contributors

Maria Ådahl Johanneberg Science Park, Göteborg, Sweden

Carolin Baedeker Wuppertal Institute for Climate Environment Energy, Wuppertal, Germany

Stella Boess Delft University of Technology, Delft, The Netherlands

Flora Bowden SustainRCA, Royal College of Art, London, UK

Clare Brass SustainRCA, Royal College of Art, London, UK

Johannes Buhl Wuppertal Institute for Climate, Environment, Energy, Wuppertal, Germany

Mike Burbridge Curtin University Sustainability Policy (CUSP) Institute, Curtin University, Perth, Australia

Jantien M. Doolaard Delft University of Technology, Delft, The Netherlands

Peter Elfstrand Tengbom Architecture, Gothenburg, Sweden

Lorenz Erdmann Fraunhofer Institute for Systems and Innovation Research ISI, Karlsruhe, Germany

Paula Femenias Department of Architecture, Chalmers University of Technology, Gothenburg, Sweden

Kamil Folta Innovation City Management GmbH, Bottrop, Germany

Justus von Geibler Wuppertal Institute for Climate Environment and Energy, Wuppertal, Germany

Rama Gheerawo Helen Hamlyn Centre for Design, Royal College of Art, London, UK

Kathrin Greiff Wuppertal Institute for Climate, Environment and Energy, Wuppertal, Germany

Viktor Grinewitschus University of Applied Science Ruhr West, Mülheim, Bottrop, Germany

Olivia Guerra-Santin Delft University of Technology, Delft, The Netherlands

Shea Hagy Building Technology, Chalmers University of Technology, Gothenburg, Sweden

Marco Hasselkuß Wuppertal Institute for Climate Environment Energy, Wuppertal, Germany

Aadjan van der Helm Delft University of Technology, Delft, The Netherlands

Natalia Romero Herrera Delft University of Technology, Delft, The Netherlands

Cecilia Holmström Tengbom Architects, Stockholm, Sweden

Bert Hooijer Rotterdam University of Applied Sciences, Rotterdam, The Netherlands

Tomasz Jaskiewicz Delft University of Technology, Delft, The Netherlands

Lina Jonsdotter Department of Architecture, Chalmers University of Technology, Gothenburg, Sweden

Floor van der Kemp Building Changes, Rotterdam, The Netherlands

David V. Keyson Delft University of Technology, Delft, The Netherlands

Jesper Knutsson Chalmers University of Technology, Civil and Environmental Engineering, Gothenburg, Sweden

Lenneke Kuijer Department of Industrial Design, Eindhoven University of Technology, Eindhoven, The Netherlands

Madeleine Larsson Department of Architecture, Chalmers University of Technology, Gothenburg, Sweden

Christa Liedtke Wuppertal Institute for Climate Environment Energy, Wuppertal, Germany

Dan Lockton Royal College of Art, London, UK

Tanja Lovrić University of Applied Science Ruhr West, Mülheim, Bottrop, Germany

Claire Matthews Independent, London, UK

Gregory M. Morrison Curtin University Sustainability Policy Institute, Curtin University, Perth, Australia

Ria van Oosterhout Rotterdam University of Applied Sciences, Rotterdam, The Netherlands

Menno van Rijn Bax and Willems, Barcelona, Spain

Holger Rohn Trifolium – Beratungsgesellschaft mbH, Friedberg, Germany

Jaap Rutten Delft University of Technology, Delft, The Netherlands

Peter Selberg Building Technology, Chalmers University of Technology, Gothenburg, Sweden

Sacha Silvester Delft University of Technology, Delft, The Netherlands

Jens Teubler Wuppertal Institute for Climate, Environment and Energy, Wuppertal, Germany

Liane Thuvander Department of Architecture, Chalmers University of Technology, Gothenburg, Sweden

Arjan van Timmeren Delft University of Technology, Delft, The Netherlands

Larry Toups Johnson Space Centre, NASA, Houston, Texas, USA; Chalmers University, Gothenburg, Sweden

Lali Virdee Institute for Sustainable Futures, London, UK

Maria Jolanta Welfens Wuppertal Institute for Climate, Environment, Energy, Wuppertal, Germany

Part I
Introduction

Chapter 1
Towards Sustainable Living

Arjan van Timmeren and David V. Keyson

Abstract The availability of technologies in our living environment offers a new approach to the study of the interaction between people and the built environment in the context of living labs. The living lab scenario can be viewed as a concertino of action as it unfolds, drawing on available material, cognitive, affective and social resources. Five phases of the translation of cognition into 'ecological rationality' can be distinguished: control, adaptation, learning, improvement (evolution/innovation), change with feedback. The overall challenge facing society today is to achieve and maintain a suitable quality of life, while reducing to a sustainable level the environmental burden to which our activities give rise.

Keywords Living lab · Sustainable living

1.1 Introduction: Toward Sustainable Living

A central challenge in the 21st century is to achieve sustainable living, while respecting the natural boundaries and resources of our planet. The increased deployment of technologies in our living environment offers a new approach to the study of the interaction between technology, the built environment and the conception and feed back loops of human centered solutions towards new ways of sustainable living. The way we describe and understand our living environments is being radically transformed, as are the tools we use to design, plan, and manage them. A new field of research and development in applied technology is emerging at the crossroads of the physical and digital sides of the built domain.

A. van Timmeren · D.V. Keyson (✉)
Delft University of Technology, Delft, The Netherlands
e-mail: d.v.keyson@tudelft.nl

A. van Timmeren
e-mail: a.vantimmeren@tudelft.nl

D.V. Keyson et al. (eds.), *Living Labs*, DOI 10.1007/978-3-319-33527-8_1

This new field focuses on the creation of unique, contemporary and vibrant shared environments for discovery and innovation. The SusLabNWE (Sustainable Labs North-Western Europe) European project featured in this book, is a front-line initiative in developing Living Lab physical infrastructures in North West Europe, as well as methods and techniques for user engagement. Initiatives such as the SusLab project focus on ways to improve the way we live, towards a more sustainable, efficient and comfortable way, and to find and support alternative, more sustainable possibilities for the design and/or the arrangement of our living environments regarding any uncertainties or potential unexpected incidents, while promoting subconscious strategies, incorporating aspects of improvisation, and gaining collaborative experience.

The living lab scenario can be viewed as a concertino of action as it unfolds, drawing on available material, cognitive, affective and social resources. Five phases of the translation of cognition into 'ecological rationality' can be distinguished: control, adaptation, learning, improvement (evolution/innovation), change with feedback.

In the living lab context, sensing and smart technologies play a crucial role. Smart technologies tend to focus on phenomena that involve easily quantifiable data, such as energy generation, use and optimization (storage and exchange), waste and water flows, and security and back-up, amongst others. In this way, ICTs can be used to elucidate the so-called 'flows' in daily use of our living environment (energy, waste, water, food, information). In most cases the resulting data can be used intelligently, as:

- Data tied to geography can provide insights in sustainable living patterns on an urban level.
- Data gives the living environment greater options for faster, more efficient change and information-based decision-making by users.
- Open data offers a connection between users and service companies, government and other stakeholders.

A lot of these smart systems are designed with the environment in mind—towards more sustainable living. The idea behind smart projects such as SusLab is that data can be used to make buildings more responsive to users, their surroundings and fluctuations in the grid and, as a second order effect, encourage households, neighborhoods and municipalities to participant in the overall production and distribution of energy to make its use more efficient, reliable and sustainable (U.S. Department of Energy 2014).

Each smart artifact found in urban space can potentially serve as a node in a data network for sensing and feedback control. This can range from home domotics, to respond to energy demand and supply and actual (real time) use of space(s), to emergency warning systems related to CO_2 and other harmful emissions, and wrong use, that are automatically forwarded to urban dwellers via e-mail or text to user platforms. While each of these technologies is useful by themselves, their combined use has the greatest potential for impact. Apart from ICT, it is forecasted that developments in cloud-based services, the Internet of Things (IoT) and

Augmented Reality (AR) will have the greatest potential of bringing smart concepts for sustainable living to the fore within the next 10 years.

Smart homes might not completely live up to the claims of corporate marketers, but their function as a testing ground for experimental technologies, like in the SusLabNWE project, offers a possible vision of what our future homes might look like: tech-enabled, hyper-efficient living environments that harness sensing technology to manifest the most seamless and automatic living experience possible.

At the same time however, the question of what we want our 'smart' future living environment, and homes in particular, to look like cannot be separated from what kind of people we aspire to be, the kinds of social relations and lifestyles we deem fruitful, or redefining our relationship with the surrounding built and natural environment.

ICTs give institutions, companies, communities, and individuals with similar goals and aspirations (e.g. resilience and sustainability) the means of sharing ideas, having conversations and organizing accordingly. It is important that any new institutional arrangements should be made in close agreement with all actors involved. If such systems are not inclusive, people might start to feel that it is useless to take action. If not, the scope that is left for them to affect their own living conditions will be reduced by the dominant technology driven culture of today. Within such an outlook, new institutional arrangements are required to cope with the use of ICT and physical environment related problems.

In putting such emphasis on the qualitative, use(r) related perspective, concepts for sustainable living can be found that provide alternatives to technological solutions and corporate 'smart' concepts, which monitor and placate citizens into passive, corporeal peripherals of technology. Such sustainable living environments are use(r)-focused, community-defined, open-source environments that harness technology to enhance the users perspective, support individual and collective autonomy, while being open to community participation on related, larger scales, and enshrine the citizen's right to privacy and protection from commodification. In this way it harnesses ICT to illuminate truths of (sustainable) living that are not absolute or self-evident in sensor-collected data, but generated and understood through the continuous physical interaction of human beings within the environment they interact. In doing so, it uses technology to reveal the unseen relations between individuals, families and/or communities and the wider technological and natural systems that support them.

An example is Natural Fuse, an art project by architect Usman Haque that promotes systems thinking and narrative imagination to encourage users to think about the effects and ethics of off-setting carbon. Natural Fuse is a collection of plants that are covered in specialized sensors, connected a computer network, and plugged into an electrical outlet. The system is effectively a buffer between electrical appliances and their power source. The plants act as a carbon sink, so essentially the energy emitted from the socket is limited to the amount of carbon the plants can sequester; this usually lasts at most a few minutes. Interestingly enough, the system is scalable, so your neighbors or anyone connected to the computer network could participate. Additionally, each individual plant can be switched to either an 'off', 'selfless', or 'selfish' mode. Once your appliance stops receiving energy, you can turn on 'selfish' mode and 'borrow' carbon allotments from other plants that are connected to the central sever granted they are in selfless mode themselves. If your plant becomes too selfish it can 'kill' other plants, causing the network to send participating users an e-mail about your heinous transgression. If your plant kills three other plants natural fuse automatically douses it in vinegar and kills it in real life.

It is one of the latest, and more apt elaborations on a more illuminated way how sensors, networks, computational power and data visualization are amalgamating in a way that could change the world forever, specifically using living environments as a metaphor to demonstrate how the power of these technologies would have to capture the encounter of technological and social complexities Real Time. These technologies and their interfaces, would create digital, bird-view level facsimile of the intricacies of interactions between human beings, the built environment and machines.

When considering the prospect of wearable smart technology and AR, this might result in a literal truncation of reality with bespoke, digitally enhanced experiences of the physical environment itself.

1.1.1 Towards Sustainable Living and Living Labs

Sustainable living implies living a lifestyle that uses as few resources as possible and causes the least amount of environmental damage impacting future generations. Shifting our lifestyles towards sustainable living implies changes to many aspects of daily life, including cooking, cleaning, washing, eating, and use of energy consuming devices.

In taking a broader view on sustainable living, perhaps the deeper challenge facing society today, is to achieve and maintain a suitable quality of life, while reducing to a sustainable level the environmental burden to which our activities give rise. This will probably also require a different kind of economy, rooted less in material throughput, being the amount of material circulating in the economy per unit of time or place. In theory such an economy is feasible. This will not only require switching over to renewable energy, but rethinking what we consider quality of life to be.

Quality of being must be derived less from matters, including goods and services embodying high environmental pressure and more from activities having little impact on the environment and nature.

Such activities do not necessarily have to be purely 'spiritual'; there are numerous more homely alternatives. Nature should be valued more as a defining factor of our wellbeing. For example, consider an inspiring work of art in the garden instead of a new kitchen. Rather than going to the tropical swimming pool, have a children's party at home. Instead of speeding down a remote ski slope with a group of friends, join the local amateur choir or enjoy a good glass of wine. Instead of buying fast food, take pleasure in "slow food" as a focal family happening.

As described by John Ehrenfeld, co-founder of the industrial ecology concept, key values in slow food are subsistence, authenticity, family, participation, the world of nature, aesthetics, and personal creativity (Ehrenfeld 2008). Thus the challenge for designers is to create products and services with core meanings and values that focus on the "being" or flourishing mode of human existence rather than the unsustainable "having" mode to which consumers cling to now.

Many aspects of sustainable living can be investigated in actual households, living labs provide an opportunity to emerge participants in a sustainable living environment, while understanding the implications of their daily routines and activities. Living Labs provide a setting for research on innovating every-day practice with an approach that facilitates an open and distributed innovation processes, engaging all relevant partners in real-life contexts (Bergvall-Kåreborn and Ståhlbröst 2009).

Given that the physical infrastructure of the home interacts with the interior systems an integral approach to studying and designing for sustainable living is required. For example, lighting fixtures, water systems, wall, ventilation equipment and floor insulation impact sustainable living. From a construction viewpoint, ideally sustainable homes should be built in such a way that they use few nonrenewable resources, building materials with a low carbon footprint, run on locally generated renewable energy, and cause little or no damage to the surrounding environment.

While this book primarily focuses on individual households, the element of community and social change should be considered as a key element in shaping sustainable living. Increasingly neighborhoods are being connected through emerging technologies and services such as locally renewable energy grids, community recycling, home cooking to order, and urban farming.

References

Bergvall-Kåreborn, B., & Ståhlbröst, A. (2009). Living lab: An open and citizen-centric approach for innovation. *International Journal of Innovation and Regional Development*, *1*(4), 356–370.

Ehrenfeld, J. R. (2008). Sustainability by design, in *A subversive strategy for transforming our consumer culture*. New Haven: Yale Univ. Press.

U.S. Department of Energy. (2014). Smart grid systems report to congress.

References

Chapter 2
The Emergence of Living Lab Methods

Natalia Romero Herrera

Abstract Innovative sustainable solutions in living and working setups need to embrace users' appropriation of technologies in their daily life practices. Successful innovation scenarios implicate adaptability in technologies for users to engage in a process in which technology and practices are adapted, and even new practices are adopted as result of the appropriation. Sustainability Living Lab (SLL) offers a socio-technical infrastructure to support user-centric innovation processes for the development and adoption of sustainable solutions. It offers a collaborative platform where professionals from different disciplines work together with future users and public and private stakeholders to generate solutions that are rooted in the dynamics of daily life practices. Future users play an active role in generating and applying contextualized practice-based knowledge in the innovation process. Central in the process is the integration of users' experiences and sustainability impact of their practices around technology appropriation. A new generation of in-situ and mixed methods is emerging to facilitate this process. This chapter introduces an integrated approach based on in-situ and mixed methods to systemize the integration of objective and subjective aspects of daily life practices at different stages of the innovation process. Three levels of integration are described with each addressing different needs and abilities of the professionals, clients and future users involved in such projects. Each level suggests specific involvement of monitoring and self-reporting activities with outcomes that varies from describing behaviours, explaining the factors that influence behaviours as well as their impact, and experimenting on alternative behaviours.

Keywords Technology appropriation · Daily life practices · In-situ methods · Mixed methods · Sustainability living lab

N. Romero Herrera (✉)
Delft University of Technology, Delft, Netherlands
e-mail: n.a.romero@tudelft.nl

D.V. Keyson et al. (eds.), *Living Labs*, DOI 10.1007/978-3-319-33527-8_2

2.1 Introduction

Sustainability Living Lab (SLL) offers a socio-technical infrastructure for sustainable innovations to emerge, be implemented and tested with and by potential users (Liedtke et al. 2012a). Three elements characterize SLL as a user-centric process:

- The design is situated in real-life (e.g. existing homes) and realistic settings (e.g. home labs)
- The focus is on behaviours and experiences of daily life practices
- The approach addresses the technical, social and temporal dimensions of practices in large scale and longitudinal setups.

Sustainability Living Lab supports a user-centric and contextualized innovation process (Schuurman et al. 2012) in the context of living and working practices. It facilitates the implementation of technical and behavioural interventions in real (-istic) contexts of use (Keyson et al. 2013).

As discussed by Krogstie (2012), Living Lab serves as a platform to combine design research with innovation praxis in which knowledge is generated through the building and deployment of designed artefacts. Sustainability Living Lab combines social, engineering and behavioural sciences with design research to unleash and manipulate the factors that sway experiences around behaviour and technology. As a user-centred process, SLL relies on future users' participation to understand practices in the presence of designed artefacts. However, existing methods fall short in supporting users in the process of identifying and articulating relevant practices and their impact when discussing the experience around designed artefacts (Krogstie et al. 2013). As practices are adopted and become part of people's routine, users need to engage in cognitive efforts to bring them to the foreground, resulting in a demanding and biased data collection process (Mulder et al. 2005).

The first generation of SLL innovations falls in two patterns: solutions designed around user behaviour (e.g. home automation) and solutions that aim to control behaviour (e.g. pervasive technologies). These solutions are characterized by a technology-centric approach failing to address the complexity of daily life practices. They assume that behaviours and needs are static elements and do not interact with other elements in social life (Scott et al. 2009). A second generation is emerging addressing the adaptability (Pallot et al. 2010) of these technologies so users can appropriate them in the complexity of their own contexts (Schwartz et al. 2014; Budweg et al. 2011). This view extends the goal of SLL, as stated by Scott et al. (2009) "beyond improving environmental product performance toward shifting lifestyles in more sustainable directions".

A prospect rises to implement in-situ and integrated design research methods that support users to capture frequent information of their daily practices integrating aspects around users' needs and values as well as sustainability impact. The knowledge generated provides an integral and contextualized view on daily life practices, encompassing:

- Description of practices (how are they implemented, what influences them, who and what is involved),
- Explanation of practices (why do they exists: what is the expected impact on people's needs, desires and experiences)
- Assessment of practices (what is the perceived and measurable impact).

The approach is based on in-situ and mixed methods to systemize the integration of objective and subjective aspects of daily life practices at different stages of the innovation process. Integration techniques are implemented at two levels: quantitative and qualitative user-centred methods are integrated to connect daily life practices, technology and user experience; and the objective and subjective aspects of practices are integrated to contextualize users' experiences and provide links to objective impacts.

In this chapter the aforementioned SLL integrated approach is presented (in Sect. 4.1: In-situ and mixed designs interventions, the in-situ tools and integrated techniques are described). The chapter starts with a brief state-of-the-art review of Living Lab's methods, the challenges and related approaches. Next, the approach is presented, illustrating three possible integration scenarios. The scenarios target different needs, resources and skills coming from stakeholders, technical facilities, design researchers and future users involved in a Sustainability Living Lab project. The chapter concludes by addressing challenges in the design and implementation of in-situ and integrated methods regarding technology, research, and participation.

2.2 User-Centric Living Lab Methods

The differentiating aspect of Living Lab Methods compared to other user-centric methods pertains to the active involvement of the users in the R&D process, entailing a collaborator role in creating new solutions (Pallot and Pawar 2012; Eriksson et al. 2005; Niitamo et al. 2006; Schuurman et al. 2012; Krogstie 2012). Users are seen as key actors in bringing the ecosystem of their everyday life central in the process of ideation, experimentation and evaluation of technological artefacts.

From the second generation of Sustainability Living Lab a shift in focus is observed, moving innovations away from addressing what technology can do to achieve sustainable outcomes, to what people can do with technology to develop sustainable practices (de Jong et al. 2008; Scott et al. 2009; Liedtke et al. 2012a; Krogstie et al. 2013; Schwartz et al. 2014). Underestimating user-bound factors like compatibility with lifestyles, aesthetics, and comfort has resulted in developing solutions that have had little to no impact on sustainability when introduced in people's life context (Scott et al. 2009; Liedtke et al. 2012b).

Therefore research is needed to develop methods and tools that encompass the complex interactions between users, technology and practices in real life context to design for the process of users' appropriation of technologies and its impact on daily life practices and sustainability. *Technology appropriation* is a user process of adopting and adapting technology so it fits into their living and working practices. Users may adapt the intended technology use and/or the technology itself to fit users' lifestyle (Dourish 2003). As consequence, the practices around a technology usage may be altered or new practices may emerge. This in reality may result in users developing new forms of using technology and appliances in the house, as for example when turning on the oven to allocate heat on a painful knee.

Two elements characterize a new generation of user-centred methods to embrace these issues:

- In-situ methods to capture the temporal and contextual nuances of users' practices (Mulder et al. 2005; de Moor et al. 2010; Hess and Ogonowski 2010).
- Mixed methods to capture the technical and social aspects of practices in a qualitative and quantitative manner (de Moor et al. 2010; Schuurman et al. 2012; Scott et al. 2009; Schwartz et al. 2014).

In-situ methods aim to capture an ecological overview of daily life practices, generating knowledge that is bounded to temporal and contextual factors. In-situ methods in Living Lab setups have been implemented as technical and non-technical instruments addressing the need for gathering insights about social practices and social networks (Hess and Ogonowski 2010), measuring user behaviour and experience (Mulder et al. 2005) and measuring quality of experience (Moor et al. 2010). As users need to engage in reporting and reflective activities, challenges related to the implementation of in-situ methods address issues on interruptibility, cognitive demand, boredom and intrusiveness (de Jong et al. 2008; Scott et al. 2009; Rek et al. 2013; Ogonowski et al. 2013). Approaches and techniques have been developed to lower burden by providing a simple structure for describing practices (Scott et al. 2009), to lower interruption by estimating appropriate times for feedback (Vastenburg and Romero 2010; de Moor et al. 2010), to provide benefit through suggestions and social support (Karaseva et al. 2015; Schwartz et al. 2014; Scott et al. 2009; Pallot et al. 2010) and by building trust, transparency and empowerment (Ogonowski et al. 2013; Rek et al. 2013).

Mixed methods extend the descriptive knowledge of practices gathered from monitoring techniques to integrate subjective aspects from a user perspective. Quantitative techniques are valuable to capture large set of objective and subjective data at a relatively low cost, that can be make easily accessible to an open network. Aggregated data provides accurate knowledge on observable behaviours (Veeckman and van der Graaf 2015). However, in the context of Living Lab quantitative methods fall short in two aspects: (a) understanding appropriation of technologies and adoption of new practices; and (b) involving user experience in ideation and evaluation of technologies. Efforts in developing mixed methods for Living Lab are still in their initial phases of conceptualization (de Moor et al. 2010; Schuurman et al. 2012; Karaseva et al. 2015; Pallot and Pawar 2012) or are presented as trials not yet formalized (Schwartz et al. 2014; Scott et al. 2009). These efforts implement integration techniques by collecting data from qualitative and quantitative sources, however they are not addressing other stages of integration and no discussion is provided on how to systemize its implementation. It is expected that a full integration in all stages of a design research project will result in adaptive innovations that are responsive to the interconnections between people's practices, their experiences and related sustainability impact.

User involvement is an ongoing challenge in the implementation of Living Lab's methods. In addition to the challenges of in-situ methods stated above, Living Lab brings other challenges that exclude user groups from participating. For instance, participation requires users to replace mature technologies with unstable or not fully functional ones, which can drastically affect practices that are well established in people's daily life (Budweg et al. 2011; Ogonowski et al. 2013). This real cost is only matched by potential benefits of user participation in contributing to innovation. These benefits in most cases fail to address the interests and needs users have when participating (Mensink et al. 2010).

From a research perspective, Living Lab poses another challenge to support large-scale and cross-national projects. On the one hand, this entails collecting data efficiently as well as ensuring consistency across cases. On the other hand, this requires flexible methods to address different needs, resources and skills from the parties involved.

2.3 Emerging Methods

In-situ methods have been proposed as a promising strategy to characterize practices from a user perspective and at different time frames. This enables comprehending practices within the complex ecosystem of users' experiences and lifestyles. State-of-the art implementations in Living Lab (de Moor et al. 2010; Mulder et al. 2005; Romero et al. 2013) refer to Experience Sampling Method (ESM) as an appropriate approach to connect user experience and practices to real contexts and for long periods of time. Daily Reconstruction Method (DRM) is an alternative in-situ strategy that characterizes practices of one day through a systematic reconstruction process on the following day.

Sensor networks are also discussed as relevant techniques to contextualize daily practices. The advantage of these two prominent strategies in Living Lab settings increases when they are integrated. Whereas integration has been mostly implemented at data analysis, integration at other stages of the design process opens up opportunities to facilitate in depth and focus insights and exploration of practices. Mixed Method Research (MMR) addresses the need for integration at different stages in a research process defining several mixed method designs that support different integration strategies.

In the following sections a brief introduction of the Mixed Methods Research, Experience Sampling Method and Daily Reconstruction Method is provided. Wireless sensor networks are out of the scope of this chapter, as they do not directly involve researcher, designer and users. For detailed information about wireless sensor networks, please refer to NRC (2001).

2.3.1 Mixed Methods Research

Mixed Methods Research (MMR) refers to the integration of qualitative and quantitative approaches to answer research questions (Creswell and Piano 2011). Methods are integrated at different stages in the research process including data collection, data analysis and data interpretation. Qualitative and quantitative data can be mixed in three different ways: *by connecting*, having one data source build on or follow up on the other; *by merging*, to compare or relate results from both data; or *by embedding*, to explain one data result by the other (see Fig. 2.1).

MMR offers a pragmatic orientation to address "practical" issues related to a research problem. For example, when dealing with the complexity of a situation, when knowledge needs to be contextualized, when individuals with different methodological orientations need to work together, when the expected impact cannot be obtained with only one type of data, or when there is an explicit need to do qualitative research.

2.3.2 Experience Sampling and Daily Reconstruction Methods

Measuring user experiences contributes to the assessment of technology appropriation. User experiences assess the interconnections between user, daily life practices and technology. It characterizes the interaction with products in different time span of usage (Roto et al. 2011): anticipated experience (before usage), momentary experience (during usage), episodic experience (after usage) and

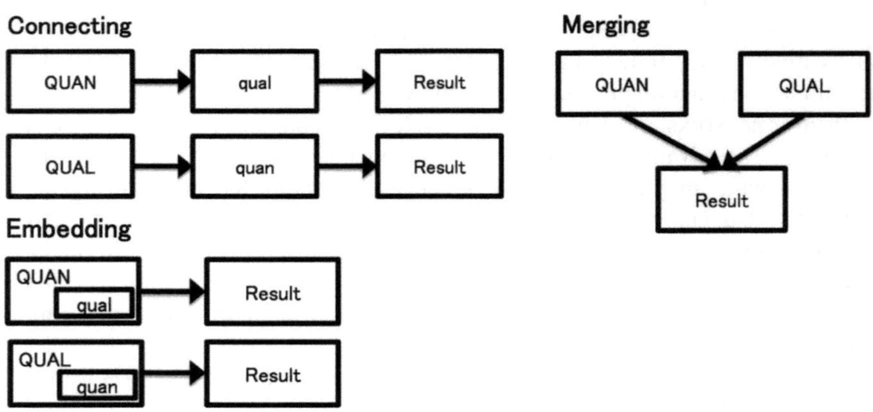

Fig. 2.1 Three ways of mixing data. Notation: a predominant method is symbolized in capitals; in the absence of a predominant method both approaches are equally represented in the results (Creswell and Piano 2011)

cumulative experience (over time). While all stages are relevant to the process of appropriation, momentary experience deserves special attention as it encapsulates the dynamics of the adaptation and adoption processes. In-situ self-reporting methods are used to capture momentary experience.

From socio-psychology research, the Experience Sampling Method (ESM) was developed in the late 60s in respond to the appearance of a technology (the pager) that could prompt people on the move allowing researchers to ask at random times questions to capture people's feelings in the moment (Barret and Barret 2001; Larson and Csikszentmihalyi 1983). ESM has evolved in the last decades, including context-aware capabilities to expand the sampling strategy from random, to time-based and to event-based. A context-aware ESM tool combines sensor networks with self-reporting techniques providing a good platform to link instances of technology use and self-reported experiences (Consolvo et al. 2006; Intille et al. 2003).

There are important considerations in the design of ESM studies. As noted by Myin-Germeys et al. (2009) and Hektner et al. (2007) ESM designs should take care of the frequency, time-demanding and cognitive effort of participants to self-report. On the long term, participants often lose their motivation to provide information every time they receive a prompt. Issues related to repetitive interruptions arise (Christensen et al. 2003), creating barriers for long-term participation, such as annoyance, burden and boredom (Scollon et al. 2003). Adaptive sampling rates aim to avoid undesired interruptions (Vastenburg and Romero 2010) while engaging strategies such as empathy, personal benefit, fun and control could keep user to self-report for longer periods (Rek et al. 2013).

Daily Reconstruction Method is an alternative method that implements users' data collection of the experience of a given day by a systematic reconstruction process conducted on the following day (Kahneman et al. 2004). Compare to ESM it reduces users' burden and captures a more complete coverage of the day, however it increases memory bias. A combination of ESM and DRM has been proposed (Khan et al. 2008) where ESM works to capture short moments in the day that are later used as memory anchors for reconstructing the experiences and practices around them.

2.4 Mixed Approach for Sustainability Living Lab

The presented approach aims to systemize longitudinal, large scale and cross cultural SLL studies by implementing in-situ methods and integration techniques at different stages of the innovation process.

Figure 2.2 illustrates the three levels of integration proposed, using a three-ring metaphor of the top view of a funnel, starting from an extended surface representing the complexity of the context under study, moving deeper into more specific and narrow areas of practices, to finally touch upon specific sustainability and human aspects of practices.

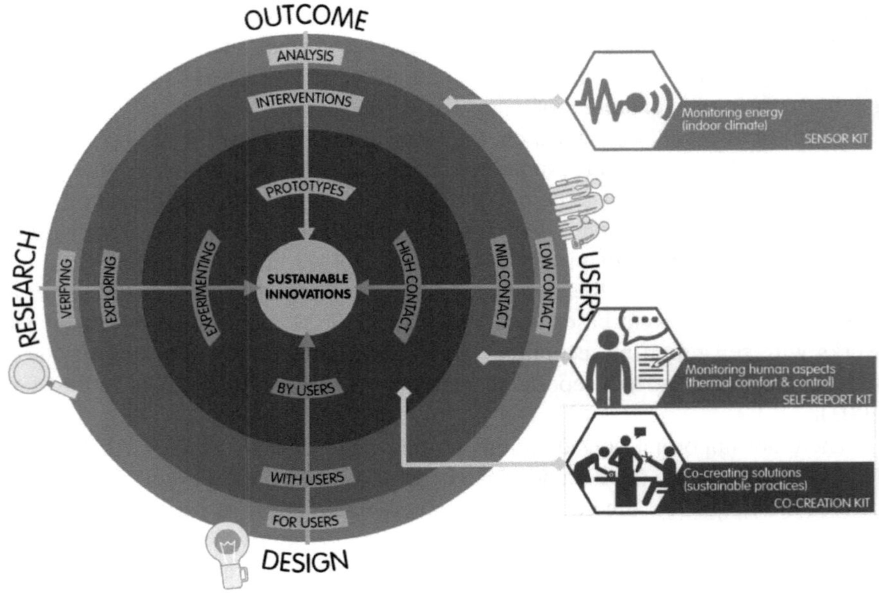

Fig. 2.2 Mixed approach for sustainability Living Lab

The four axes showed in Fig. 2.2 represent different types of user involvement (Users), design research approaches (Design), research goals (Research) and innovation outcomes (Outcome).

- **Users**: users' involvement defines different roles of users as collaborators in the design research activities. The outer ring represents sporadic and passive involvement of users in collecting data. The middle ring indicates an active role in generating and interpreting insights. In the inner ring users are active collaborators in ideating, prototyping and evaluating solutions.
- **Design**: design research approaches incorporate users' needs by means of different activities that result in solutions addressing different levels of complexity. The outer ring provides solutions that address general user needs in context, by means of surveys, interviews and monitoring sensor networks. The middle ring offers solutions that involve deeper insights, which are generated together with users, by means of in-situ self-reporting tools. The inner ring brings solutions that are developed by users, by means of co-design and prototyping sessions.
- **Research**: the depth and richness of knowledge that can be generated varies at the different levels of integration. The outer ring characterizes knowledge based on the validation and verification of current practices. Knowledge in the middle ring moves deeper into explaining and exploring existing and alternative practices; and the inner ring relates to knowledge generated by the experimentation and evaluation of sustainable practices.

- **Outcome**: different levels of knowledge of practices can be obtained at the different levels of integration. The outer ring offers long-term analysis and description of existing situations and their impact on sustainability. The middle ring implements interventions exploring deeper into more specific and narrow areas of practices. The inner ring involves prototypes which touch upon complex and specific sustainability and human aspects of practices.

These axes form quartets of [User, Design, Research, and Outcome] setups to characterize each level of integration. For example:

- **The outer ring** is characterized by a sporadic and passive user involvement in collecting data (User), where activities address generic users' needs (Design), the generated knowledge describes the sustainability of current practices (Research), and innovation is informed by identifying directions and potential impact (Outcomes).
- **The middle ring** is described by an intensive and active involvement of users in collecting and generating insights (User), activities involved deeper and complex users needs (Design), knowledge explores alternative sustainable practices (Research), and innovation develops contextual interventions on specific areas of practices (Outcomes).
- **The inner ring** incorporates users as active collaborators in ideating, prototyping and evaluating solutions (User), activities materialize expectations and desires of users (Design), knowledge projects the impact of new practices (Research), and innovation develops prototypes with validated impact on sustainable practices (Outcomes).

2.4.1 Integration Techniques Based on Mixed Methods

The integration levels are implemented by integration techniques based on Mixed Methods Research (Creswell and Piano 2011). The overall integration gives priority to qualitative methods in understanding, experimenting and evaluating user appropriation of technologies and emergent practices. Quantitative methods are *embedded* offering an objective and subjective layer to measure impact. The interaction techniques support the development and application of *mixed tools*. In this section, the integration techniques and the tools for each level are described. In Sect. 4.1: In-situ and mixed designs interventions an implementation of these mixed tools is presented in the context of home energy use.

First level of integration (outer ring): This level is characterized by *merging* techniques for data analysis. Quantitative data from sensors and other objective data sources are merged to describe baseline impact on sustainability. Merging of quantitative data is also implemented with qualitative data from interviews conducted at the beginning and end of the study to describe sustainable and non-sustainable practices.

Second level of integration (middle ring): *Connecting* and *merging* techniques are applied in data collection, analysis and interpretation. In data collection, quantitative and qualitative self-report tools are connected to support users inform qualitative and reflectively on daily practices. For instance, by connecting quantitative reports as inputs for qualitative self-report tools, so the earlier work as memory anchors to facilitate reflections in the latter (ESM supporting DRM). For data analysis, quantitative data from sensors and quantitative and qualitative data from self-report tools are merged to describe the impact of a specific context on sustainability and on user experience. The outcomes are contextual users insights on daily practices and *mixed probes*: visualizations of the integration of objective and subjective data around practices.

For data interpretation, this level also supports user research design sessions (e.g. contextual interviews and user re-enactment). Mixed probes are connected to these sessions to get deeper and focus explanations of the phenomena described in the first level.

Third level of integration (inner ring): after analysis, *connecting* techniques are applied between the resulted mixed probes and co-design and co-prototyping sessions to interpret the results and generate requirements for the design of artefacts/prototypes.

An interactive setup of user experimentation and evaluation applies real-time *merging* of sensor networks and self-reporting data to enrich data collection. Through, in-situ interventions and in-situ experiments, merged data of sensors and self-reports is used to provoke reflective insights from users as well as to evoke experiences by guided interactions with artefacts, respectively.

2.4.2 Choosing the Appropriate Level of Integration

An ideal innovation process includes all levels of integration to address in its full extent the complexity of users' appropriation of technology and adoption of sustainable practices. However, the approach offers alternative setups to address specific configuration in SLL projects. The criteria for selecting the appropriate level of integration consider:

- The setup and scope of the project
- Project resources (technical infrastructure)
- Collaborators (researchers, designers and users) skills, experience and availability.

A SLL project may encompass several studies with different setups and scopes. Setups involving prominently quantitative methods and with a descriptive goal are placed in the outer ring—first level of interaction. For instance, the first level of integration supports the implementation of long-term monitoring studies and pre and post user interviews with the purpose to define a baseline of practices and their sustainable performance. When the outcome is aimed to go beyond a descriptive baseline, qualitative methods are needed to support the involvement of users and therefore a deeper level of integration is suggested.

Project resources also affect the integration depth in the study. The deeper the layer of integration the higher the need for a robust technical infrastructure to support the development and application of mixed tools.

Finally, the availability, experience and skills of collaborators define to what extend mixed tools and design research sessions can be developed and applied. On the one hand, the intensity and frequency of the sessions depends on the time availability of collaborators. On the other hand, skills are needed for richer and deeper use of the mixed tools. For instance, design researchers' skills in ethnographic and co-design may result in better practices to incorporate mixed tools in the sessions. Similarly, cognitive skills are needed from users in generating and applying mixed probes as well as in participating in in-situ interventions and experiments. Therefore different users' needs and abilities require different variations of in-situ self-reporting tools and integration techniques.

2.5 Conclusion and Challenges

This chapter introduces a methodological approach for Sustainability Living Lab that stages an innovation process based on user-driven in-situ methods, sensor networks and integration techniques. The integration techniques intends to empower collaborators in connecting and contextualizing daily life practices, technologies and user experiences in the process of developing sustainable innovations. By incorporating tools based on quantitative and qualitative methods and by mixing objective and subjective data, the integration techniques elicit and trigger descriptive and reflective insights at different stages of the innovation process.

The central and active role of users as collaborators is supported by means of mixed tools that are developed and applied by them at different stages. Different levels of integration are proposed by setting up different research activities and user involvement (see Fig. 2.2):

- **First level—outer ring** offers verification of daily practices in context by means of low contact user research and monitoring tools. This level is implemented by merging integration techniques for data analysis.
- **Second level—middle ring** explores and analyses opportunities for sustainable practices by means of design interventions. Merging and connecting techniques are implemented at all stages of data collection, analysis and interpretation.
- **Third level—inner ring** supports users in the development and evaluation of their own solutions for sustainable behaviour by means of co-creation and self-experimenting sessions. Real-time merging and connecting techniques are implemented at all stages of collection, analysis and interpretation.

There are two main impacts on innovation that are expected by using this approach. First of all, the resulted innovations address the complexity of technology appropriation in daily life practices. Secondly, and as consequence, such innovations enable dynamic processes of adoption of sustainable practices.

The promising aspects of this approach still require further research to address issues with respect to technology dependency and methodological scope.

The integration techniques rely strongly on high-end and expensive technical infrastructure based on wireless sensor networks and big data analysis. Stability, reliability and scalability of this infrastructure are required to guarantee successful implementations in real life contexts, for long periods of time, and while capturing, analysing and visualizing continuous streams of contextual and behavioural data. When resources are not sufficient to ensure these requirements, cheaper alternatives will result in unstable, less reliable and less scalable setups and higher efforts from collaborators.

The implementation of large-scale and cross-national projects requires that the application of methods and techniques is replicable and comparable. Despite the effort to systematize the proposed approach, as reported earlier, the action of conducting the methods and techniques is vulnerable to contextual and subjective factors. This may result in knowledge generated by data gathered at different frequency and depth (quality). This limitation opens the discussion in the Sustainability Living Lab agenda with regard to the comparability of cases and a user-driven process.

Acknowledgment The development of the Mixed approach has been funded by the SusLabNWE Interreg European program.

References

Barrett, L., & Barrett, D. (2001). An introduction to computerized experience sampling in psychology. *Social Science Computer Review, 19*, 175–185 (Sage Publications).

Budweg, S., Schaffers, H., Ruland, R., Kristensen, K., & Prinz, W. (2011). Enhancing collaboration in communities of professionals using a Living Lab approach. *Production Planning and Control, 22*(5–6), 594–609.

Christensen, T. C., Barrett, L. F., & Bliss-Moreau, E. (2003). A practical guide to experience-sampling procedures. *Journal of Happiness Studies, 4*, 53–78.

Consolvo, S., Harrison, B., Smith, I., Chen, M., Everitt, K., Froehlich, J., & Landay, J. (2006). Conducting in situ evaluations for and with ubiquitous computing technologies. *Journal of Human Computer Interactions, 22*, 107–122.

Creswell, J., & Piano, V. (2011). *Designing and conducting mixed methods research.* Sage Publications.

de Jong, A. M., Bakker, C. A., & Scott, K. (2008). The Living Lab project; user centered sustainable design. *Sustainable Innovation 8.*

de Moor, K., Ketyko, I., Joseph, W., & Deryckere, T. (2010). Proposed framework for evaluating quality of experience in a mobile, testbed-oriented living lab setting. *Mobile Networks and Applications, 15*(3), 378–391.

Dourish, P. (2003). The appropriation of interactive technologies: Some lessons from placeless documents. *Computer Supported Cooperative Work (CSCW), 12*(4), 465–490.

Eriksson, M., Niitamo, V. P., & Kulkki, S. (2005). State-of-the-art in utilizing Living Labs approach to user-centric ICT innovation-a European approach. *Lulea: Center for Distance-spanning Technology.* Lulea: Lulea University of Technology Sweden.

Hektner, J. M., Schmidt, J.A., & Czikszentmihalyi, M. (2007). *Experience sampling method: measuring the quality of everyday life.* Sage Publications.

Hess, J., & Ogonowski, C. (2010). Steps toward a living lab for socialmedia concept evaluation and continuous user-involvement. *Proceedings of the 8th international interactive conference on Interactive TV and Video* (pp. 171–174). ACM.

Intille, S. S., Tapia, E. M., Rondoni, J., Beaudin, J., Kukla, C., & Agarwal, S., et al. (2003). Tools for studying behavior and technology in natural settings. *Proceedings of Ubicomp'03*. (pp. 738–739). ACM Press.

Kahneman, D., Krueger, A. B., Schkade, D. A., Schwarz, N., & Stone, A. A. (2004). A survey method for characterizing daily life experience: The day reconstruction method. *Science, 306*(5702), 1776–1780.

Karaseva, V., Seffah, A., & Porras, J. (2015). A social-media-based living lab: an incubator for human-centric software engineering and innovation. *Presented at the ICSSP 2015: Proceedings of the 2015 International Conference on Software and System Process* (pp. 194–198). ACM.

Keyson, D., Almahmud, A., & Romero, N. (2013) Living Lab and research on sustainability: Practical approaches on sustainable interaction design. *Proceedings of International Conference of Ambient Intelligence Ami'13*. Springer.

Khan, V. J., Markopoulos, P., Eggen, B., IJsselsteijn, W., & de Ruyter, B. (2008). Reconexp: A way to reduce the data loss of the experiencing sampling method. *Proceedings of the 10th international conference MobileHCI* (pp. 471–476). ACM.

Krogstie, J., Stålbrøst, A., Holst, M., & Gudmundsdottir, A. (2013). Using a Living Lab Methodology for Developing Energy Savings Solutions. *Proceedings of AMCIS 2013*. AISEL.

Krogstie, J. (2012). Bridging Research and Innovation by Applying Living Labs for Design Science Research. *Nordic Contributions in IS Research* (pp. 161–176). Berlin Heidelberg: Springer.

Larson, R., & Csikszentmihalyi, M. (1983). The experience sampling method. *New directions for methodology of social and behavioral science, 15*, 41–56.

Liedtke, C., von Geibler, J., and Baedeker, C. (2012a). The sustainability living lab as a reflective user-integrating research infrastructure. In *Proceedings of* the *3rd International Conference on Sustainability Transitions* (pp. 206–222).

Liedtke, C., Welfens, M. J., Rohn, H., & Nordmann, J. (2012b). LIVING LAB: User-driven innovation for sustainability. *International Journal of Sustainability in Higher Education, 13*(2), 106–118.

Mensink, W., Birrer, F., & Dutilleul, B. (2010). Unpacking European Living Labs: analysing innovation's social dimensions. *Central European journal of public policy, 4*(1), 60–85.

Mulder, I., Hofte, Ter, G. H., & Kort, J. (2005). SocioXensor: Measuring user behaviour and user eXperience in conteXt with mobile devices. In *Proceedings of Measuring Behavior* (pp. 355–358).

Myin-Germeys, I., Oorschot, M., Collip, D., Lataster, J., Deles-paul, P., & Van Os, J. (2009). Experience sampling research in psychopathology: Opening the black box of daily life. *Psychological Medicine, 39*, 1533–1547.

National Research Council. (2001) *Embedded, everywhere: A research agenda for networked systems of embedded computers*. Washington, DC: The National Academies Press, 2001.

Niitamo, V. P., Kulkki, S., & Eriksson, M. (2006). State-of-the-art and good practice in the field of living labs. In *Proceedings of the 12th International Conference on Concurrent Enterprising: Innovative Products and Services through Collaborative Networks* (pp. 26–28).

Ogonowski, C., Ley, B., Hess, J., Wan, L., & Wulf, V. (2013). Designing for the living room: long-term user involvement in a living lab. *Proceedings of CHI'13* (pp. 1539–1548). ACM.

Pallot, M., and Pawar, K. (2012). A holistic model of user experience for living lab experiential design. *Proceedings of the 18th International ICE Conference on Engineering, Technology and Innovation* (pp. 1–15). IEEE.

Pallot, M., Trousse, B., Senach, B., Scapin, D. (2010). Living Lab research land-scape: From user centred design and user experience towards user co-creation. *First European Summer School "Living Labs", Aug 2010*, Paris, France.

Rek, M., Romero, N., Jimenez Garcia, J., & van Boeijen, A. (2013) Motivation to Self-report: Capturing user experiences in field studies. *Proceedings of CLIHC'13*. Springer.

Romero, N., Almahmud, A., Beella, S. & Keyson, D. (2013) Towards an Integrated Methodology to Design Sustainable Living Practices. *Proceedings of International Conference of Ambient Intelligence Ami'13*. Springer.

Roto, V., Law, E., Vermeeren, A., & Hoonhout, J. (2011). White paper (Eds): User experience white paper. *Outcome of the Dagstuhl Seminar on Demarcating User Experience, 39*, 1161–117.

Schuurman, D., Lievens, B., De Marez, L., & Ballon, P. (2012). *Towards optimal user involvement in innovation processes: A panel-centered Living Lab-approach* (pp. 2046–2054). IEEE.

Schwartz, T., Stevens, G., & Jakobi, T. (2014). What people do with consumption feedback: A long-term living lab study of a home energy management system. *Interacting with Computers, 27*(6), 551–576.

Scollon, C., Kim Prieto, C., & Diener, E. (2003). Experience sampling: Promises and pitfalls, strengths and weaknesses. *Journal of Happiness Studies, 4*(2003), 5–34.

Scott, K., Quist, J., & Bakker, C. (2009). Co-design, social practices and sustainable innovation: involving users in a living lab exploratory study on bathing. *Proceedings of Joint Actions on Climate Change Conference* (pp. 8–9).

Vastenburg, N., & Romero, N. (2010). Adaptive experience sampling: Addressing the dynamic nature of in-situ user studies. *Proceedings of ISAmI International Symposium on Ambient Intelligence* (Vol. 72/2010, pp. 197–200). Springer Advances in Soft Computing.

Veeckman, C., & van der Graaf, S. (2015). The city as Living Laboratory: Empowering citizens with the citadel toolkit. *Technology Innovation Management Review, 5*(3), 6–17.

Chapter 3
Social Practices as a Main Focus in Living Lab Research

Marco Hasselkuß, Carolin Baedeker and Christa Liedtke

Abstract This chapter introduces the theoretical background of social practice theories as a main focus in sustainable living lab research as well as its connection to the living lab approach and links to desired broader socio-technical transition paths. Applications of practice theories to sustainable consumption and in the field of heating are introduced and conclusions drawn for using practice theories in living lab research and experiments.

Keywords Social practices · Living lab · Heating

3.1 Introduction

Against the background of environmental problems arising from the growing extraction of natural resources and resource depletion, achieving a sustainable development is an indispensable challenge in the twenty-first century. As Rockström et al. (2009) show the current ecological impact, caused by economic action, is already exceeding the ecosystem's capabilities to compensate for interventions in some areas. However, projections of current economic and human development still show that without radical changes the use of natural resources will even increase in the next decades. Bringezu and Bleischwitz (2009) i.e. predict that the global extraction of raw materials might double or triple until 2030, if western consumption patterns are also adopted in now developing countries. A societal transformation (WBGU 2011) towards more sustainable patterns of production and consumption is required, most

M. Hasselkuß · C. Baedeker (✉) · C. Liedtke
Wuppertal Institute for Climate Environment Energy, Wuppertal, Germany
e-mail: carolin.baedeker@wupperinst.org

M. Hasselkuß
e-mail: marco.hasselkuss@wupperinst.org

C. Liedtke
e-mail: christa.liedtke@wupperinst.org

D.V. Keyson et al. (eds.), *Living Labs*, DOI 10.1007/978-3-319-33527-8_3

urgently in the fields of living, mobility and food (EEA 2013). Transition research (i.e. Geels 2002) has recently gained much attention as a theoretical framework and blueprint for empirical work on sustainable development. Transition is thereby understood as a "radical, structural change of a societal (sub)system that is the result of a co-evolution of economic, cultural, technological, ecological and institutional developments at different scale-levels" (Rotmans and Loorbach 2010).

Focusing solely on technical solutions and individual behaviour to achieve such transition towards more sustainable production and consumption systems while assuming people's needs as fixed entities, however, disregards the dynamics of everyday life in which technologies themselves create needs—and, thus, often leads to unsustainable outcomes. Therefore, the consumer's active position should be strengthened through user-driven innovation and the interdependencies of everyday life routines need to be taken into account. To conceptualise such daily routines and their change towards more sustainable patterns of action theories capable to capture such dynamics are required: social practice theories (Shove et al. 2012) and transition research (Geels 2002) are integrated in order to achieve an appropriate understanding.

LivingLabs are combined lab-/household systems, which put the user, i.e. the home occupant, and value chain related actors (producer, handicraft, etc.) on centre stage in the innovation process. Practice theory emphasises the everyday doings as entities, consisting of the elements meaning, material and competence (Shove et al. 2012), which are similarly (re)produced by 'practioners' (those who engage in a social practice) in performances (e.g. practices like cooking, skiing, heating) and routinely involve practical knowledge and use of materials. 'Even though routines of practices are emphasised by theory, it can be used to examine change as well. We take social practices as the basis to design sustainable PSS and aim to reduce negative rebound effects from wrong applications. The LivingLab infrastructure provides means to observe practices involving technical artefacts and opportunities for users to experiment with such practices. learn and appropriate them in the process of everyday use. The potentials of new prototypes to change current practices and, thereby, potentially change rules and resources of systems of provision bottom-up, can be analysed in real-life settings. The potentials of new prototypes for product-service-systems (transformational products, advice and information, feedback systems, etc.) to change current heating practices can be analysed in real-life settings by involving users and stakeholders through action research methods and transdisciplinary research designs. Social practices in using the new product have to be regarded during the innovation process, since all too often products designed for environmental efficiency under given circumstances are misused or overused, resulting in unintended and generally less sustainable outcomes ('the rebound effect') (Liedtke et al. 2012b; Buhl 2014). Studies in failed innovations have shown that the benefits of eco-designed products, technologies or infrastructures are hardly realised if designed without reference to user practices (Spaargaren 2011). Therefore, the Sustainable LivingLab (SLL) approach refers to Social Practice Theory to conceptualise environmental behaviour and awareness and design sustainable product-service-systems around the home (Baedeker et al. 2014; Liedtke et al. 2013, 2015).

As already proposed in earlier phases of building up the SLL research infrastructure (Scott et al. 2009), the approach builds on social practice theory. In SusLabNWE this focus was maintained for two reasons:

(i) Design processes with regard to user practices may lead to greater user acceptance, making it easier to spread sustainable PSS novelties. Studies in failed innovations shown that the benefits of eco-designed products, technologies or infrastructures are hardly realised if designed without reference to user practices. At the same time, the key role of technological innovation for a transition towards more sustainable lifestyles is in no way neglected (Spaargaren 2011).
(ii) Social practice theories are beneficial in analysing routine behaviour related to a specific case study, e.g. heating.

3.2 Theories of Social Practices

In recent years, theories of social practices have gained a lot of attention in the analysis of consumption (i.e. Reckwitz 2002; Warde 2005; Brand 2010; Shove et al. 2012; Røpke 2015). The body of these theories emerged from sociological theories by Bourdieu (1977), Giddens (1984) and Schatzki (1996), who worked on practices as a central category of analysis. Reckwitz later (2002) aimed to provide a cohesive theory of practice as a sociological approach on its own. As one of the baselines, social practice theories draw on the idea of duality of structure, as proposed by e.g. Giddens (1984). Criticising both the determination of action by social structure in many sociological theories as well as the overemphasis of individual interpretation and agency in many theories of action, Giddens introduced his theory of structuration as an attempt to overcome the micro-macro-dualism in sociology. According to this framework, actors are knowledgeable and reflexive, both enabled and constrained by social structure (as virtual sets of rules and resources) in their actions. Structure at the same time is only (re)produced by actions, consisting of rules (of legitimation and signification) and of resources, subdivided into allocative resources (i.e. financial means, technology) and authoritative resources (as a source of power over other actors). Actors know about rules as memory traces on different levels of consciousness. Structures are (re)produced in a recursive process through (mostly routinised) social practices. Reckwitz (2002) accordingly identifies social practices as the location of the social, where action and structure are mediated. He defines practices as "a routinised type of behaviour which consists of several elements, interconnected to one another: forms of bodily activities, forms of mental activities, 'things', and their use, a background knowledge in the form of understanding, know-how, states of emotion and motivational knowledge" (Reckwitz 2002, p. 249). All activities can be considered a practice from this perspective. Actors need specific skills, largely implicit knowledge, and control both over their bodies and over material artefacts. In Giddens' terms, for routine practices actors largely draw on the practical consciousness; meanings and legitimation associated with a practice are usually not reflected.

Yet, practices are not always enacted in exactly the same way but actions can vary. According to Shove et al., another important differentiation is between 'practices as performance', emphasized by Reckwitz (2002), at the level of carriers: concrete, observable actions and competences as different skills and knowledge, which actors need in order to engage in practices. At the level of 'practices as entities', as highlighted by Schatzki (1996), practices represent the concept of social structure related to Giddens' idea of shared rules and resources. This also allows to account for individual deviation in practice performance without any effect on practice as entity, as Reckwitz also describes.

Thus, the interdependency between routines, technology and social norms can be scrutinised (Reckwitz 2002; Warde 2005). "The social is neither reduced to rational actions of individuals (homo economicus) nor to value-based normative rules (homo sociologicus) or to symbolic structures 'inside' or 'outside' of the individual mind [...]" (Brand 2010, p. 220). At the same time, the key role of technological innovation for a transition towards more sustainable lifestyles is in no way neglected (Spaargaren 2011). Spaargaren et al. (2006) argue that the adoption of sustainability innovations can neither be understood only as a conscious choice of individuals (such as early movers with a well-developed environmental awareness) nor just as a passive, involuntary or mechanistic result of changes technological systems.

Accordingly, it is practices, rather than individual desires, that drives consumption (Warde 2005). In this sense new technologies can be said to trigger new demands and, thus, possibly unintended resource and energy consumption. We take social practices as the basis to design sustainable PSS and aim to reduce negative rebound effects from wrong applications. Unexpected user behaviour or wrong application of potentially sustainable innovations is an important cause for rebound effects (Liedtke et al. 2012a). Through a break of on-going sense-making in practices, reproduced practices can be changed. Warde (2005) showed that consumption is not a practice itself but rather engaging in many practices requires a certain level of consumption of goods or services. This also means that consumption of many goods and services and the according resource consumption result from non-reflexive, routine enactment of practices rather than from individual desires. Styles of consumption are interwoven with social practices of certain activities but also with daily routines i.e. in households (Brand 2010). Consumers, then, combine a number of different practices related to nutrition, mobility etc. and form them into lifestyles (Spaargaren 2003, 2011). Further condensing the theoretical work, Shove et al. (2012) identify three elements of practices: meanings, competences and materials (Table 3.1).

The 'materials' element shows how practices are directly or indirectly related to resource use for objects or infrastructures needed to engage in a given practice.

Table 3.1 Elements of social practices (based on Shove et al. 2012)

Elements of social practices
Meanings (mental activities, emotions, motivational knowledge)
Materials (objects, infrastructures, tools, hardware, body)
Competences (understanding, practical knowledgeability)

Table 3.2 Current and sustainable level of resource consumption in the housing sector, calculated for Finnish data (Lettenmeier et al. 2014, p. 497)

Housing—from 10.8 to 1.6 tons/(person a)				
Reduction required by	**Factor 6.8**	Direct consumption amount	Present	**38 m^2/capital** (house)[1] **11500 kWh** (heat and electricity)[1]
			Future	**20 m^2/capital** (zero energy house) **1000 kWh** (electricity)
Share in households' material footprint	Present **27 %**	Material intensity	Present	**65 kg/m^2/a**(house, unheated/uncooled)[2] **0.6 kg/kWh** (Finnish heat and electricity)[2]
	Future **20 %**		Future	**65 kg/m^2/a** (house, heated/cooled) **0.3 kg/kWh** (European electricity)

[1]*Source* Lähteenoja et al. 2007
[2]*Source* Kotakorpi et al. 2008, p. 43

Continuous engagement in a practice, rising numbers of practioners and differentiation thus keeps the engine running for continued extraction of natural resources (for a structuration theory based view on the connection between practices in social systems and the natural system see e.g. von Geibler et al. 2010, p. 155).

Given sustainability challenges we propose that analysis through a practice theoretical lens should be merged with the concept of environmental space (Spangenberg and Lorek 2002) to analyse relevant practices in terms of material use: how e.g. heating is actually performed can be carved out by practice theory but is not a goal in itself; rather, it is to be matched with sustainability strategies and resource saving potentials. We hypothesise that these result in high resource consumption as many practices with positive meaning involve large material requirements. Examples for changes of social practices in relation to sustainable levels of resource use are shown in Lettenmeier et al. (2014). Table 3.2 illustrates the required reduction in resource use induced by housing (tons/person per anno).

A decrease of living space as an example of consumer preference is an example for reduction of heating requirements: "Shared space use is an option for increasing individual living space. Co-housing is seen as a promising practice emerging in the context of sustainable living" (Lettenemeier et al. 2014, p. 505).

3.3 Using Social Practice Theories in Research on Sustainable Consumption

To apply social practice theories in the SLL infrastructure it is helpful to take a look at some methodological strategies developed in previous empirical work making use of this theoretical framework. Within the environmental social sciences, theories of practices are used by an increasing number of authors to analyse the greening of consumption in the new, global order of reflexive modernity. The use of practices as key methodological units for research and governance is suggested as a way to avoid the pitfalls of the individualist and systemic paradigms that dominated the field of sustainable consumption studies for some decades (Spaargaren 2011). Previous empirical work on consumption and social practices (e.g. Evans 2011a, b; Halkier and Jensen 2011; Hargreaves 2011) has shown analytical affordances and methodological challenges of such approaches. Halkier and Jensen (2011) point out the affordances of a constructivist approach to social practices of (1) understanding consumption as entangled in webs of social reproduction and changes, rather than focusing individual consumer choices, and (2), viewing ways of consuming as continuous relational accomplishments in "intersectings of multiple practices" (Halkier and Jensen 2011, p. 117). Here, the concept of duality of structure and agency inherent in practice theory offers important insights. Using such affordances, Evans (2011a, b) e.g. shows that food waste is not a consequence of immoderate consumer choices but of managing the multiplicity of different everyday practices and contingencies. A problem of a purely constructivist perspective, as promoted by Halkier and Jensen, however is to foreground the discourse and negotiation of normative elements in consumption (here: healthy food) among practioners at the expense of downplaying factual knowledge about boundaries or indicators—the same applies to e.g. knowledge about planetary boundaries (Rockström et al. 2009) in considering ecological consumption. While it is a clear affordance of practice theoretical research on (sustainable) consumption to highlight the social embeddedness of consumption and the negotiation within social networks about normative elements and acceptability of practices, power relations as well as intersections of the different daily practices, many studies so far lack a clear concept of sustainability or analysis of practices in the most relevant fields of activity (housing, food, mobility; see: EEA 2013).

3.4 Heating and Social Practices

As Røpke (2015) shows, the enlarged role of consumers beyond simple choices for "greener" products is illustrated—among other factors such as prosumer roles—when considering it from the perspective of systems of provision. One of the improvements of practice theories is to understand domestic consumption as preconfigured by socio-material infrastructures with implicit cultural and policy regimes, e.g. for heating, cooling or lighting at home (Spaargaren 2011). At the "very moment of turning the tap" (Spaargaren 2011, p. 816) such systems of provision (as rules and resources) shape practices of use, but at the same time systems are reproduced in their current form. The consumer becomes co-manager of systems of provision and plays a key role in system change, being in control of e.g. radiators as part of wired and piped systems (Røpke 2015).

Research by Boulanger et al. (2013) on heating practices indicates that practices have changed in the last decades. In the 1960s activities like cooking, washing were focused around a stove in usually one heated room while other rooms were poorly insulated and cold. Today all rooms are heated to comfortable temperatures for light clothes automatically with little effort (leading to less reflected heating routines) and activities are dispersed through the entire house. Gram-Hansen (2010) showed variance in heating energy consumption of families living in similar buildings due to their heating practices.

The role of user practices for system change, illustrated in heating energy systems of provision, points to the need for a theoretical account. This can be achieved by searching for interconnections to transition research as argued in the next section.

3.5 Transition Research and Practice Theories

The concept of transition offers a broader perspective beyond technological solutions by embracing cultural and institutional change. At the Wuppertal Institute (Liedtke et al. 2013) an integrative model of sustainable practices was developed and proposed, integrating the multi-level perspective of transition research and practice theories. It focuses on reconfiguration of social practices involving consumption of some kind towards a lower level of consumption, re-use/longer use or design and implementation of low resources product-service-systems. We hypothesise that a change of social practices, as routine patterns of action, plays a crucial role in transition processes. An important field is the transition of current (mainly western) consumption (and production) patterns (Jackson 2005; Røpke 2015; Spangenberg and Lorek 2002; Baedeker et al. 2008; Liedtke et al. 2013; Schneidewind and Palzkill 2011; Stengel 2011).

As indicated above the argument for this link is twofold: (1) Regimes and transition can usefully be explained more informatively using practice theories and (2) as Warde (2005) points out, consumption is an element in many social practices to competently engage in. Social practices are the appropriate level of transition analysis also because research shows that for many products, the use phase (users' social practices) is highly relevant for its environmental impact, i.e. clothing (Paulitsch and Rohn 2004) or heating/space heating (Liedtke et al. 2012a, b). Private households can directly or indirectly influence part of the environmental impact through their consumption patterns in many of these fields (Spangenberg and Lorek 2002) and through ways of usage.

In spite of mutual criticisms between the two approaches recently calls for developing links have been proposed (cf. Geels et al. 2015). Transition can in terms of practice theories be seen as a circumscribed process or trajectories of change, within the time-space bound reproduction of social practices (Spaargaren et al. 2006). Transition research studies complex socio-technical change processes

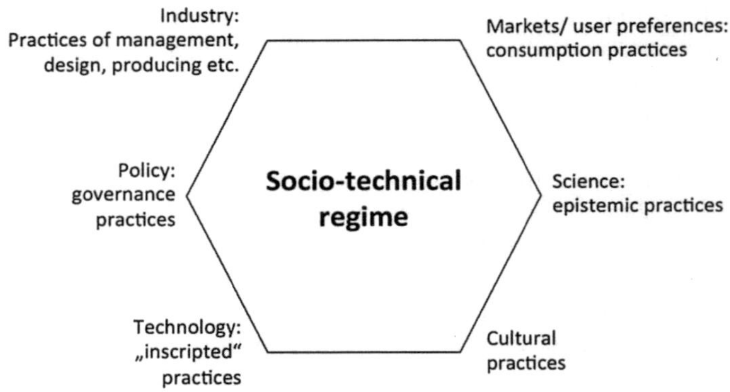

Fig. 3.1 Socio-technical regime elements (own depiction, adapted from Geels 2002)

to meet challenges of sustainable development, aiming to identify underlying patterns and dynamics (Geels and Schot 2007). The multi-level perspective (Geels 2002) analyses change as an interplay of developments on the three functional levels 'landscape', 'regime' and 'niche'. Transition means a deep change of 'regime' of a specific system, which refers to the currently predominant structure of a social system in terms of culture, dominant values and patterns of action. A process of change in material infrastructures, organisational structures, values and norms to establish new patterns of perception and action (transition) can be induced by combining several different developments (Kemp and Loorbach 2006).

We argue for conceptualising regime as a system of interrelated social practices (Watson 2012), drawing on the structuration theory (Giddens 1984) and social practice theories (e.g. Shove et al. 2012; Kuijer 2014). Watson argues: "[…] practices (and therefore what people do) are partly constituted by the socio-technical systems of which they are a part; and those socio-technical systems are constituted and sustained by the continued performance of the practices which comprise them" (Watson 2012, p. 2). These regime elements as social practices are depicted in Fig. 3.1.

The institutions of a given socio-technical regime can usefully be understood as interrelated social practices, which are systematically (re-)produced over time and space and, thus, can be considered institutionalised (cf. Giddens' concept of institutions as (re)produced practices showing the largest extent over time and space).

Transformation to sustainable production and consumption systems can be analysed from a practice theory point of view as change in social practices, i.e. in the area "market/user preferences", which then puts other regime-practices under pressure (Liedtke et al. 2013)—practices can e.g. be changed through transformational objects (Hassenzahl and Laschke 2015), sustainable product-service-systems (Liedkte et al. 2015) or educational strategies (Bliesner et al. 2014).

3.6 Conclusions for SLL Research

Even though social practice theories emphasise routine actions it can be used to examine change as well. For change to take place, a break of on-going sense-making in practices must occur or can be induced by certain events or governance strategies. The opportunities for co-design and user integration offered by the approach were pointed out in the earlier Living Lab design study (Scott et al. 2009) and consequently used in the SusLabNWE project.

For example, transformational products appear promising to disturb routines of practices (Hassenzahl and Laschke 2015). The idea behind these is instead of automating processes, transformational objects intervene and disturb acting on impulse (i.e. leaving the heating on while a window is open). As Liedtke et al. (2015) argue in contrast to more subtle "nudging" approaches, transformational objects remind users of habits and a conscious need to take action—this is thought to support learning processes.

The responsibility for more sustainable production and consumption should however not only be put on the shoulders of individual consumers (Welfens et al. 2010; Walker 2015)—a perspective strengthened by social practice theories by conceptualising the entangled routines and conjunctures of practices. Much more, changes in structures and processes in economy are needed in parallel. As we have shown the developed SLL infrastructure provides an adequate setting for research and real-life experiments with focus on social practices that can integrate experimenting with new practices, assisting technologies and learning opportunities for users beyond individualized responsibility.

References

Baedeker, C., Bahn-Walkowiak, B., Bleischwitz, R., Kolberg, S., Mont, O., & Stengel, O., et al. (2008). Survey on consumption behaviour and its driving forces, European Environment Agency.

Baedeker, C., Greiff, K., Grinewitschus, V., Hasselkuß, D., Keyson, D., & Knutsson, J., et al. (2014). *Transition through sustainable Product and Service Innovations in Sustainable Living Labs: application of user-centred research methodology within four Living Labs in Northern Europe*. Paper for presentation at the 5th International Sustainable Transitions (IST) Conference, August 27–29, 2014 Utrecht, The Netherlands. Retrieved January 12, 2016 from http://suslab.eu/fileadmin/suslab/Images/IST-5thconference_SusLabNWE.pdf.

Bliesner, A., Liedtke, C., Welfens, M. J., Baedeker, C., Hasselkuß, M., & Rohn, H. (2014). 'Norm-Oriented Interpretation Learning' and Resource Use: The concept of 'open-didactic exploration' as a contribution to raising awareness of a responsible resource use. *Resources, 3*, 1–30.

Boulanger, P.-M. Couder, J. Marenne, Y. Nemoz, S., Vanhaverbeke, J., Verbruggen, A., & Wallenborn, G. (2013). Household energy consumption and rebound effect. Science for a sustainable development, Final Report, Brussels.

Bourdieu, P. (1977). *Outline of a theory of practice*. Cambridge: Cambridge University Press.

Brand, K. W. (2010). Social practices and sustainable consumption: Benefits and limitations of a new theoretical approach. In M. Gross & H. Heinrichs (Eds.), *Environmental sociology: european perspectives and interdisciplinary challenges* (pp. 217–234). Dordrecht/Heidelberg/London/New York: Springer.

Bringezu, S., & Bleischwitz, R. (2009). *Sustainable resource management: Global trends, visions and policies*. Sheffield, UK: Greenleaf.
Buhl, J. (2014). Revisiting rebound effects from material resource use. indications for Germany considering social heterogeneity. *Resources, 3*, 106–122.
EEA. (2013). Environmental pressures from European consumption and production. A study in integrated environmental and economic analysis. Technical Report 2. European Environment Agency, Copenhagen. Retrieved January 12, 2016 from http://www.eea.europa.eu/publications/environmental-pressures-from-european-consumption/at_download/file.
Evans, D. (2011a). Beyond the throwaway society: ordinary domestic practice and a sociological approach to household food waste. *Sociology*, 0038038511416150.
Evans, D. (2011b). Blaming the consumer—once again: the social and material contexts of everyday food waste practices in some English households. *Critical Public Health,*. doi:10.1080/09581596.2011.608797.
Geels, F. W. (2002). Technological transitions as evolutionary reconfiguration processes: A multi-level perspective and a case-study. *Research Policy, 31*(8), 1257–1274.
Geels, F. W., & Schot, J. W. (2007). Typology of sociotechnical transition pathways. *Research Policy, 36*(3), 399–417.
Geels, F. W., McMeekin, A., Mylan, J., & Southerton, D. (2015). A critical appraisal of Sustainable Consumption and Production research: The reformist, revolutionary and reconfiguration positions. *Global Environmental Change, 34*, 1–12.
Giddens, A. (1984). *The constitution of society: Outline of the theory of structuration*. Berkeley/Los Angeles: University of California Press.
Gram-Hanssen, K. (2010). Residential heat comfort practices: Understanding users. *Building Research and Information, 38*(2), 175–186.
Halkier, B., & Jensen, I. (2011). Methodological challenges in using practice theory in consumption research. Examples from a study on handling nutritional contestations of food consumption. *Journal of Consumer Culture, 11*(1), 101–123.
Hassenzahl, M., & Laschke, M. (2015). Pleasurable Troublemakers. In S. Walz & S. Deterding (Eds.), *The gameful world: Approaches, issues, applications* (pp. 167–196). Cambridge, MA: MIT Press.
Hargreaves, T. (2011). Practice-ing behaviour change: Applying social practice theory to pro-environmental behaviour change. *Journal of Consumer Culture, 11*(1), 79–99. doi:10.1177/1469540510390500.
Jackson, T. (2005). Motivating Sustainable Consumption. A Review of Evidence on Consumer Behavior and Behavioral Change. Report to the Sustainable Development Network. Retrieved January 12, 2016 from https://www.c2p2online.com/documents/MotivatingSC.pdf.
Kemp, R., & Loorbach, D. (2006). Transition management: a reflexive governance approach. In: J.-P. Voss, D. Bauknecht & R. Kemp (Eds.), *Reflexive governance for sustainable development* (pp. 103–130). Cheltenham: Edward Elgar.
Kotakorpi, E., Lähteenoja, S., & Lettenmeier, M. (2008). Household MIPS. Natural resource consumption of Finnish households and its reduction. The Finnish Environment; Ministry of the Environment: Helsinki, Finland.
Kuijer, S. C. (2014). Implications of Social Practice Theory for Sustainable Design. Delft. Retrieved January 12, 2016 from http://studiolab.ide.tudelft.nl/studiolab/kuijer/files/2014/01/Kuijer-Lenneke_2014_Implications-of-Social-Practice-Theory-for-Sustainable-Design_PhD-thesis.pdf.
Lähteenoja, S., Lettenmeier, M., Kauppinen, T., Luoto, K., Moisio, T., & Salo, M., et al. (2007). Natural resource consumption caused by finnish households. *Proceedings of the Nordic Consumer Policy Research Conference*, Helsinki, Finland, 3–5 October 2007; Nordic Council of Ministers, Ministry of Trade and Industry, National Consumer Research Centre: Helsinki, Finland.
Lettenmeier, M., Liedtke, C., & Rohn, H. (2014). Eight tons of material footprint—Suggestion for a resource cap for household consumption in Finland. *Resources, 3*(3), 488–515.

Liedtke, C., Baedeker, C., von Geibler, J., Hasselkuß, M. (2012a). User-integrated Innovation: Sustainable LivingLabs. In: V. Fricke, U. Schrader & V. W. Thoresen (Eds.), *2nd PERL International Conference "Beyond Consumption—Pathways to Responsible Living": Conference Proceedings* (pp. 203–219). March 19th–20th 2012, Technical University Berlin.

Liedtke, C., Welfens, M. J., Rohn, H., & Nordmann, J. (2012b). Living Lab: User-Driven Innovation for Sustainability. *International Journal of Sustainability in Higher Education, 13*(2), 106–118.

Liedtke, C., Hasselkuß, M., Welfens, M. J., Nordmann, J., & Baedeker, C. (2013). Transformation towards sustainable consumption: Changing consumption patterns through meaning in social practices. Paper for presentation at the 4th IST-conference, Zurich.

Liedtke, C., Baedeker, C., Hasselkuß, M., Rohn, H., & Grinewitschus, V. (2015). User-integrated innovation in Sustainable Living Labs: An experimental infrastructure for researching and developing sustainable product service systems. *Journal of Cleaner Production, 97*, 106–116. doi:10.1016/j.jclepro.2014.04.070.

Paulitsch, K., & Rohn, H. (2004). Nackte Tatsachen: Warum das T-Shirt ökologisch schlechter ist als sein Ruf. In F. Schmidt-Bleek (Ed.), Der ökologische Rucksack: Wirtschaft für eine Zukunft mit Zukunft (pp. 68–78). Stuttgart.

Reckwitz, A. (2002). Toward a theory of social practices. *European Journal of Sociology, 5*(2), 243–263.

Rockström, J., et al. (2009). A safe operating space for humanity. *Nature, 461*, 461–472.

Rotmans, J., & Loorbach, D. (2010). Towards a better understanding of transitions and their governance—A systemic and reflexive approach. In: J. Grin, J. Rotmans & J. Schot (Eds.), *Transitions to sustainable development—new directions in the study of long term transformation change* (pp. 105–220). New York. Routledge.

Røpke, I. (2015). Sustainable consumption: Transitions, systems and practices. In J. Martínez-Alier & R. Muradian (Eds.), *Handbook of Ecological Economics* (pp. 332–359). Cheltenham/Northhampton: Edward Elgar Publishing.

Schatzki, T. (1996). *Social practices*. Cambridge.

Schneidewind, U., & Palzkill, A. (2011). Suffizienz als Business Case – Nachhaltiges Ressourcenmanagement als Gegenstand einer transdisziplinären Betriebswirtschaftslehre. Impulse zur WachstumsWende Nr. 2. Wuppertal Institut für Klima, Umwelt, Energie GmbH, Wuppertal. Retrieved January 12, 2016 from http://epub.wupperinst.org/frontdoor/index/index/docId/3955.

Scott, K., Quist, J., & Bakker, C. (2009). Co-design, social practices and sustainable innovation: Involving users in a living lab exploratory study on bathing. *Proceedings of Paper for the "Joint Actions on Climate Change" Conference*, Aalborg, Denmark.

Shove, E., Pantzar, M., & Watson, M. (2012). *The dynamics of social practice: Everyday life and how it changes*. Los Angeles/London: SAGE.

Spangenberg, J. H., & Lorek, S. (2002). Environmentally sustainable household consumption: from aggregate environmental pressures to priority fields of action. *Ecological Economics, 43*(2–3), 127–140.

Spaargaren, G. (2003). Sustainable consumption: A theoretical and environmental policy perspective. *Society & Natural Resources, 16*, 687–701.

Spaargaren, G. (2011). Theories of practice: Agency, technology, and culture. Exploring the relevance of practice theories for the governance of sustainable consumption practices in the new world-order. *Global Environmental Change, 21*, 813–822.

Spaargaren, G., Martens, S., & Beckers, T. A. M. (2006). Sustainable technologies and everyday life. In P.-P. Verbeek & A. Slob (Eds.), *User behavior and technology development: Shaping sustainable relations between consumers and technologies* (pp. 107–118). Dordrecht: Springer.

Stengel, O. (2011). *Suffizienz: die Konsumgesellschaft in der ökologischen Krise*. München: Oekom Verlag.

von Geibler, J., Kristof, K., & Bienge, K. (2010). Sustainability assessment of entire forest value chains: Integrating stakeholder perspectives and indicators in decision support tools. *Ecological Modelling, 221*(18), 2206–2214.

Welfens, M. J., Liedtke, C., Rohn, H., & Nordmann, J. (2010). Living Lab: Research and development of sustainable products and services through user-driven innovation in experimental-oriented environments. Paper presented at the ERSCP-EMSU Conference "Knowledge Collaboration and Learning for Sustainable Innovation", Delft, The Netherlands, October 25–29, 2010.

Walker, G. (2015). Beyond individual responsibility: Social practice, capabilities and the right to environmentally sustainable ways of living. In Y. Strengers & C. Maller (Eds.), *Social pracitces, intervention and sustainability: Beyond behaviour change* (pp. 45–59). New York: Routledge.

Warde, A. (2005). Consumption and theories of practice. *Journal of Consumer Culture, 5*(2), 131–153.

Watson, M. (2012). How theories of practice can inform transition to a decarbonised transport system. *Journal of Transport Geography, 24*, 488–496.

WBGU (2011). World in Transition—A Social Contract for Sustainability. German Advisory Council on Global Change (WBGU). Retrieved January 12 2016 from http://www.wbgu.de/fileadmin/templates/dateien/veroeffentlichungen/hauptgutachten/jg2011/wbgu_jg2011_en.pdf.

Chapter 4
Green Economy as a Framework for Product-Service Systems Development: The Role of Sustainable Living Labs

Carolin Baedeker, Christa Liedtke and Maria Jolanta Welfens

Abstract This chapter focuses on the role of Sustainable Living Labs for the implementation and diffusion of low resource and sustainable Product-Service Systems (SPSS) in a "Green Economy". In recent years the concept of "Green Economy" has emerged as a strategic priority for governments and intergovernmental organisations. Several governmental and industrial strategies reflect aspects of a variety of its definition searching for implementation of its principles. Therefore, we will start off by giving an overview of different definitions of the green economy showing its relevance for micro-oriented approaches. Subsequently, eight key areas of intervention in a Green Economy, which set up the frame and orientation of our sustainable Living Lab (SLL) approach (Sect. 4.1), are presented. The Sect. 4.2 deals with the relevance of Living Labs for the transformation process of the socio-economic regime. This process consists of changing the production-consumption systems towards sustainability through modifying processes and SPSS on the micro level. SLL focuses on sustainability innovations and offers a number of new characteristics reflecting the intervention arenas, which are described in Sect. 4.3. The Sustainable Living Lab approach offers fundamental sustainability-oriented research infrastructure, in which relevant actors are actively integrated into the development, design and testing of new PSS aiming for the transition of our sociotechnical regime towards sustainability. SLL use a three-phase-model as their methodological framework, a description of which is also provided in Sect. 4.3. The actor integrated innovation process is described in a case study from Germany dealing with one of the most relevant

C. Baedeker (✉) · C. Liedtke · M.J. Welfens
Wuppertal Institute for Climate, Environment, Energy, Wuppertal, Germany
e-mail: carolin.baedeker@wupperinst.org

C. Liedtke
e-mail: christa.liedtke@wupperinst.org

M.J. Welfens
e-mail: jola.welfens@wupperinst.org

D.V. Keyson et al. (eds.), *Living Labs*, DOI 10.1007/978-3-319-33527-8_4

areas of needs: heating and airing. It shows a high potential of social innovation for sustainable development. This could foster a more human need oriented and low investment technical development of low-resource SPSS. (see Sects. 4.4 and 4.5). The chapter concludes with an outlook (Sect. 4.6).

Keywords Green economy · Product-service systems · Livinglabs · Sustainable livinglabs sust production and consumption · Transition to sust · Transition path

4.1 Sustainable Product Service Systems—Quality of Life Within Ecological Boundaries

The World Energy Outlook and the UN Global Environment Outlook have recently shown that despite of hundreds of internationally agreed goals and objectives the world's development still remains unsustainable: The accumulation of greenhouse gases continues to be unabated. Taken into account the current and intended efforts of China, Japan, the USA, and the EU to mitigate carbon emissions, the global energy demand will grow by more than one-third by 2035 corresponding to a long-term average global temperature increase of 3.6 °C. The UN assessed 90 of the most-important environmental goals and objectives in the Global Environment Outlook and found that significant progress had only been made in four of them. A few years prior to these studies, Rockström (2009) estimated that at least three planetary boundaries had already been crossed: climate change, loss of biodiversity and the nitrogen cycle (UNEP 2012; IEA 2012).

In order to keep up the services provided by nature, the following global goals have been suggested in literature for the target year 2050:

- The ecological footprint per person should not exceed 1.2 ha.
- The worldwide per capita consumption of non-renewable resources should be less than 5–6 tons per year. This goal implies a tremendous increase in resource efficiency and a reduction in demand (sufficiency) in industrialized countries. In Germany and other industrialized countries, for instance, it means a factor 10 increase, requiring an absolute annual improvement in resource productivity of almost 5 %, starting now in order to preserve ecosystem services. Globally, resource consumption needs to be reduced by factor 2 (Schmidt-Bleek 1994; 1998; 2007).

Especially the consumption of industrialised nations currently exceeds sustainable levels by far. Direct material consumption of private households e.g. calculated in a study in Finland currently amounts to 40 tons per capita and year—which is approximately five times higher than the sustainable level (Lettenmeier et al. 2014). The largest share of this amount is induced in the fields of demand linked to housing, mobility and food (EEA 2013; Lettenmeier et al. 2014).

Our welfare production has to happen within the natural system boundaries. It must be ensured that the ecosystem services provided by nature are not reduced.

Currently we are consuming more ecosystem services and more natural resources than nature is providing on a sustainable basis.

Change can only take place through the transformation of production and consumption systems. This involves the modification of products, services and business models towards resource-efficient production patterns and lifestyles. Digitalisation and Industry 4.0, as well as social trends and movements (e.g. Sharing Economy) account for an increasing relevance and economic recognition of value chains and prosumer approaches—whether or not these are environmentally friendly will not be discussed here. This makes it even more important to direct social and technical innovations towards using them to design sustainable production and consumption patterns in an actor integrated way (Liedtke et al. 2013a, b; 2015b). However, further interactions and structures for the development, implementation and diffusion of these sociotechnical innovations are required. One suitable concept is that of Living Labs (Almirall and Wareham 2008; Mulder and Stappers 2009; Pallot et al. 2010) and especially SLL (Bergvall-Kåreborn 2009; Baedeker et al. 2014; Liedtke et al. 2015c). These concepts develop new prospects of change for the production and consumption system based on the existing sociotechnical system, which are then sampled in real, day-to-day situations involving both the group of people affected and supporting actors. This approach needs i.a. an economically sustainable and stable model such as that of Green Economy. Green Economy requires sociotechnical innovations—which should provide the foundation for appropriate business models—and the development and promotion of a low-resource society and lifestyles (Liedtke et al. 2015a, b).

4.2 Green Economy—A Possible Approach for Sustainable Socio-Economic Systems

In recent years the concept of 'Green Economy' has emerged as a strategic priority for governments and intergovernmental organisations (e.g. UNEP 2011; OECD 2011). In Europe, it prominently features in a range of medium- and long-term EU programmes and strategies, including the Europe 2020 Strategy, the 7th Environment Action Programme, the EU Framework Programme for Research and Innovation (Horizon 2020) and sectorial policies in areas such as transport and energy.

A Green Economy essentially is one in which socio-economic systems are organised in ways that enable society to live well within planetary boundaries. The concept therefore has several dimensions.

There is no internationally agreed definition of the term *Green Economy* but the existing definitions by different relevant international organisations such as UNEP, UNDP, OECD, World Bank and EEA are broadly characterised by three main objectives (see EEA 2014):

- Improving resource use efficiency: a Green Economy is one that is efficient in its use of energy, water and other material inputs,
- ensuring ecosystem resilience: it protects the natural environment, its ecosystems' structures and flows of ecosystem services, and
- enhancing social equity: it promotes human wellbeing and fair burden sharing across societies.

These objectives are addressed in different depth and scope by the definitions of relevant international and national organisations.

Green Economy is clearly defined by **UNEP** (2011, p. 19) as a vehicle that facilitates the transition to sustainable development: "moving towards a Green Economy must become a strategic economic policy agenda for achieving sustainable development. A Green Economy recognises that the goal of sustainable development is improving the quality of human life within the constraints of the environment, which include combating global climate change, energy insecurity, and ecological scarcity. However, a Green Economy cannot be focused exclusively on eliminating environmental problems and scarcity. It must also address the concerns of sustainable development with intergenerational equity and eradicating poverty."

The **OECD** (2011, p. 19) focuses the discussion on Green Economy on "green growth" and points out that green growth means fostering economic growth and development while ensuring that natural assets continue to provide the resources and environmental services on which our wellbeing relies. To do this it must catalyse investment and innovation, which will underpin sustained growth and give rise to new economic opportunities.

The **WWF** (2012, p. 7) underlines that "green economies should improve people's wellbeing, and restore, maintain and enhance the healthy natural environment that people and other species need to survive and thrive. Green economies are a means to achieving sustainable development and should therefore be based on the principle of equity within and between generations. Global sustainable development goals are required to build a shared understanding of the outcomes that economies should achieve, in terms improving human wellbeing and maintaining natural systems."

Green Economy Coalition—**GEC** (a group of NGO´s, trade union and other organisations) defines a Green Economy as "a resilient economy that provides a better quality of life for all within the ecological limits of the planet"(GEC 2015).

The **European Commission (EC** 2011, p. 4) notes that Green Economy is "an economy that generates growth, creates jobs and eradicates poverty by investing in and preserving the natural capital upon which the long term survival of our planet depends. It can also launch the necessary reform of international sustainable development governance."

The German Ministry of Education and Research (German BMBF) states that "The Green Economy's vision is of an internationally competitive, environmentally and socially sustainable economy. The concept combines ecology and economy positively with each other. Green Economy increases social welfare, fights poverty and aims at social justice. Against the background of recognized ecological limits, an environmentally friendly, high-quality and thus sustainable growth based on a comprehensive understanding of the relationships in business, finance and politics, should be made possible. The aim is to develop modified, sustainable production and consumption patterns, in order to ensure prosperity and a high quality of life worldwide and especially to future generations. Besides the positive impacts on environment and society, the competitiveness and resilience of Germany's position will also be strengthened." (BMBF 2014, p. 3).

In Table 4.1 an overview of the main ecological, social and economic objectives of the definitions above are presented.

The usually concerted balance of the target definition between the three dimensions, which follows a Green Economy, is surprising. It is obvious that, in general, they have a nearly common understanding of Green Economy. Beside the ecological balance orientation the definitions seek social justice and wellbeing. It remains rather unclear which structural changes in the economic field will be pursued to support the balance of ecosystems and social development. At micro level

Table 4.1 Ecological, social and economic characteristics of a Green Economy addressed in the relevant definitions, *source* Wuppertal Institute 2015

Ecological	Social	Economic
Environmentally sustainable (BMBF)	Socially sustainable (BMBF)	Internationally competitive and resilient economy (BMBF)
	Improving social equity (UNEP)	
	Socially inclusive (UNEP)	
	Ensuring social welfare (BMBF)	Resource efficient (UNEP)
Recognized ecological limits (BMBF)	Poverty eradication (BMBF)	Catalysing investment and innovation (OECD)
Ensuring ecosystem resilience (EEA)	Supporting social innovations (EEA)	Generating high-quality sustainable growth (OECD, EC, BMBF)
Preserving the natural capital (EC)	Increasing social justice (BMBF)	Creating and securing jobs (OECD, EC)
Maintaining and enhancing the healthy natural environment (WWF)	Improving quality of human life (UNEP)	Eradicating poverty by investing in and preserving the natural capital (OECD)
Low carbon (UNEP)	Based on the principle of equity within and between generation (WWF)	Enhancing technological progress and innovation
Significantly reducing environmental risks and ecological scarcities (UNEP)	Supporting social justice (BMBF)	Increasing social welfare (BMBF)
Providing a better quality of life for all within the ecological limits (GEC)	Improved human wellbeing (UNEP)	Encouraging sustainable production and consumption patterns (BMBF)
Prevention of the loss of biodiversity and ecosystem services (UNEP)	Social equity (UNEP)	Resilient economy (GEC) that improve people's wellbeing (GEC, WWF)
Combating global climate change (UNEP)	Social resilience (BMBF)	Growth in income and employment driven by public and private investments (UNEP)

the approach of Sustainable Living Labs gives the possibility to develop a business model, under the premises of the environmental, social and individual sustainable development, to gradually help achieving structural changes in the economic structure. This will be exemplified in a case study from the SusLabNWE project (see Sect. 4.5). The contribution of Sustainable Living Labs to implement such visions and goals of Green Economy will be reflected in the outlook of this section.

To bend the frame in which the Sustainable Living Labs develop this, eight core components for a sustainable transition towards sustainability are named, where Sustainable Living Labs can develop and test a transition role and transition paths. These are included once more in the Outlook and connected to the Case Study and the lessons learned.

4.2.1 Transition to a Green Economy—Eight Core Intervention Arenas Relevant for Sustainable Living Labs

Sustainable consumption and production within the Green Economy constitute a future market whose potential for our society's development has yet to be fully exploited. Actor interaction in the context of sustainable production and consumption is complex (see Fig. 4.1). Eight points of reference demarcate the arenas in

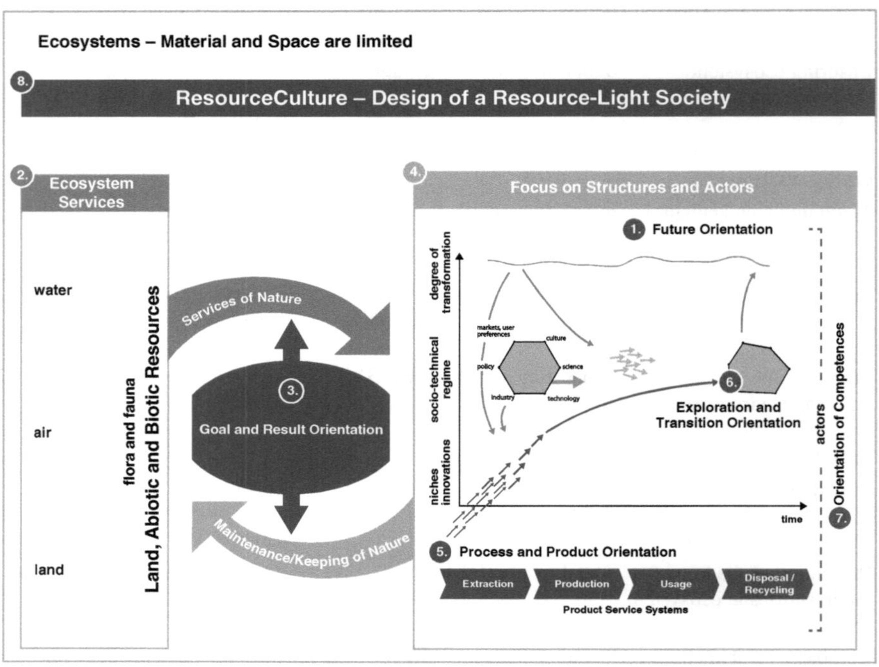

Fig. 4.1 Transition to a green economy—eight core points of reference relevant for sustainable Living Labs, *source* Liedtke et al. 2015b based on Geels 2004, 2010, Geels and Schot 2007

which this interaction must take place: a socially accepted value is attached to available material and immaterial resources, and this Resource Culture (8) forms the requisite basis for the implementation of sustainable production and consumption systems with a future focus (1). Conservation of ecosystem services (2) is the prerequisite for life on Earth and economic activity. In order to stay within planetary boundaries, goals must be defined as guard rails (3), compliance with these guard rails helps to ensure that our economic systems are sustainable and low-resource. This approach to resource management requires infrastructures and governance structures, which provide a frame of reference and guidance (4) for users and stakeholders as they develop low-resource technologies, processes, products and services (5). Users construct their environment themselves, working together. Through open-didactic exploration (6), which aims at supporting the transition to sustainability, users are empowered to play an active role in the design of complex systems (7).

This creates the framework for developing low-resource product-service systems, which are products and services that are highly serviceable to users while at the same time consuming less resources from cradle to grave or cradle, meaning from resource extraction, production, transport and usage to disposal or recycling. Such properties can and have to be considered already at the stage of developing and designing PSS (Schmidt-Bleek 1994, 2007; Schmidt-Bleek and Tischner 1995; Liedtke et al. 2013a, b; Irwin 2011; Manzini 2015; Walker 2006, 2014).

Low-resource SPSS (balanced use of ecosystem services) can help to act more sustainable in everyday life by promoting more corresponding social practices (Shove et al. 2012). To do so, Sustainable Living Labs offer a relevant sustainability-oriented research and development infrastructure, in which relevant actors are actively integrated into development, design and testing of new future SPSS (future orientation, Liedtke et al. 2015c).

Sustainable production and consumption are fundamental strategies for a broad social transition towards a more sustainable sociotechnical regime and Green Economy. In essence, all products, services and infrastructures on which we rely should be sustainable and low-resource (goal orientation). This could be achieved through the development of user-centred product-service systems, which integrate efficiency, consistency and sufficiency strategies (process and product-service orientation) into production-consumption systems (actor and structure orientation—changing the sociotechnical regime). Diverse and creative solutions and innovations help to ensure that sustainability gains are not simply cancelled out by additional consumption (rebound effects). Instead, positive rebound effects should be fostered by mainstreaming sociotechnical niche innovations, thereby reducing resource consumption in absolute terms while enhancing quality of life and wellbeing (implementing a resource culture—designing a low-resource society). There is potential for action by all user and stakeholder groups here, which need a complex frame structure for their training, experience and competence development in a complex, exploration oriented environment. The development of user-integrated sustainable product-service-systems can shape the transition to a Green Economy in a more sustainable society (Liedtke et al. 2015b). This eight core intervention arena has to be addressed for a transition towards a sustainable and resilient economy—SLL includes this challenge.

4.3 Living Labs in a Green Economy—An Approach for Transforming the Production and Consumption System

Since industrialization the production systems have been focused on an important part of the economy responsible for producing welfare and socio-economic progress. Households and the interaction of the production and consumption systems were only the "black box" on the demand side. The *Homo economicus* has been perceived as an actor with a high and well-known deciding and behaviour structure. However, social sciences, evolutionary economic[1] and market research as well as the latest neuromedicine have shown that deciding and behaviour processes are relatively unknown. They are of high complexity and diversity (Strünk et al. 2012; Liedtke et al. 2008a).

The sustainability research shows the need for a holistic approach taking into account and transforming the production-consumption-system. Only sustainability-oriented efficiency and sufficiency strategies will be able to solve the future tasks, challenges and problems. Thus the future potentials for resource efficiency, climate change, poverty reduction and broadening welfare lie in the fields of individual decision processes and behaviour combined with organizational learning processes recognizing social context situations. Therefore, we need more information and research results about these processes, we need more knowledge about what people want and how they use products and services in their living environment at home, on the road and on their job. Such an interactive, scientific and stakeholder integrative research approach needs real-life test beds, kits and framework conditions (Liedtke et al. 2008b).

Living Labs should give the possibility of using a technical and social flexible framework for analysing the production-consumption system interaction. The conditions of the test bed have to be configured based upon the investigated social context to enable the optimisation of production-consumption systems. Hence it is necessary to analyse people in their real social context and day-to-day situations, preferably lifelike. The brand of the Living Lab design is a combined and integrated technological-socioeconomic approach to enable optimised interaction of production and consumption. Technological and social innovations can only be developed interactively (Almirall 2009; Almirall and Wareham 2008; Mulder et al. 2008; Mulder and Stappers 2009; Bergvall-Kåreborn 2009; Pallot et al. 2010; Geibler et al. 2014; Liedtke et al. 2008b; Liedtke et al. 2015c).

The Living Lab can be a strategic research instrument for science, companies and society. It should help to generate competitive advantages by matching

[1]Evolutionary economics is a part of economics that is inspired by evolutionary biology. Much like mainstream economics it stresses complex interdependencies, competition, growth, structural change, and resource constraints but differs in the approaches which are used to analyze these phenomena (Friedmann 1998).

sustainable requirements and contributions. It should give the possibility of arranging a competition of product-service systems for the highest resource efficiency solution (including benchmarking processes). A Living Lab should support both testing newly designed and developed product-service-systems and support the basic research agenda for analysing the developing processes of existing production and consumption patterns. For public awareness it is necessary to communicate the results and to position the product-service systems in a global economic system. Users should be actively integrated in research for the best sustainability solution (Bergvall-Kåreborn 2009; Liedtke et al. 2008b, 2015c).

Research results also show that product mixes, perceptions, valuations and behaviour patterns inside of a single social milieu and between different milieus vary significantly. The resource input can be different by a factor of seven in the same social milieu (Kleinhückelkotten 2005). If this is a realistic result, people use a comparable product mix very differently. Living Labs must allow to test and arrange verified research concepts i.e. focusing on custom-fit product and service mixes in different types of households (i.e. an average basket of goods of a specified type of household) or phases of life with their conditions (families, young singles, young old people, old- or middle-aged people, disabled people, etc.). Indicators for resource efficiency—from cradle to cradle—and sustainability should be measurable so that potential contributions to different national, European and company's sustainability strategies can be evaluated (Liedtke et al. 2008b).

4.4 From Living Labs Towards Sustainable Living Labs

The Green Economy concept promotes investments in sustainability innovations as leverage for a low-carbon, resource-efficient and socially inclusive economy. Technological or social innovations alone cannot achieve the necessary system changes. Innovative sociotechnical strategies and their experimental testing are needed to offer steps towards the named sustainability goals through product-service-systems (PSS).

The Sustainable Living Lab approach is based on four points of departure. The most fundamental consideration is the assumption that a Living Lab infrastructure could be used to foster sustainable development and, thereby, meet the challenges in climate change, scarcity of resources and reduced richness in species that pose threats to human economic activities and human existence in general. The second consideration puts user-centred innovation processes in the focus of this strategy. The fact that Living Lab enters globally interwoven production-consumption systems argues for the third point of departure, namely the need for a holistic approach to allow for resource saving innovations on a systemic level. The fourth cornerstone of our Living Lab infrastructure is formed by the focus on the home as the most significant manifestation of a private household and as a hub in the system of production and consumption (Jong et al. 2010).

Sustainable patterns cannot be achieved solely through technological efficiency innovations. Many product service innovations with a high sustainability potential fail because they are rejected by consumers or create negative rebound effects (Sorrell 2007; Druckman et al. 2011). Another important factor is unexpected user behaviour or the incorrect application of potentially sustainable efficiency innovations (Liedtke et al. 2012a, b). Living Lab can be defined as "a user-centric innovation milieu built on every-day practice and research, with an approach that facilitates user and actor influence in open and distributed innovation processes engaging all relevant partners in real-life contexts, aiming to create sustainable values" (Bergvall-Kåreborn et al. 2009, p. 3). The Sustainable Living Lab approach is gaining more and more importance on different political scales (international, national, regional, local). The approach of real-life laboratories combined with user/actor-integrated innovation research has found ample support in recent years in different research programmes of international and national research policy and funding agendas (e.g. Horizon 2020, INTERREG, Climate KIC Initiative, German Ministry of Education and Research).

Sustainable Living Labs (SLL) are a locally based regional, national and international infrastructure set up to enable innovation processes in which users and value chain-relevant actors actively participate in development, testing and marketing phases. Interactive innovation processes take place gradually in users' real life surroundings (user observation, field tests) and user interaction laboratories (e.g. for prototyping). An SLL, led by sustainability criteria, aims to contribute to global and universally applicable patterns of production and consumption, including the actor-integrated development of business cases, enabling transition processes to be marketed to companies and users (Liedtke et al. 2015c; Geibler et al. 2012).

Compared to existing Living Lab approaches the Sustainable Living Lab infrastructure offers a number of new qualities. In addition to our clear focus on sustainability innovations, the systematic connection of product-service-systems development to Living Lab is not yet an established field of research. Furthermore, SLL offers the unique combination of laboratory situations in a Living Lab with real-life experiments in urban districts, in which households are asked to become involved in the development process on a voluntary basis. Thus, the product-service-systems developed in a user-integrated innovation process will have a greater chance of being distributed successfully (Liedtke et al. 2015c).

The European infrastructure and methodological framework (three-phase model of research) developed with several research and business partners in the Living Lab Design Study (2008–2010) led by the TU Delft is unique. The three-phases model (Insight Research, Prototyping, Field Testing, see 2) has been adopted and further developed in the project "SusLabNWE—Creation of a Networked Infrastructure for Innovation on Sustainability in the Home Environment" (2012–2015), financed by INTERREG. In the first phase, the Insight Research, the existing status quo in a field of interest is explored and the required level of change is analysed in the real-life environment of actual households. In the second phase, during Prototyping, actual product-service-innovations are developed and tested in the Living Lab facilities. In the last phase, Field

Testing, developments are evaluated and redesigned if necessary. Throughout the development resource efficiency and saving potentials of the new prototype will be evaluated (Bakker et al. 2010; Welfens et al. 2010).

The European SusLabNWE project sets up an international infrastructure of Sustainable Living Lab test facilities (Fraunhofer inHaus, Living Lab-Container at Hochschule Ruhr-West) and real-life experiment settings at different locations, which co-operate in user-centred development of sustainability innovations around the home. The focus area of the German consortium is located in the Ruhr area in North Rhine-Westphalia (NRW). The real-life experiments are set up in the model region, which is involved in SusLabNWE as the regional SusLabNRW sub-project.

4.5 SusLabNWE in Germany: Implementation of the Three-Phases Model as a Framework for PSS Development Towards a Green Economy

The above-described methodology (see Fig. 4.2) is concretised and applied in the German focus region Innovation City Ruhr, Model Town Bottrop. All actions are aligned to the three-phases model of research (see Fig. 4.3 Case study design).

During **Insight Research** a pre-analysis of heating energy consumption in the building stock of Innovation City was carried out and monitoring continued for the whole length of the project. Until today around 1200 interviews have been conducted through energy consulting by Innovation City Management GmbH. With an associated pre-analysis of building characteristics the comparison of heating energy consumption has been possible, depending on different types of buildings in Innovation City Ruhr. Further insight research has been conducted by analysing heating and cooling behaviour in recruited households in the Model Town Bottrop. Sensor technology (mobile data loggers) has been applied in 80 households in the

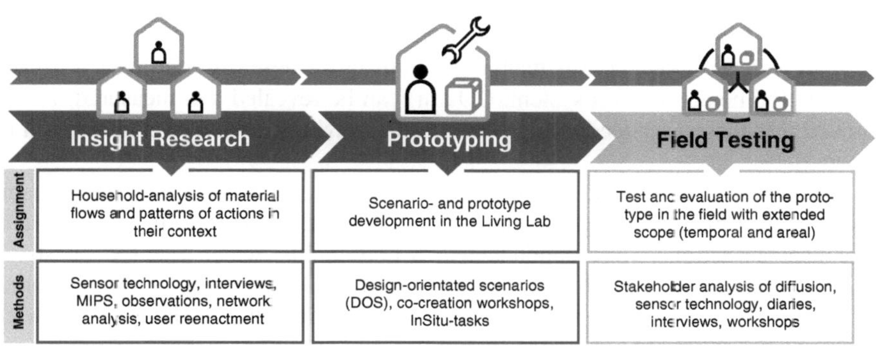

Fig. 4.2 The sustainable Living Labs three–phases model of research, *source* Liedtke et al. 2015c, p. 111

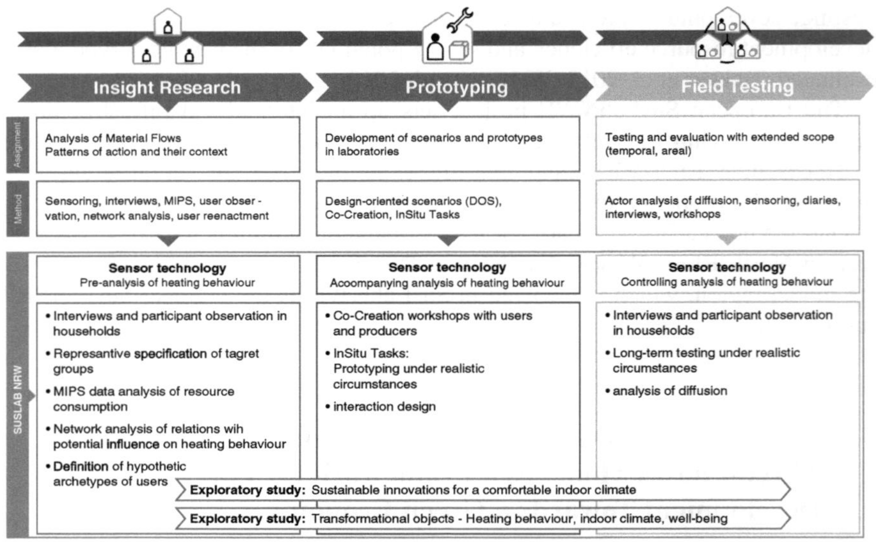

Fig. 4.3 Case study design: methods conducted in the three-phases model in SusLabNWE in Germany, *source* Liedtke et al. 2015c, p. 112

winter pilot 1 (Winter 12/13), measuring room temperature, CO2-concentration and humidity for a period of one week. Participants were asked to write down when they were at home, opened a window and for how long. If possible, energy consumption was monitored through the entire time or at least once a day (Grinewitschus et al. 2013). In the winter pilot 2 (Winter 13/14) the Prototyping Phase started. Within the Prototyping 40–50 households were equipped with the data loggers mentioned above for two weeks each. All companies involved in this project participated during this prototyping phase either by providing home-automation systems or CO2 ms, which visualize indoor air-quality (measured by CO2 level) using the colours of a traffic light. Within the first week a 'basis measuring' was conducted, whereas in the second week the home-automation systems and CO2 m were installed. In this manner the impact of these different user feedback advices (home automation system, CO2 m) can be revealed and measured. It was proved that these user feedback advices lead to savings of energy consumption of up to 25 %.

Furthermore, within the period of winter pilot 1, qualitative interviews, combined with a participant observation to also consider nonverbal information and unconscious actions, were conducted in 6 households, which focused on social practices related to warmth, comfort, room airing or welcoming guests at home. Within the winter pilot 2 a network analysis on the influence of household advice relations i.e. to neighbours, friends but also heating maintenance personnel was conducted resulting in network maps (see Chap.15 Actor and network analysis). In the phase of Insight Research also the connection of energy and material consumption was analysed. Further, we assume that different household types exist with characteristic attributes in (1) energy consumption and (2) resource consumption and that these attributes are connected with (3) ownership and purchase behaviour and (4) time management patterns. To analyse these relations we implemented an experiment design comprising classification of households within social milieus (see Chap. 19 Household Typology and social milieus), Material Input Per Service unit (MIPS) analysis as well as characterisation of ownership, purchase and time management behaviour (see Chap. 21 Material Footprint of household activities).

4.6 SusLabNWE in Germany: Results and Further Developments

The assessments of the described measurements and questionnaires confirm the assumption that the deciding point in energy efficiency is not the technical equipment and rehabilitation measures but the behaviour, knowledge and personal sensation of the user. We found that the user influence especially outshines investment-heavy measures. For example zero energy homes become dissipative homes when a terrace door is always open. In contrast simple and investment-low measures (like smart home systems, CO2 meter) can have much larger effects on energy efficiency. It can be shown that technical improvements used to calculate saving potentials are annulled by user behaviour. The predicted savings of thermal insulation or more efficient heating systems are often not reached in practice. Next to user behaviour we found maladjustments of heating systems responsible for high energy demand. Solutions for this must be found with user, producer and maintenance group combined. The results in the phases Insight research and Prototyping proved the necessity of an user and stakeholder integrated development of products-service systems.

The implementation of the three-phases model in Germany in close interaction with households, companies, energy consultants and researchers led to the conclusion that a custom fit energy consultation like it is intended in the City of Bottrop requires the development of an innovative product-service-system. Therefore, a concept for user and household centred energy efficiency consulting for tenants and homeowners in the field of heating within SusLabNWE in Germany was developed. The concept is based on a pre-analysis of the energy efficiency potential by using the evaluation methods developed in the SusLab project (e.g. user- and household-centred sensoring, feedback systems). The "product-service-system" is based on the idea to receive in-depth insight into the room-climate of a client using a portable measurement technology. The collected data during a fixed period of time will serve as a basis foundation for further analyses by the energy consultant. The analysis of room climate delivers a first impression on social practices and user behaviour, climatically facts (like humidity, temperature), heating system functionality and building insulation characteristics. Based on the results a suitable, economically optimized proposal of measurements (product service solutions) for increasing energy efficiency can be given to tenants and homeowners. The distinguishing feature of the customized energy consultation on the one hand roots in its foundation of individual sensoring data (humidity, temperature, energy consumption) within the households over a longer period of time. On the other hand it roots in tailoring advice specifically to the individual results of sensor measurements and focusing on low investment measures leading to behavioural changes.

The Innovation City Management GmbH in Bottrop developed in cooperation with the Hochschule Ruhr West and the Wuppertal Institute the concept "Heating plus"—a product-service-system (PSS) for Energy Consultation.

Focus of the PSS "Heating plus" is the activation of energy efficiency potentials in private households through low-investment measures in a product-service-system: it combines energy consultation for households with individual measurement of indoor climate (temperature, humidity, CO2-level), which is easily applicable at a low price, and low-investment measures to feedback users with information on ventilation behaviour. Accompanying this, a qualification for energy consultants and handicraft service suppliers, active in the field of heating, will be developed for the application of the PSS. It is central to not only develop a PSS that focuses on technical influence on the user but to establish which technology supports the user to become aware of and shape routines of heating behaviour, experiment with them in everyday life and train them—in the sense of evolving from action to knowledge. The aim is to enable learning by experience and open up a creative space on the interaction of social practices in households and not to denounce "wrong" user behaviour.

The PSS "Heating plus" can be summarized in the three main steps:

- Collecting data: Metering in households
- Pre-Consulting: optimized individual/user- and household-centred consultation.
- Implementation of user-fit products and services
- Qualification of energy consultants and handicraft service suppliers

Innovativeness lies in the development of a PSS by strongly integrating users and stakeholders along the value chain of heating through Sustainable Living Labs. Additionally, the form of energy consultation in the PSS by combining it with measurement and user feedback, which enables to identify individual needs and advices very specifically, is an innovation, which can activate considerable energy savings in private households. Based on the state of the art in research and own prior results saving potentials of heating energy of 15–20 % are possible by applying the concept of "Heating plus". Thus, between 1,050 and 1,400 kWh/a of heating energy, between 288.12 and 384.16 kg CO2-equiv./a and up to 90 € per household can be saved each year.[2] Accumulated to 40.2 million households in Germany (2014) this would mean a reduction of 15.4 million tonnes CO2-equiv. per year.

[2]Assumptions for average households: Gas heating, 70 m^2 living space with a heating energy demand of 100 $kWh/m^2/a$ and emissions of 0.2744 kg CO2-equiv. per kWh as well as costs of 6.52 Euro CT/kWh.

Within SusLabNWE the concept of PSS "Heating plus" was developed. To finalize this new concept additional funding is necessary. Therefore new capital should be acquired in future projects to enforce the realization of the concepts.

4.7 Outlook: Sustainable Living Labs as a Transition Path to Green Economy

The article shows that Sustainable Living Labs can play an important role in the operationalisation and implementation of a Green Economy based on sustainability innovations. They represent an experimental setting for implementing Green Economy, where technological and social innovation could be developed and tested.

Key dimensions for Sustainable Living Labs in the Green Economy are (based on Geibler et al. 2015 forthcoming):

- Innovations for a Green Economy have the objective of contributing to transform production and consumption systems towards sustainability. This includes a progressive structural and sociotechnical change of business models, value chains and lifestyles that can be supported by the development of low-resource and sustainable product-service-systems.
- This change can only be successful if system-relevant actors are involved in innovation and development processes. Next to science and business, consumers such as households and governance-shaping actors are important. They should be integrated into Living Labs for a Green Economy towards sustainability.
- The current focus of innovations for a Green Economy, other than for sustainable development, lies on the ecological and economical dimensions of sustainability—although the definitions of Green Economy fundamentally emphasize the social and societal aspects. Accordingly, the innovation processes in Living Labs for a Green Economy should be aimed at developing low-resource product-service-innovations that implicitly co-address these aspects in the development or modification phases.

Therefore, in order to measure a contribution of innovations for Green Economy to sustainable development (integration of the aspects mentioned in the definition), a good methodology for evaluation, a solid data base and a meaningful set of indicators are needed, which should be taken into account from the innovation process and dynamically be developed further.

Looking at SusLabNWE innovations for Green Economy could be operationalized in the case of the product-service-system (PSS) concept "Heating plus". This case shows explicitly that within such a research infrastructure low-resource and sustainable product-service-systems could be developed and tested effective by integrating all actors of the value chain such as producers, installers, households, consultants and municipality.

In conclusion it can be said that in the Sustainable Living Lab research with its inter- and transdisciplinary design the eight components of sustainable transition described above are relevant and interacting (see Fig. 4.1). The implementation of SLL in SusLabNWE in Germany in relation to the "Heating Plus" case shows that at the micro level the approach of Sustainable Living Labs gives the possibility to develop business models, under the premises of environmental, social and individual sustainable development helping to achieve structural changes in the economic structure due to product-service-systems. Most relevant for the SLL based research are social practices and needs that derive from social contexts, embedded into cultural, societal, environmental and political frameworks. Therefore SLL can lead to sustainable production and consumption patterns focusing on sociotechnical innovation, cooperation, participation and interaction in value chains. They are a necessary vehicle for the transition path to Green Economy and they will play an important role on its implementation.

References

Almirall, E. (2009). *Understanding innovation as a collaborative*. Ramon Llull: Co-Evolutionary Process.

Almirall, E., & Wareham, J. (2008). Living Labs and open innovation: Roles and applicability. *The Electronic Journal for Virtual Organizations and Networks, 10*(3), 21–46.

Bakker, C., Eijk, D. v., Silvester, S., Reitenbach, M., Jong, A. d., & Keyson, D. (2010). *Understanding and modelling user behaviour in relation to sustainable innovations: The LIVING LAB method*. Paper presented at the TMCE, Ancona, Italy.

Baedeker, C., Greiff, K., Grinewitschus, V., Hasselkuß, M., Keyson, D., Knutsson, J., et al. (2014). *Transition through sustainable Product and Service Innovations in Sustainable Living Labs: application of user-centred research methodology within four Living Labs in Northern Europe*. Paper presented at the 5th International Sustainable Transitions (IST) Conference, August 27–29, 2014, Utrecht, Netherlands.

Bergvall-Kåreborn, B., Ihlström Eriksson, C., Ståhlbröst, A., & Svensson, J. (2009). *A milieu for innovation–defining Living Labs*. 2nd ISPIM Innovation Symposium, New York, United States.

BMBF (2014) *Forschungsagenda Green Economy. Die neue Hightech-Strategie. Innovationen für Deutschland*. Berlin: Bundesministerium für Bildung und Forschung (BMBF).

Druckman, A., Chitnis, M., Sorrell, S., & Jackson, T. (2011). Missing carbon reductions? Exploring rebound and backfire effects in UK households. *Energy Policy, 39*(6), 3572–3581.

EC (2011). Rio +20: Towards the Green Economy and better governance. Brussels, Belgium: European Commission (EC).

EEA (2014). Resource-efficient Green Economy and EU policies. European Environment Agency (EEA). EEA Report 2/2014, 1–107.

EEA (2013). SOER 2015—The European environment—state and outlook 2015.

Friedman, D. (1998). Evolutionary economics goes mainstream: A review of the theory of learning in games. *Journal of Evolutionary Economics, 8*(4), 423–432.

GEC (2015). *Who we are. Green Economy Coalition (GEC)*. Retrieved October 29, 2015, from http://www.greeneconomycoalition.org/about.

Geels, F. (2004). From sectoral systems of innovation to socio-technical systems: Insights about dynamics and change from sociology and institutional theory. *Research Policy, 33*, 897–920.

Geels, F. (2010). Ontologies, socio-technical transitions (to sustainability), and the multi-level perspective. *Research Policy, 39*, 495–510.

Geels, F., & Schot, J. (2007). Typology of sociotechnical transition pathways. *Research Policy, 36*, 399–417.

Geibler, J., Baedeker, C., Hasselkuß, M., Jordan, N.D., & Rohn, H. (2012). Living Labs as a reflexive user-driven research infrastructure to promote sustainable consumption and production patterns: Experiences from exploratory research. In: *18th Annual Conference of the International Sustainable Development Research Society (ISDRS): Conference Proceedings*. Hull.

Geibler, J., et al. (2015). *Analyse von Visionen einer Green Economy*. Wuppertal (upcoming): Schlüsseldimensionen für Living Labs in der Green Economy.

Geibler, v. J., Erdmann, L., Liedtke, C., Rohn, H., Stabe, M., Berner, S., et al. (2014). Exploring the potential of a German Living Lab research infrastructure for the development of low resource products and services. *Resources 2013*, 575–598.

Grinewitschus, V., Lovric, T., Rumler, N. (2013). Influence of user behavior and home automation on energy consumption. In: Proceedings of the 7th International Conference on Energy Efficiency in Domestic Appliances and Lighting (EEDAL13), Coimbra, Portugal, pp. 11–13.

IEA. (2012). *World energy outlook 2012*. International Energy Agency (IEA): Paris, France.

Irwin, Terry (2011). Design for a Sustainabel future. In J. C. Hershauer, G. Basile & S. G. McNall (Eds.), (pp. 41–60). Santa Barbara: The Business of Sustainability.

Jong, A. d., Krekeler, M., Lettenmeier, M., Liedtke, C., Pohling, A., Welfens, J. M. (2010). Overall research agenda for the European LIVING LAB network. Deliverable D 4.5, internal paper.

Kleinhückelkotten, S. (2005). *Suffizienz und Lebensstile*. Berlin, Germany: Ansätze für eine milieuorientierte Nachhaltigkeitskommunikation. Berliner Wissenschafts-Verlag.

Lettenmeier, M., Liedtke, C., & Rohn, H. (2014). Eight tons of material footprint—suggestion for a resource cap for household consumption in Finland. *Resources, 2014*(3), 488–515.

Liedtke, C., Baedeker, C., & Borrelli, L. M. (2015a). Stellschrauben für Nachhaltigkeit: Trends und Verantwortungen in Produktion und Konsum. In: M. Roth, C. Ulbert & T. Debiel (Eds.), *Stiftung Entwicklung und Frieden, Institut für Entwicklung und Frieden, Käte Hamburger Kolleg/Centre for Global Cooperation Research: Globale Trends 2015* (pp. 299–314). Frankfurt, Germany: S. Fischer Verlag.

Liedtke, C., Baedeker, C., & Borrelli, L. M. (2015b). Transformation towards a sustainable society—key intervention area. *Innovative Energy and Research, 4*(2), 1–10.

Liedtke, C., Baedeker, C., Hasselkuß, M., Rohn, H., & Grinewitschus, V. (2015c). User-integrated innovation in Sustainable Living Labs: an experimental intrastructure for researching and developing sustainable product service systems. *Journal of Cleaner Production, 97*, 106–116.

Liedtke, C., Buhl, J. & Ameli, N. (2013a). Designing value through less by integrating sustainability strategies into lifestyles. *International Journal of Sustainable Design, 2*(2), 167–180.

Liedtke, C., Buhl, J., & Ameli, N. (2013b). Microfoundations for sustainable growth with eco-intelligent product service arrangements. *Sustainability, 5*(3), 1141–1160.

Liedtke, C., Baedeker, C., von Geibler, J., Hasselkuß, M. (2012a). User-integrated innovation: Sustainable Living Labs. In V. Fricke, U. Schrader & V. W. Thoresen (Eds.), *2nd PERL International Conference "Beyond Consumption—Pathways to Responsible Living": Conference Proceedings*, March 19–20, 2012 (pp. 203–219). Berlin, Germany: Technical University.

Liedtke, C., Welfens, M. J., Rohn, H., & Nordmann, J. (2012b). LIVING LAB: User-driven innovation for sustainability. In *International Journal of Sustainability in Higher Education*, Vol. 13, 2/2012, pp. 106—118.

Liedtke, C., Welfens, M. J., Schaefer, I., Stengel, O., & Onischka, M. (2008a). Didaktische Module "Vom Wissen zum Handeln", Modul "Wirtschaft/Neue Wirtschaftsordnung, (WIN)". Stiftung Forum für Verantwortung, ASKO EUROPA-STIFTUNG, Europäische Akademie Otzenhausen. Wuppertal Institute: Wuppertal, Germany.

Liedtke, C. Lettenmeier, M., & Welfens, M. J (2008b). Major results of social science research concerning consumer behaviour, sustainability and innovations, draft paper for internal Workshop in the EU-project: Design Study for the LIVING LAB Research Infrastructure, to research human interaction with, and stimulate the adoption of, sustainable, smart and healthy innovations around the home, Wuppertal September 2008.
Manzini, E. (2015). *Design when everybody designs: An introduction to design for social innovation*. Cambridge, MA: MIT Press.
Mulder, I., & Stappers, P. J. (2009). Co-creating in practice: Results and challenges. In *Collaborative Innovation: Emerging Technologies, Environments and Communities (Proceedings of the 15th International Conference on Concurrent Enterprising: ICE 2009*, (pp. 1–8).
Mulder, I., Velthausz, D., & Kriens, M. (2008). The Living Labs harmonization cube: communicating Living Labs' ESSENTIALS. eJOV Executive *The Electronic Journal for Virtual Organization and Networks, 10*, 14.
OECD. (2011). *Towards green growth*. Paris, France: OECD Green Growth Studies. OECD Publishing.
Pallot, M. et al. (2010). Living Lab research landscape: From user centred design and user experience towards user co-creation. In *First European Summer School "Living Labs"*.
Rockström, J., Steffen, W., Noone, K., Persson, A., Chapin, F. S. III, & Lambin, E. (2009). Planetary boundaries: Exploring the safe operating space for humanity. *Ecology and Society, 14*(2), 1–32.
Schmidt-Bleek, F. (1994). Wieviel Umwelt braucht der Mensch? MIPS, das Maß für ökologisches Wirtschaften, Berlin/Basel/Boston.
Schmidt -Bleek, F., & Tischner, U. (1995). Produktentwicklung, Nutzen gestalten—Natur schonen, Wirtschaftskammer Österreich (WIFI Broschüre Nr. 270), Wien.
Schmidt-Bleek, F. (1998). *Das MIPS-Konzept. Weniger Naturverbrauch, mehr Lebensqualität durch Faktor 10*. München.
Schmidt-Bleek, F. (2007). Nutzen wir die Erde richtig? Frankfurt am Main.
Shove, E., Pantzar, M., & Watson, M. (2012). *The dynamics of social practices*. SAGE: London, United Kingdom.
Sorrell, S. (2007). *The rebound effect: An assessment of the evidence for economywide energy savings from improved energy efficiency*. London: UK Energy Research Centre.
Strünck, C., Arens-Azevêdo, U., Brönneke, T., Hagen, K., Jaquemoth, M., & Kenning, P. (2012). Ist der "mündige Verbraucher" ein Mythos? Auf dem Weg zu einer realistischen Verbraucherpolitik. Stellungnahme des Wissenschaftlichen Beirats Verbraucher- und Ernährungspolitik beim Bundesministerium für Ernährung, Landwirtschaft und Verbraucherschutz (BMELV).
UNEP. (2011). *Towards a green economy: Pathways to sustainable development and poverty eradication*. United Nations Environment Programme (UNEP): Nairobi, Kenya.
UNEP (2012). GEO5. Global environment outlook: Environment for the future we want. United Nations Environment Programme (UNEP): Nairobi, Kenya.
Walker, S. (2006). *Sustainable by design: Explorations in theory and practice*. London: Earthscan.
Walker, S. (2014). The dynamics of energy demand: Change, rhythm and synchronicity. *Energy Research and Social Science, 1*(2014), 49–55.
Welfens, M. J., Liedtke, C., Rohn, H., & Nordmann, J. (2010). Living Lab: Research and development of sustainable products and services through user-driven innovation in experimental-oriented environments. In *ERSCP-EMSU Conference "Knowledge Collaboration and Learning for Sustainable Innovation", Delft*.
WWF. (2012). *Building green economies: Creating prosperity for people and planet*. WWF-UK: Godalming, United Kingdom.

Part II
Research Methods for Living Labs

Chapter 5
Living Labs to Accelerate Innovation

David V. Keyson, Gregory M. Morrison, Carolin Baedeker and Christa Liedtke

Abstract A key goal of LivingLabs is to provide industry, including large companies and SMEs, knowledge institutes and policymakers with a unique new infrastructure in testing and the co-development of sustainable products, services, legislation and combinations of these, takes place directly with users. The three-Tier Model of living lab research which aims to connect industry, academic, and public stakeholders, while facilitating user-centred studies, is presented in this section consisting of: (a) insight research involving the study of current practices in existing homes, (b) studies in prototype houses equipped with innovative products and services focused on sustainable living, and (c) field testing, in which research prototypes are up-scaled such that existing homes can be equipped with innovative sustainable technologies.

Keywords Living labs · Research through design · Three-tier model · Innovation

5.1 The Three-Tier Model

Living labs as real-world test beds offer the potential to accelerate the innovation processes by which users, designers, and other stakeholders actively participate in the development, testing, and diffusion of new products and services.

D.V. Keyson (✉)
Delft University of Technology, Delft, The Netherlands
e-mail: D.V.Keyson@tudelft.nl

G.M. Morrison
Curtin University Sustainability Policy (CUSP) Institute Curtin University, Bentley, Australia
e-mail: Greg.morrison@curtin.edu.au

C. Baedeker · C. Liedtke
Wuppertal Institute for Climate, Environment, Energy, Wuppertal, Germany
e-mail: carolin.baedeker@wupperinst.org

D.V. Keyson et al. (eds.), *Living Labs*, DOI 10.1007/978-3-319-33527-8_5

The innovation processes are examined using observational techniques and prototypes positioned in existing homes as well as in living laboratories for short-term studies on sustainable living. The living labs are designed to provide a context in which users can interact with and report on sustainable innovations and share experiences on living routines. Central to living labs is the development of user-centered design research methodologies and measures for in situ studies which can provide insights into the usability and adoption of sustainable innovations for industry, public and academic stakeholders. Living Labs can contribute to global and universally applicable patterns of production and consumption in compliance with measures of efficiency, sufficiency, and consistency. Design is led by the application and development of criteria for sustainability from an integral perspective.

The ability to be creative and innovative in a society influenced by stress, lack of time, and other factors, while harnessing the technology available in today's knowledge society is key towards increasing the overall productivity of future societies. While technology enables new value-chains, product and service development, increasingly there is need to focus on true user requirements and social innovation in order to be successful in today's increasingly global and competitive market. New R&D and innovation methodologies need to be developed to meet the challenges.

The Living Labs concept can be generally understood as the ability to bring user, technology and business into an open innovative development process that establishes real life environments. The concept supports long-term cooperation, co-creative research and development by involving at an early stage the user in the innovation process for 'sensing, prototyping, validating and refining complex solutions in multiple and evolving real life contexts'. The long-term cooperation between researchers, companies and end users distinguish this concept from other approaches, which revert to traditional methods. In this regard heterogeneous empirical methods have been applied to studying behavior and media usage (Eriksson et al. 2005).

SusLabNWE (Sustainable Lab North-West Europe) aimed to resolve territorial challenges related to industrial competitiveness and sustainability in North West Europe. In reports from the European Commission ('Design as a driver of user-centered innovation') and policymakers such as DTI in the UK ('Creativity, Design and Business Performance') and by the Dutch government ('Value of Creation') user-centered design and development is highlighted as a key driver to increase innovation and competitiveness.

As Europe moves from an advanced industrial economy to one that is knowledge-based, the region must work together to change output from industrial goods to tailor-made products and services taking into account consumer needs while ensuring a sustainable society. Janez Potocnik, European Commissioner for Environment, noted its impact on sustainability when stating that to meet EU2020 targets for resource efficiency, it is critical to change the behavior of European consumers, to work on people's awareness, and to influence their habits (ESDN Quarterly Report 2015).

Many companies, especially those involved in fast-moving consumer goods have recognised the need for user insight in product development and have adapted their development approaches accordingly. Companies, such as those involved in designing complex building integrated products (heating systems, building materials, solar panels, etc.), in many cases do not sell directly to the consumer, and are only now starting to follow a user-centric design approach. However, in many cases the expertise and facilities to enable such an approach are lacking.

5.1.1 Co-development

A key goal of LivingLabs is to provide industry, including large companies and SMEs, knowledge institutes and policymakers with a unique new infrastructure in testing and the co-development of sustainable products, services, legislation and combinations of these, takes place directly with users (Fig. 5.1).

5.1.2 Approach

Traditional methods for generating insights on consumers rarely make it possible to experience the full benefits of new or hypothetical products, and often fail to predict accurately whether consumers will understand the technologies that underpin truly innovative products. As a result, new products and innovations often fail in the market, and companies have increasingly poor returns on their investment in product innovation.

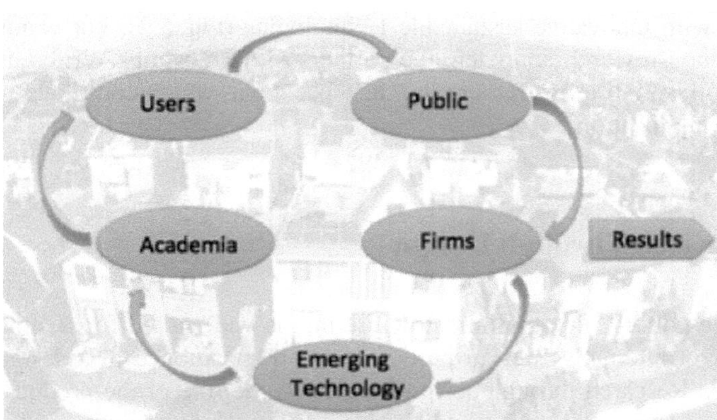

Fig. 5.1 Key actors in market and enabling technology in co-design process

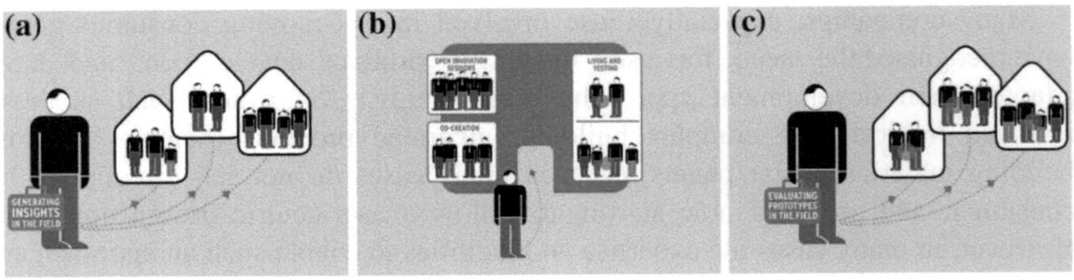

Fig. 5.2 Three-tier model of research in the LivingLab design study

The involvement of users in the design and evaluation process of sustainable technologies is fundamental to developing usable and acceptable products and services. Central to involving users is the living lab R&D methodology where innovations such as services, products or application enhancements are created and validated in collaborative multi-contextual empirical real-world environments.

5.1.3 LivingLabs Model

With the goal of setting up a living lab infrastructure, including use of existing homes and a network of research houses, SusLab NWE built on the EU Living Lab project. The project was a 2-year design study completed in January 2010 that focused on the role of Living Labs as a platform for observing and influencing sustainable user behavior in real-world environments via innovations in sustainable products and services. Within the Living Lab project, a Three-Tier Model of research was developed, consisting of (a) Insight research involving the study of current practices in existing homes, (b) studies in prototype houses equipped with innovative products and services focused on sustainable living, and (c) field testing, in which research prototypes are up-scaled such that existing homes can be equipped with innovative sustainable technologies (Fig. 5.2). For example based on a common research approach, the partners in the SusLabNWE configured the testing toolkits, lab infrastructure and ICT backbone which contribute to building up the testing locations.

5.2 Research Through Design

Living Labs offer a flexible infrastructure that is well suited for design research. A number of alternative design proposals can be positioned in a design research framework. Research through design focuses on the role of the product prototype as an instrument of design knowledge enquiry. The prototype can evolve in degrees of granularity, from interactive mockups to fully functional prototypes, as a means to formulate, develop and validate design knowledge. The designer-researcher can begin to explore complex product interaction issues in a realistic user context and reflect back on the design process and decisions made based on actual user-interaction with the test prototype. Observations of how the prototype was experienced may be used to guide research through design as an iterative process, helping to evolve the product prototype.

Given the design challenge of developing products that may be complex in terms of function and context of use, research through design can provide a means to increase the external validity of a given design concept. However, though the research product prototype may develop into a highly desirable design through a series of rapid design iterations and validations, it is often difficult for the designer to reverse engineer the resulting final prototype and form a substantial contribution to design knowledge in terms of generalizable results. On the other hand controlled lab studies exploring a certain aspect of interaction, may lead to high internal validity but findings may be difficult to generalize and apply in context when not integrated as part of the total product experience. Empirical Research Through Design as a method builds on the notion of Living Labs in which controlled empirical studies can be integrated into real-world working prototypes as a means to gain deeper insights into interaction issues and to contribute to design science. In this manner data-driven field findings are fed back into the theoretical model of interaction behind the prototype concepts (Keyson and Bruns 2009).

In developing a field prototype it is important not to confound the research variables. For example, if one is trying to study voice versus touch interfaces it is important that the visual interface be the same in both variants. Cumulative designs that vary across multiple dimensions for each test may confound statistical analyses and create difficulties sorting out dependent and independent variables. Furthermore the field prototype should be robust enough so as to function in longitudinal studies without requiring dedicated support and be intuitive to operate.

As detailed in Fig. 5.3 below the primary process of ERDM is based on the following phases:

- Hypothesis forming. This step is so critical, many designers jump into a design process without any working hypothesis.
- Iteration 1 to n
- Design
- Participant Study
- Findings
- Conclusions—feed forward into next Design Iteration
- Final Design
- Final Evaluation

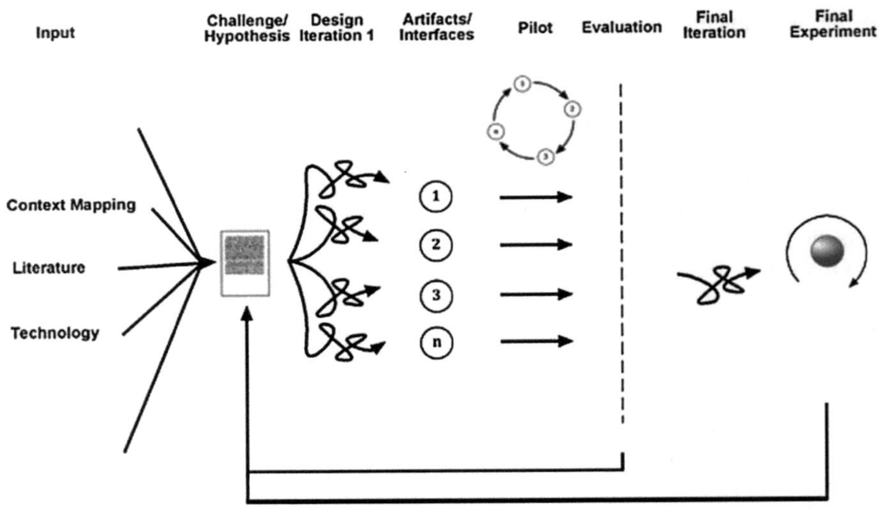

Fig. 5.3 The empirical research through design method process (ERDM). Multiple designs with embedded research variables are rapidly prototyped and piloted, leading to a final experiment involving a limited set of design variants

5.3 Continuous Involvement of User Communities

The challenge in increasing acceptance of technologies which can foster sustainable living, such as energy feedback systems, lie not in the introduction of more and more product and service interfaces, rather in focused studies on usability issues and user needs and aspirations in relation to their daily living routines.

To understand the full scope of user demands and needs in relation to the acceptance and eventual adoption of sustainable technologies, it is important to involve users in the design process as well as in the experience of new technical artefacts. Sleeswijk Visser (2009) argues that the same users should participate at all the different stages of the design process. Such 'returning participants' can more effectively give feedback, as they already have a relatively deep understanding of the application's concepts.

The three-tear approach enables local communities to take part in design driven studies via early insights studies in their homes, experience directly innovations in dedicated research houses, and evaluate the introduction of new products and services in their own dwellings. In this manner localized comparative studies can be conducted in which research houses serve as an incubator and catalyst for increasing the chance of adoption of sustainable innovations in the homes of the users.

References

Eriksson, M., Niitamo, V. P., & Kulkki, S. (2005). State-of-the-art in utilizing Living Labs approach to user-centric ICT innovation—a European approach.

ESDN. (2015). The SDGs and their impact on the European SD governance framework ESDN Quarterly Report No.35. 2015.

Keyson, D. V., & Bruns, M. (2009). Empirical research through design: ISRD, Seoul.

Sleeswijk Visser, F. (2009). *Bringing the everyday life of people into design Doctoral thesis Delft*. Rotterdam: Sleeswijk Visser. ISBN 978-90-9024244-6.

Chapter 6
Splashing: The Iterative Development of a Novel Type of Personal Washing

Lenneke Kuijer

Abstract This chapter describes a case study on personal washing that was developed in association with two subsequent Living Lab projects. Drawing on theories of practice, the case study explored the application of a practices-oriented approach to reducing household resource consumption. Personal washing was taken as a target practice because of its high and growing water and energy consumption. The case study used an iterative process to develop a feasible, but highly less resource intensive alternative to the dominant practice of showering in the Netherlands. Splashing emerged as a promising proto-practice from subsequent performances, both in the lab and the field.

Keywords Practices-oriented design · Personal washing · Splashing · Energy consumption · Proto-practice

6.1 Introduction

This section describes a research and design method that was developed in two subsequent Living Lab projects[1] using a research through design process that had both methodological and empirical aims (Zimmerman et al. 2010). The methodological aim was to explore the implications of taking a practices-oriented approach

[1]The FP7 LivingLab project and the SusLabNWE project related to a PhD research conducted at the Department of Industral Design at Delft University of Technology.

L. Kuijer (✉)
Department of Industrial Design, Eindhoven University of Technology, Eindhoven, The Netherlands
e-mail: s.c.kuijer@tue.nl

D.V. Keyson et al. (eds.), *Living Labs*, DOI 10.1007/978-3-319-33527-8_6

instead of a user-centred one. A practices-oriented approach has been argued to have potential to overcome various limitations of user-centred sustainable design approaches, such as a risk of missing opportunities for change on larger scales (Kuijer and Bakker 2015). The approach builds on the three-tier methodology developed in the LivingLab projects of insight generation, prototyping and field testing (Bakker et al. 2010). The empirical aim, which this section focuses on, was to develop new ideas for achieving the strongly less resource intensive ways of life required to achieve the EU targets for carbon emission reductions.[2] This section describes a case that focuses on the practice of personal washing, which, particularly due to its high and increasing demand for warm water is an increasingly resource intensive target practice. More detailed outcomes of the research can be found in Kuijer (2014).

Current personal washing practices in Western countries are resource intensive and moving into directions that are increasingly so. According to a 2010 study conducted among 1,200 Dutch households, average water use for personal washing per person per day in the Netherlands is 51.4 l, of which 48.6 l is used for showering (Foekema and Thiel 2011). An average shower lasts 8 min and requires 62 l of water. Water use for personal washing—including energy use to warm the water[3]—has increased strongly over the past century. Average weekly amounts increased from approximately 105 l[4] a 100 years ago, which were presumably not or marginally heated, to an average of 340 l of warm water per week today. To place this number in perspective, the United Nations use a recommended minimum of 15 l per day (105 l per week) for personal washing to achieve acceptable standards of living.

The now widely available shower timers, which in all kinds of ways aim to motivate people to take shorter showers, are having some effect in combatting high water consumption, but they work only in certain situations and tend to lead to small reductions, for short periods of time (Kuijer and Bakker 2015). This is in part because showering is, for many, a very pleasant activity that besides getting clean, offers functions like caring for one's body, waking up, relaxing and getting warm or refreshed. Moreover, with some exceptions in the form of water saving shower heads, current developments in bathroom design do not offer much prospect of improvement; power showers, massage showers and rain showers discharge ever growing flows of water while shower frequency and duration are increasing (Foekema and Thiel 2011; Kuijer et al. 2010). To address these problems, the *splash* project developed a strongly less resource intensive, pleasurable way of washing called "splashing" through an iterative design process.

[2]http://ec.europa.eu/energy/en/topics/energy-efficiency.

[3]It has to be noted that the environmental impact of a shower derives primarily from the energy required to heat the water, but in this study the amount of water is used as a proxy for the combined resource use of water and energy to heat the water.

[4]This is an estimation assuming a daily flannel wash using the 5 l 'lampet' (jug and basin) and a weekly bath of 50 l.

The methodology used in this project draws on practice theory, a group of theories from sociology that takes socially shared practices—such as bathing, cooking or commuting—as the fundamental unit of analysis and design. Aim of the methodology is to reconfigure everyday practices like bathing into less resource intensive directions. Inspired by practice theory, the method centralizes bodily performances—'doings with things', in the lab and in the field, as the primary place where the design of such reconfigurations happens.

The starting point for the iterative design process described in this section was the idea of washing from a bucket. The idea of a bucket wash resulted from an elaborate analysis of Dutch bathing practices, including analysis of its history and forms of bathing in other cultures. Aim of the design process was to develop this idea into a pleasurable, less resource intensive alternative for showering. This section briefly describes the identification of the bucket wash as a promising concept. It then moves on to the three main design cycles through which this rough idea was fleshed out into the so called 'proto-practice' of splashing and closes by briefly reflecting on the concept of splashing and its potential and challenges.

6.2 The 'Bucket Wash'

Important for understanding the splashing concept and the origins of the idea of the bucket wash are the practice theory concepts of practices-as-performance and practices-as-entity.

6.2.1 Practice-as-Performance and Practice-as-Entity

In the strands of practice theory that have so far been picked up most in design research,[5] two 'ideal-types' of practice are distinguished. One is the practice-as-performance, which is basically all human action, and can be observed directly. Shove et al. (2012) view performances as an integration of elements, which they group into three broad categories: meanings, competences and materials. In a particular instance of showering, water, soap, shampoo, shower nozzles, a human body—with its sebum, hair and dirt, combines with skills of removing dirt and certain ideas of what it means to properly clean and take care of the body, and to enjoy a 'good' shower. The practice-as-performance varies in each situation, because different elements are available and different configurations make sense. For example, a shower at the gym is different from a shower at home, a shower to

[5]Notably the work of Schatzki (1996), Schatzki et al. (2001), Reckwitz (2002a, b) and Shove et al. (Ingram et al. 2007; Shove et al. 2007, 2012; Shove and Watson 2006).

wake-up is different from a shower to wind down, and my shower last night was slightly different from the one the night before.

Although more or less different from each other, all these performances are forms of showering. Viewed as such, showering, or more broadly, personal washing, exists as an entity that can be traced over space and time beyond individual performances or performers. Importantly, there is a recursive relation between individual performances of showering, in which water and energy are consumed, and the practice of personal washing as an entity. To understand why personal washing, in many European countries today, takes the form of daily showering, and subsequently, what may be less resource intensive ways of personal washing, it is helpful to trace the practice-as-entity back in time and to study variety in its occurrence over space. Subsequently, it requires understanding of how this practice is configured in relation to the amount of resources it currently requires. This is what the case study set out to do initially.

6.2.2 Analysing Showering

The search for the origins of showering and examples of less resource intensive ways of personal washing, which corresponded with the 'insight generation' tier of the LivingLab methodology, included a variety of studies. Most information about the history of personal washing was found in literature. In the Netherlands, showering has only fairly recently replaced the 'lampet'—a set of a jug and bowl used in the bedroom—for daily washing, and the bathtub for the weekly full body washing. The spread of the shower went hand in hand with the connection of households to mains water, which happed in The Netherlands between 1853 and 1970 (Moel et al. 2006) and natural gas in the 1960s and 70s (Correlje and Verbong 2004). While showering is experienced as relaxing and pleasurable today, this has not always been the case. Kira, a prominent ergonomics researcher who has made detailed analysis of bathrooms and bathing postures explains how in the 1960s, when his book was first published, the shower was experienced as 'spartan, distinctly masculine, business-like, and even uncontrollable, destructive and rough' (Kira 1976, p. 37). Similarly, Hand et al. (2005) find that in the UK, showering has only become a serious alternative to the bath since the 1980s. Personal washing is a practice that has not always been promoted by the authorities: Stuller (1991) reports instances in the Middle Ages where 'over bathing', meaning more than once a month, was punished by law. However, the connections made between hygiene and disease in the 19th century, and the increasing standards of bodily cleanliness to rise on the social ladder (Bushman and Bushman 1988) have made daily washing a subject of concern for national policy. This is reflected, for example, in the once highly influential household manuals and contemporary educational programmes, which have contributed to the spread and uptake of (daily) showering.

Supplementing this historic analysis, a detailed cultural inquiry about bathing in India, Japan and The Netherlands (Matsuhashi et al. 2009) revealed insights about the details of alternative, less resource intensive forms of personal washing. Blogger Chris Chopp explains that in India the common way of bathing is a bucket bath, entailing a bucket containing around 19 l of water and a mug to pour the water over your body: "Begin by pouring one or more mugs of water over your body from the head down until the entire body is wet. Then apply shampoo and rinse, making sure the soapy water does not enter the bucket but ends up on the floor' (Chopp 2012). From analysis of these different forms of bathing, a clear relation between postures and the use of tools was detected. For example, most of the actions using water from a reservoir were done when participants were sitting on a stool".

Diving into the history of personal washing and analysing washing in other cultures taught the design team to see showering from a different perspective. Although near-to-daily showering is the norm in Western Europe, it has become so only during the past 50 years and is not so common in other cultures.

After assessing the resource requirements of different ways of bathing, a concept that is here referred to as 'bucket wash' emerged as a promising direction. The 'bucket wash' represents a way of bathing in which water for washing is contained in a bucket sized container. It is clear that the bucket wash is not immediately appealing to people accustomed to showering. However, early experiments by 'showerers' improvising a bucket wash showed that aspects of it were experienced as rewarding, effective and relaxing (Kuijer and de Jong 2009; Scott et al. 2009). These results, combined with insights into bathing in Dutch history, India and Japan indicated that a shift away from showering, with its constantly flow of warm water going down the drain still warm and practically clean, is possible. Starting from the rough idea of a bucket wash, the aim of the design process was to iteratively develop ways of personal washing that had potential to work in the Dutch context. The following sections describe its three cycles, which combine the prototyping and field study tiers of the LivingLab methodology.

6.3 From Bucket Wash to Splashing

In an early design exploration of the bucket wash idea, in a master graduation project, a student had renamed it 'splash'. Reason for this renaming was that it was the splashing of water involved in the bucket bath that early experimenters with the concept particularly mentioned that they liked (Karakat 2009). From these experiments, it also became clear that Dutch mainstream bathroom design did not facilitate pleasurable splashing: new materials were required to make splashing work. Cycle 1 set out to develop these in a highly participatory manner.

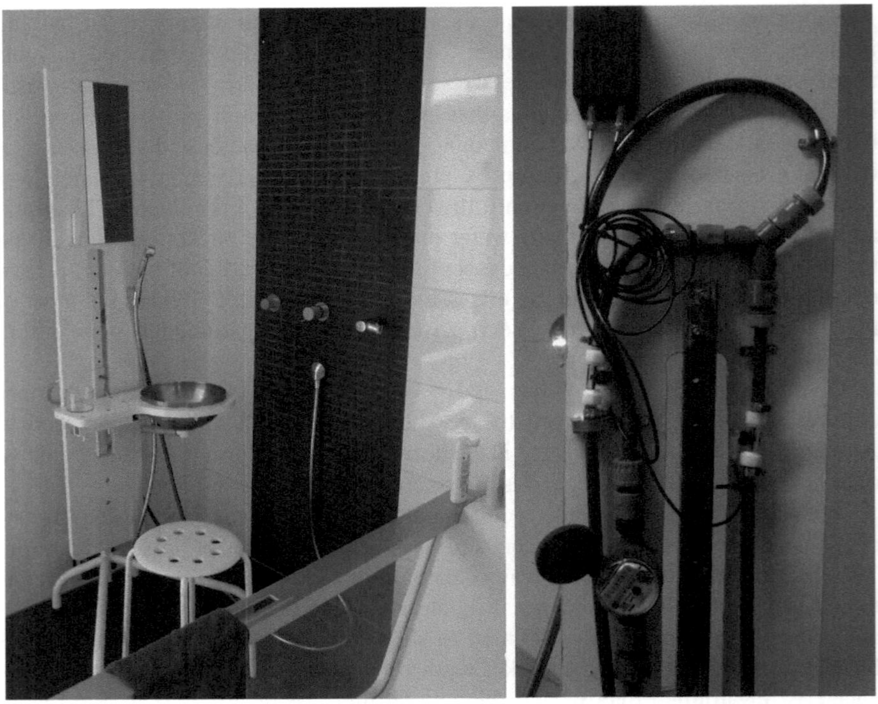

Fig. 6.1 The splash prototype installed in a shower during a field trial (*left*) and the back of the prototype with water meter and logger device (*right*)

6.3.1 Cycle 1: Generative Improv Performances

The goal of this first cycle was to flesh out the rough idea of washing from a bucket into a variety of more detailed ways of washing. The set-up of the study was designed to explicitly 'break with the rigidity of existing structures in existing household situations and to find participants with the required creativity, courage, and ability to envision something strongly different' (Kuijer et al. 2013). The tailored method developed for this aim, named 'Generative Improv Performances' (GIP) can be described in terms of a stage, a scene and its players. For this particular study, the stage involved a simulated bathroom containing rough, open prototypes of a basin on a pedestal and a movable seat, and additional objects such as cups, sponges, soaps bottles and towels. The 'stage' was deliberately not in a bathroom to create a situation where people could 'act out of the normal' (ibid). The scene was inspired by the type of instructions used in improv performances, which are open assignments with which the actors engage creatively. Participants got the assignment to 'perform splashing as if it is your normal way of washing with which you are satisfied' and were not allowed to include the use of continuously flowing water. The 'players' were trained improv actors, with the idea that they are skilled to improvise in a challenging situation and to act outside of what is currently considered normal (Vera and Crossan 2004).

The study resulted in 25 varied performances of splashing, which, next to ideas for new bathing artefacts included new skills and meanings, such as skills of pouring and sponging and meanings of enjoying deliberate and elaborate soaping of the body. The performances were analysed in detail and summarized in graphical overviews using a pictogram library. From these performances, a list of dimensions of variety on which the performances differed from each other was composed. These included for example a variety of ways of wetting, soaping and rinsing, different postures taken and different likes and dislikes, which formed important input for the next design cycle. Through the study, splashing was fleshed out and could now be described as

> an active, flexible way of washing the body with water from a basin, involving sitting and standing postures, a range of ways of applying water and soap, involving scoops, sponges and hands, in varying sequences. Rather than rinsing with constantly flowing water, soap plays a central role in cleaning the body. Splashing can be quick and functional, washing selective parts of the body, but also a relaxing, time taking ritual with a focus on scent and deliberate body care (Kuijer et al. 2013).

This fleshed out idea of a variety of ways of splashing and the dimensions of variety were used to develop a refined concept of splashing, including objects that allow for a wide variety of types of splashing to be performed. The resulting prototype, developed and made by master graduation student Knupfer (2011) was used in a series of field studies.

6.3.2 Cycle 2: Evaluative Field Studies

The goal of the field studies was to evaluate whether the refined concept of splashing developed in response to the results of the GIP study worked in an everyday life setting and had the expected reduced level of water and energy use.

Looking in particular at the dimensions of variety, the new design allowed for sitting and standing postures, and contained a central, 2-litre basin and a hand held shower. This splash fixture, designed for easy installation in existing shower cubicles came with a scoop and a sponge. Importantly, its deployment in the field was accompanied by a set of use suggestions and explanations of what others enjoyed about splashing, which were also based on the GIP study. Referring back to the practices as a constellation of elements, this set of objects, use suggestions and interpretations formed the proto-practice of splashing. To evaluate levels of water (and energy) use involved in splashing, the prototype was equipped with a water meter and in later studies with a water logger that recorded water use for the basin and hand shower separately (Fig. 6.1).

In order to evaluate the integration of splashing into everyday life settings and to study its effects over the course of repeated performances, participants were recruited who were willing to use the splash prototype in their own homes, instead of their shower for period of 1 week up to 1 month. Participants were interviewed about their existing routines for personal washing, asked to keep a diary of their splashing experiences and interviewed again after the period in which they trialled splashing.

Five households of varied composition took part, with a total of eleven participants performing splashing at least once, and up to fifteen times. Four out of these eleven participants preferred splashing over showering, three mostly enjoyed it, but preferred showering because it is warmer, and four said it just didn't work for them for various reasons, which mainly came down it being too much of a hassle, or too cold. For those enjoying splashing, water use per splash varied between 10 and 20 l of warm water.

Again, splashing was fleshed out further through performance. Bas, for example, who very much enjoyed splashing, developed a fixed procedure starting with filling the basin, wetting his body 'from hair to toes' with one sponge and then soaping his body with another. In this process he washed his hair in one go with his body. Then he got fresh water and rinsed his hair and body with the scoop. Altogether, his 'splash' would take no more than 5 min. In terms of postures, he first sits down, but when washing his lower body and while rinsing he stands up. Bas liked splashing for its speed, but also because he thought it felt really good. To quote him 'you clearly use less water than in a shower, but it feels like it is more'. Similar to Bas, Astrid really enjoyed splashing. Her routine, however, was quite different, as were her reasons for liking it. Astrid started with her toes and worked up to her head. This way she prevented getting cold. For her, washing her body and washing her hair were separate routines. When washing her body, she used only the basin, while for washing her hair, she would place the hand-shower in the holder and use that for wetting and rinsing it. While Bas really liked splashing for its speed, Astrid was so enthusiastic about it because it allowed her to really take her time and wash herself deliberately.

Through these performances, splashing was starting to take shape as something that could work and be pleasurable for people in a variety of ways. While viewing splashing as too much of a hassle can easily be attributed to the fact that some of the participants were not open enough to trying something new, an important remaining challenge for splashing was that people felt cold in their performances. The final cycle that formed part of the splash project therefore focused on integrating a low energy form of body heating in the prototype. It was therefore more technology focused than earlier cycles.

6.3.3 Cycle 3: Technical Refinement

In this third iteration, results of the field studies, together with other data collected on bathing and splashing formed the basis for a redesign of splashing. A master graduation project was executed as part of the SusLabNWE project. The project was focused on developing a design and working prototype of a splash fixture including integrated heating (Henny 2013). After exploring several possibilities,

Fig. 6.2 Refined splash fixture model with outer band of wall mounted fixture functioning as a radiator capable of warming up the user in a closed shower cubicle within minutes (image by Henny 2013)

the student choose a radiator system, which effectively extends the hot water supply of the splash into a tube radiator integrated into the wall mounted vertical element of the appliance (Fig. 6.2.). Domestic hot water supply to showers is at least 65 °C due to legionella regulations. This hot water is first led through the radiator tube to heat the person and shower cubicle and then used for washing, which requires water of around 38 °C.

The prototype of the design built by Henny was tested both for technical performance of the heater and for use experience through one-time uses in a shower facility on campus. The technical tests, using a thermal imaging camera, showed that the radiator heats up to its maximum capacity of around 900 W/m^2 (appr. 70 °C) in less than 25 s. After 13 min, the temperature of the 4.1 l of water in the radiator was still above 40 °C, which is warm enough to bathe with. While mainly based on radiation (meant to warm the body of the user directly), the heater also warmed up the space. After about 10 min, the temperature in the relatively large space the prototype was installed in rose 1.6 °C.

In the lab-based user test with 11 participants, an average of 18 l of water was used, ranging from 7 to 27 l. Again, a variety of postures and procedures was identified. Two of the participants used only the hand shower. All others used both the basin and the hand shower. While getting chilly or cold was still an issue for some participants, others felt comfortable or even comfortably warm. To shape and evaluate the integration of the heater in the proto-practice, further tests are required.

In the current tests, participants were for example not informed that there was a heater in the product. Explaining about the radiator and the way it works may affect their thermal experience, just like Rohles (1980) showed that merely changing the interior design of a room without adjusting its temperature could make people feel warmer. Moreover, the testing space was relatively large for a shower cubicle. Longer term tests with the prototype in a setting such as the SusLab Concept House were envisioned as a next step where technical performance, usability and use experience could be tested and tweaked over longer periods of time with a variety of users.

6.4 Conclusions

Showering is by far the most popular form of personal washing in the Netherlands. It is also highly resource intensive. A main culprit for this resource intensity is the paradigm of continuously flowing water. This observation, in combination with the identification of other forms of washing with lower resource intensities that are based on relatively small reservoirs of contained water, led to the selection of the bucket wash as a promising direction for less resource intensive personal washing.

From this rough idea that emerged from insight generation studies, splashing was developed through three iterative cycles of prototyping and field studies into a fleshed out, coherent set of objects (including a height adjustable basin, a seat, a scoop, a hand shower, an integrated radiator, sponges and soaps), actions (such as scooping, pouring, rinsing, sitting and splashing) and meanings (including deliberate body care, control over what is washed when, taking time, and flexibility). Together, they form a way of washing that has potential to work as an attractive alternative for showering with strongly lower resource requirements: an average of 62 l versus 20 l of warm water.

While this all sounds promising, it has to be noted that reaching large scale reductions in water and energy consumption would require widespread uptake of splashing. Judging from the speed with which the shower has overtaken the bath, this is not entirely impossible. When reaching a certain momentum, splashing might be able to reposition showering and thus catalyse the shift. In such a view, showering requires large amounts of water, can be tiring because you have to stand all the time making it difficult to wash your feet, is inflexible because you have to wet your entire body without being able to direct what is wet or rinsed when, and soap is rinsed off before you've had time to appreciate it (Kuijer et al. 2013).

However, so far splashing remains a concept that has only been evaluated on a relatively small scale. It is therefore difficult to predict the larger scale, longer term effects of its introduction. The splash studies show for example that splashing, being a more flexible form of washing may increase frequencies of washing. Future development of splashing should explore the effects of the heater and longer-term effects of splashing on washing frequencies and water requirements, while at the same time making available its specific bathroom fixtures, skills and images of the pleasures of bathing to enable it to spread into society. A living lab infrastructure with facilities such as the Concept House and larger scale field trials could form an ideal incubation space to further facilitate this process.

In terms of methodology, the practice-oriented approach taken in the splash case led to the addition of historic and cross cultural exploration to the insight generation tier. Moreover, in the prototyping tier it explicitly took practices as a unit of design, developing proto-practices rather than merely product prototypes. Finally, by viewing performances as the creative integration of elements into new practices, performances of splashing, both in the lab and the field, were treated as the places where these proto-practices were designed.

Acknowledgments I would like to thank all those innumerable advisors, students and participants that have been involved in the splash projects, Delft University of Technology for hosting it and the FP7 and NWE frameworks for funding it.

References

Bakker, C., van Eijk, D., Silvester, S., Reitenbach, M., de Jong, A., & Keyson, D. (2010). *Understanding and modelling user behaviour in relation to sustainable innovations: The LIVING LAB method.* Ancona, Italy: Paper presented at the TMCE.

Bushman, R., & Bushman, C. (1988). The early history of cleanliness in America. *Journal of American History, 74*(4), 1213–1238.

Chopp, C. (2012). A bucket a day, keeps the dirt away: A guide to bucket baths. *Full Stop India, Tips* and *Advise From a Western Traveler,* 2012. http://www.fullstopindia.com/a-bucket-a-day-keeps-the-dirt-away-a-guide-to-bucket-baths.

Correlje, A., & Verbong, G. (2004). The transition from coal to gas: radical change of the Dutch gas system. *System innovation and the transition to sustainability: theory, evidence and policy* (pp. 114–134). Cheltenham: Edward Elgar.

Foekema, H., & van Thiel, L. (2011). *Watergebruik thuis 2010.* Amsterdam: TNS NIPO, in opdracht van VEWIN.

Henny, F. (2013). *Implementation of a new bathing practice.* (Master). Delft: Delft University of Technology.

Hand, M., Shove, E., & Southerton, D. (2005). Explaining showering: A discussion of the material, conventional, and temporal dimensions of practice. *Sociological Research Online, 10*(2).

Ingram, J., Shove, E., & Watson, M. (2007). Products and practices: Selected concepts from science and technology studies and from social theories of consumption and practice. *Design Issues, 23*(2), 3–16.

Karakat, H. (2009). *Designing for alternative (sustainable) bathing practices.* (MSc). Delft: Delft University of Technology. Retrieved from http://repository.tudelft.nl/view/ir/uuid:bda7a61d-2a45-4bd7-9979-25fc611ce55d/.

Kira, A. (1976). *The bathroom.* New York: Viking press.

Knupfer, L. (2011). *An alternative bathing practice for sustainable renovation.* (MSc thesis). Delft: Delft University of Technology. Retrieved from http://repository.tudelft.nl/view/ir/uuid%3Ab9db7836-66b2-4745-bdab-c3aef0f1a2dd/.

Kuijer, L. (2014). Implications of Social Practice Theory for Sustainable Design.

Kuijer, L., & Bakker, C. (2015). Of chalk and cheese: behaviour change and practice theory in sustainable design. *International Journal of Sustainable Engineering* (ahead-of-print), 1–12.

Kuijer, L., de Jong, A., & Van Eijk, D. (2013). Practices as a unit of design: An exploration of theoretical guidelines in a study on bathing. *Transactions on Computer-Human Interaction, 20*(4), 22.

Kuijer, L., & de Jong, A. (2009). A practice oriented approach to user centred sustainable design. In *Paper presented at the Sixth International Symposium on Environmentally Conscious Design and Inverse Manufacturing Sapporo, Japan*.

Kuijer, L., McHardy, J., & Scott, K. (2010). The challenge of the bucket wash: creating desirable sustainable practices. In *Paper presented at the 7th International Conference on Design and Emotion, Chicago*.

Matsuhashi, N., Kuijer, L., & de Jong, A. (2009). A culture-inspired approach to gaining insights for designing sustainable practices. In *Paper presented at the EcoDesign 2009: Sixth International Symposium on Environmentally Conscious Design and Inverse Manufacturing Sapporo, Japan*.

Moel, P. J., Verberk, J. Q., & Van Dijk, J. (2006). *Drinking water: principles and practices*. Singapore: World Scientific.

Reckwitz, A. (2002a). The status of the "material" in theories of culture: From "social structure" to "artefacts". *Journal for the Theory of Social B, 32*(2), 195–217.

Reckwitz, A. (2002b). Toward a theory of social practices: A development in culturalist theorizing. *European Journal of Social Theory, 5*(2), 243–263. doi:10.1177/13684310222225432.

Rohles, H. (1980). Temperature or temperament: A psychologist looks at thermal comfort. *ASHRAE Transactions, 86*(1), 5–13.

Schatzki, T. (1996). *Social practices: A wittgensteinian aproach to human activity and the social*. Cambridge: Cambridge University Press.

Schatzki, T., Cetina, C. K., & von Savigny, E. (Eds.). (2001). *The practice turn in contemporary theory*. London and New York: Routledge.

Scott, K., Quist, J., & Bakker, C. (2009). *Co-design, social practices and sustainable innovation: involving users in a living lab exploratory study on bathing*. Aalborg, Denmark: Paper presented at the joint actions on climate change conference.

Shove, E., Pantzar, M., & Watson, M. (2012). *The dynamics of social practice: Everyday life and how it changes*. London: Sage.

Shove, E., & Watson, M. (2006). Manifesto of practice oriented product design. http://www.lancaster.ac.uk/fass/projects/dnc/papers.html. Retrieved 06 Aug 2008.

Shove, E., Watson, M., Hand, M., & Ingram, J. (2007). *The design of everyday life*. Berg.

Stuller, J. (1991). Cleanliness has only recently become a virtue + the history of bathing and hygiene. *Smithsonian, 21*(11), 126–&.

Vera, D., & Crossan, M. (2004). Theatrical improvisation: Lessons for organizations. *Organization Studies, 25*(5), 727–749.

Zimmerman, J., Stolterman, E., & Forlizzi, J. (2010). An analysis and critique of research through design: Towards a formalization of a research approach. In *Paper presented at the DIS2010*, Aarhus Denmark.

Chapter 7
Design with Intent and the Field of Design for Sustainable Behaviour

Dan Lockton

Abstract Design for sustainable behaviour necessarily involves a multidisciplinary perspective, drawing on insights around human action from multiple fields, and making them relevant to designers. This chapter explores some considerations which build on these multidisciplinary concepts, around questioning assumptions and understanding people's lives better, and introduces the Design with Intent toolkit, a design pattern collection which aims to facilitate reflective exploration of problem-solution spaces in 'behaviour change' contexts, with a brief exploratory example of its application to provoke discussion with householders as part of SusLabNWE.

Keywords Design patterns · Sustainable behaviour · Behaviour change

7.1 Introduction

Both design and sustainability are about the future—bringing into being a world where humanity and other forms of life will "flourish on the planet forever" (Ehrenfeld 2008) or where we can "go about our daily affairs… [knowing] that our activities as civilised beings are expanding our future options and improving our current situation" (Sterling 2005). Design might be one of the mechanisms by which much of our current predicament has come about (Papanek 1971), but perhaps "the future with a future for 'us' can only be reached by design" (Fry 2015).

Thus, design and sustainability are deeply enmeshed (Schmidt-Bleek and Tischner 1993)—as we see throughout this book and the variety of examples within the SusLabNWE project. A major component of this is design which relates

D. Lockton (✉)
Royal College of Art, London, England
e-mail: dan@danlockton.co.uk

D.V. Keyson et al. (eds.), *Living Labs*, DOI 10.1007/978-3-319-33527-8_7

to people's actions: what has become known as design for behaviour change, behavioural design, or in the case of specific focus on sustainability, *design for sustainable behaviour* (e.g. Lilley 2009; Wever 2012; Daae and Boks 2014; Strömberg et al. 2015). In this section, we will examine some of the issues and dimensions of the topic, and introduce the *Design with Intent* toolkit (Lockton et al. 2010), a design pattern collection for environmental and social behaviour change, which aims to enable exploration of assumptions and problem-solution spaces in 'behaviour change' contexts, which was used in an exploratory way as part of householder interviews around energy use and heating.

7.2 Design and Behaviour

Behaviour change is currently fashionable in design—commercially, politically, and academically—addressing everything from fitness tracking to compliance with tax return procedures to getting passengers to board trains more efficiently. There is an increased policy focus in a number of countries on applying principles from behavioural economics to complex social and sociotechnical issues through initiatives such as the UK's quasi-privatised Behavioural Insights Team, stemming from the popular *Nudge* (Thaler and Sunstein 2008), while advertising agencies—no strangers historically to applying psychology to influence behaviour (Bernays 1928; Packard 1957; Schwartz 1966)—have embraced this burgeoning interest, for example through Ogilvy and Mather's creation of the specialist 'behavioural practice' #ogilvychange [sic]. Others such as Payne (2012) have taken a wider perspective, looking at how to 'inspire' more sustainable behaviour.

Nevertheless, this approach is usually centred on quite small changes in current behaviour, with short time-frames, rather than long-term futures. In this context, different understandings and definitions of 'sustainability', and degrees of ambition for change, complicate the notion of design for sustainable behaviour. What kind of behaviour? Whose behaviour? What kind of sustainability? Is the intended change in behaviour a reduction in some unsustainable behaviour, or a shift to something very different? As Ehrenfeld (2008, p. 20) puts it, "Reducing unsustainability will not create sustainability"; the degree of intervention varies with the boundary of how the 'problem' is considered, whether it is at the level of individual interaction with products, or part of a more systemic societal transition (Irwin et al. 2015). Large-scale changes in human behaviour are central to many visions of more sustainable futures, often going hand-in-hand with scientific and technological advances.

In many cases, technological advances would require people to change the way they act for the potential environmental impact to be achieved, whether through adoption of new ways of doing things at home and at work, decisions about capital investment or purchases, and fundamental changes in assumptions, attitudes and political norms. But design is also often about how broader social influences affect what people do, and how people's relations, and actions individually and together help construct what society, community and culture are.

The sociological concept of practices (Reckwitz 2002) as units of analysis (rather than behaviour) can offer a useful way of exploring issues for designers. Practices are "the mundane activities that make up most of what people do in their daily lives, such as bathing, cooking, laundering and cleaning… socially shared entities with a certain persistence over time and space," as Kuijer et al. (2013) describe them. By considering people's actions in larger, 'supraindividual' socially and culturally situated contexts, designers can potentially develop a better understanding of design's role in much larger—potentially systemic—change. This is an approach taken in a number of studies within the SusLabNWE project.

By enabling closer co-creation with people ('users') themselves in contexts approaching the complexity of real life, the Living Lab approach, as discussed throughout this book in relation to SusLabNWE, enables sustainability transitions to be explored in a variety of different ways, but all through "living change" (Scott et al. 2011) rather than designing in the abstract.

7.3 Multidisciplinarity and Complexity

In academia, work on design for sustainable behaviour has grown significantly in recent years, with the aim of reducing the unintended environmental and social impacts of products and services in use—or increasing the intended impacts—through design which concentrates on *understanding* and *influencing* user behaviour. Design 'interventions' largely involve redesign of products and services, changing the affordances and constraints available to users, or the design of interfaces (often digital) which give users information—and feedback—on use or the impacts of behaviour, for example energy use, waste generation or transportation choices (e.g. Lockton et al. 2008; Wever et al. 2008). Negotiating the large field of possible design techniques can be a challenge for designers briefed with 'changing behaviour', and so a number of toolkits and guides have been developed (e.g. Selvefors et al. 2014; Daae and Boks 2014) which aim to provide designers with a more structured process.

Design for sustainable behaviour is inherently multidisciplinary (Niedderer et al. 2014), drawing on knowledge and models from other fields relating to human action. These include social and cognitive psychology, behavioural economics, human-computer interaction, ethnography, science and technology studies, ergonomics, cybernetics, ethics, and architecture, as well as other facets of design for behaviour change, such as *social implication design* (Tromp et al. 2011), *persuasive technology* (Fogg 2009), *practice-oriented design* (Kuijer et al. 2013; Scott et al. 2011; Pettersen 2015), *product experience* (Desmet and Hekkert 2007) and *transformational products* (Laschke et al. 2011). Many fields, in both research and practice, both within and without what are termed the 'behavioural sciences', have insights or frameworks to contribute, and each works with particular *models* seeking to explain human behaviour in different ways—even if those models are sometimes mutually incompatible (Gintis 2007). Being aware of, and attentive to, the models we are using as designers, is an important part of a reflective approach.

7.3.1 Questioning Assumptions

One valuable contribution that a multidisciplinary approach can make is to enable multiple perspectives on situations, paying attention to issues which designers might otherwise not consider, or might not be afforded the luxury of considering due to the way that briefs are framed. At the very least, a multidisciplinary approach can trigger us to question our assumptions and reflect on the models we are using. Because—much as many designers might like to discover a 'formula' for behaviour change—the complexity and interconnectedness of real-life behaviour and practices is deeply enmeshed with social and cultural contexts, power structures, and other people's actions, and more nuanced than any singular vision can ever capture, which highlights the naïvety of very determinist stances (Lockton 2012), appealing as they might be to a "planning" mindset (Watson et al. 2015). We should question the assumptions embedded in work that presumes one-to-one mappings between design features and resulting 'behaviours' (Broady 1966). People will not always behave how designers intend or expect them to (Kanis 1998; Stanton and Baber 2002); even as designers attempt to 'script' behaviour (Akrich 1992; Jelsma and Knot 2002). As Brand (1994) puts it, in reference to the built environment, "All buildings are predictions. All predictions are wrong". Assumptions about people—how they live, how they make decisions, and what affects their actions—are integral to design, while designers are engaged generally in "act[ing] to change the actuality of the world" (Dilnot 2015).

Although these assumptions and issues around them are not necessarily always explicit stances taken by designers or researchers, they embody tensions that arise when a new approach touches on areas that have previously been the preserve of other disciplines with different traditions, expectations and aims. We cannot avoid having models of people (Dubberly and Pangaro 2007) but the question of how these models and assumptions are applied in design is of practical relevance—how those models can be translated, tested, questioned and improved through use in the real world, rather than in laboratory studies (here the Living Lab approach offers a valuable intermediate step).

Looking at the differences between how designers themselves model 'their' intended users in relation to sustainability or other issues (Lockton et al. 2012), and how users themselves understand and think about the situation can be instructive here in understanding how design for sustainable behaviour techniques are applied in practice, and how they could be improved. This suggests the need for a structured way of exploring the assumptions and implications inherent in design which seeks to influence behaviour, both to negotiate the large field of possible design techniques from different disciplinary backgrounds and traditions—and their appropriateness for different situations—and to enable a more reflective design approach. The ability to question and reframe the assumptions inherent in a brief, as part of a *problem-framing* (Dorst 2015) or even *problem-worrying* (Anderson 1966) approach potentially requires the designer to have a much greater awareness of the problem-solution space (Maher et al. 1996), including both deeper contextual enquiry, through researching the situation in the field, a knowledge of the repertoire of design approaches which might be applicable (Lawson 2004), and deeper knowledge of sustainability in context (Liedtke et al. 2013).

7.3.2 Understanding People's Lives

Taking a subject central to SusLabNWE—domestic energy use behaviour change—Strengers (2011) considers it "alarming" that the model of individual householders as "micro-resource managers", and the language of 'demand management', continue to dominate the design of feedback systems. Brynjarsdóttir et al. (2012) describe persuasive design for sustainability as "a modernist enterprise", focusing both on individuals at the expense of broader social considerations, and on narrowing the broad scope of sustainability into "the more manageable problem of 'resource minimisation'," drawing on Scott's (1999) conception of how states have attempted to make populations 'legible' through reducing their variety (of behaviour as of other characteristics). In other areas of design for behaviour change, Fantini van Ditmar and Lockton (2016) explore the ways in which simplistic models of motivation underlie much of the quantified self technology arising from Silicon Valley, while Whitson (2014) draws parallels between the increasing use of quantified 'gamification' in design for behaviour change (employing game elements, such as earning points, in non-game contexts) and forms of governance and normalisation drawing on Foucault (1977).

The criticism links well with approaches highlighting the potential value of considering social practices (Wilhite 2013; Shove et al. 2012) in this area, rather than 'behaviour'—specifically because social practice theory's emphasis on shared activities and ways of meeting daily needs can "lift understandings of resource consumption to [a] supraindividual level" (Kuijer et al. 2013). Scott et al. (2011) call for "a more comprehensive understanding of 'users' as social creatures, and the role of consumption in everyday life, than has ever been undertaken through design".

What a lot of these issues perhaps come down to, is something around *inclusion*: to what extent are real people, in real contexts, included in design processes around sustainability (Wiek et al. 2014)? Many design for sustainable behaviour interventions—often arising from work in human-computer interaction (HCI)—are not necessarily designed inclusively, in the sense of considering all users' needs and abilities, including older people, people with disabilities, and even people on lower incomes (Eikhaug and Gheerawo 2010). As Langdon and Thimbleby

(2010) argue, HCI can (and must) learn from inclusive design research, and embrace opportunities to involve a much wider range of users in the development of new interfaces, and the same applies in behaviour change. One prescription might thus be for more participatory processes, including co-design, hackathons and participatory prototyping.

7.4 The Design with Intent Toolkit

The *Design with Intent toolkit* (Lockton et al. 2010) aims to help designers and other stakeholders explore the space of behaviourally relevant design concepts, through presenting examples and insights from different disciplines using a *design pattern* format (Fig. 7.1). This could lead to idea generation, through use as a 'suggestion tool' to help a form of directed brainstorming, or serve as an exploratory, reflective or teaching tool. The toolkit was developed via an iterative, participatory process, running workshops with students and designers throughout its development (Fig. 7.2) to understand how it is being used and how to improve its structure and content. The patterns were extracted—and abstracted—from a literature review of treatments of human behaviour in a range of disciplines. While the toolkit has been applied in sustainability contexts (Lockton et al. 2013a), it has also been developed as a tool for interaction designers more widely (Lockton 2017).

Fig. 7.1 Selection of Design with Intent toolkit cards

Fig. 7.2 A workshop using the cards at Philips Research, Eindhoven

In the toolkit, 101 design patterns for influencing behaviour are described and illustrated (Fig. 7.2), grouped into eight 'lenses'—categories which provide different disciplinary 'worldviews' on behaviour change, challenging designers to think outside the immediate frame of reference suggested by the brief (or the client), and helping with transposing ideas between domains. The lenses (described in Table 7.1) are not intended to be ontologically rigorous, but primarily a way of triggering multiple viewpoints within an ideation session. The patterns are essentially recurring problem-solution instances, described in a way which can be easily referenced, to enable practitioners to recognise the situation. The pattern form can help a designer recognise that a 'new' problem situation is similar or analogous to one encountered previously elsewhere, even in a different context. This makes them a useful format for cross-disciplinary transfer. However, the classification is not perfect: there are many ways to view certain concepts, depending on disciplinary perspective. The intention is that all apply to multiple fields; examples from (for instance) software, can often be translated into the physical world, and vice versa.

7.4.1 Example: Applying the Toolkit to Explore Householders' Perspectives

Within the SusLabNWE project, a preliminary exploration was made of how a subset of the Design with Intent patterns could be used for research with householders, about how they saw possible sustainable behaviour-related interventions fitting into their daily lives. We worked with four households in Dartford, Kent, in south-east England.

For each of the eight lenses, one pattern was applied to an energy-related issue within the home, to generate a plausible concept (or adapt an existing one) and a simple 'provocation' card created (Fig. 7.3). The energy issues centred on heating

Table 7.1 The Design with Intent toolkit lenses and patterns

Lenses	Patterns
Architectural The Architectural lens draws on techniques used to influence user behaviour in architecture, urban planning and related disciplines such as traffic management and crime prevention through environmental design	Angles; converging and diverging; conveyor belts; feature deletion; hiding things; material properties; mazes; pave the cowpaths; positioning; roadblock; segmentation and spacing; simplicity
Errorproofing The Errorproofing lens represents a worldview treating deviations from the target behaviour as 'errors' which design can help avoid, either by making it easier for users to work without making errors, or by making errors impossible in the first place	Are you sure?; Choice editing; conditional warnings; defaults; did you mean?; Interlock; matched affordances; opt-outs; portions; task lock-in/out
Interaction All the patterns are really about interaction design in one form or another, but the interaction lens brings together some of the most common design elements of interfaces where users' interactions with the system affect how their behaviour is influenced, including from the field of persuasive technology (Fogg 2009)	Feedback through form; *kairos*; partial completion; peer feedback; progress bar; real-time feedback; simulation and feedforward; summary feedback; tailoring; tunnelling and wizards
Ludic Games are great at engaging people for long periods of time, influencing people's behaviour through their very design. The Ludic lens includes a number of 'gamification' techniques for influencing user behaviour that can be derived from games and other 'playful' interactions, ranging from basic social psychology mechanisms such as goal-setting, to common game elements such as scores and levels	Challenges and targets; collections; leave gaps to fill; levels; make it a meme; playfulness; rewards; role-playing; scores; storytelling; unpredictable reinforcement
Perceptual The Perceptual lens combines ideas from product semantics, ecological psychology and Gestalt psychology about how users perceive patterns and meanings as they interact with the systems around them	(A)symmetry; colour associations; contrast; fake affordances; implied sequences; metaphors; mimicry and mirroring; mood; nakedness; perceived affordances; possibility trees; prominence; proximity and grouping; seductive atmospherics; similarity; transparency; watermarking
Cognitive The Cognitive lens draws on research in behavioural economics and cognitive psychology looking at how people make decisions, and how this is affected by 'heuristics' and 'biases'. If designers understand how users make interaction decisions, that knowledge can be used to influence interaction behaviour. Equally, where users often make poor decisions, design can help counter this	Assuaging guilt; commitment and consistency; decoys; desire for order; do as you're told; emotional engagement; expert choice; framing; habits; personality; provoke empathy; reciprocation; rephrasing and renaming; scarcity; social proof

(continued)

Table 7.1 (continued)

Lenses	Patterns
Machiavellian The Machiavellian lens comprises design patterns which, while diverse, all embody an 'end justifies the means' approach. This may be unethical, but is nevertheless commonly used to control and influence consumers through advertising, pricing structures, planned obsolescence and lock-ins	Anchoring; antifeatures and crippleware; bundling; degrading performance; first one free; forced dichotomy; format lock-in/out, functional obsolescence; i cut, you choose; poison pill; serving suggestion; slow/no response; style obsolescence; worry resolution
Security The Security lens represents a 'security' worldview, i.e. that undesired user behaviour is something to deter and/or prevent though 'countermeasures' designed into products, systems and environments, both physically and online, with examples such as digital rights management	Coercive atmospherics; peerveillance; sousveillance; surveillance; threat of injury; threat to property; what you can do; what you have; what you know; what you've done; where you are; who or what you are

system control and feedback on overall energy use, two areas which had emerged from earlier work with households in London (Lockton et al. 2013b) as part of the initial UK phase of SusLabNWE.

After a discussion about energy-related issues in their homes, their daily routines and decision-making, and their priorities for change, householders (five in total—three on their own, and one couple) were asked to 'think aloud' with the eight cards, talking through whether they believed they would find each idea desirable (and why) and whether they believed it would 'work' in their context (and why), and grouping them accordingly (Figs. 7.4 and 7.5).

While this was a small, exploratory use of these cards, it revealed some interesting details about the differences between householders' views towards interventions around energy use. For example, one householder did not *like* the idea of the self-turning thermostat, but said she thought it would nevertheless work in her house in terms of reducing energy use, because of other household members who often turned the thermostat up and forgot about it. Another said that she thought an energy-saving game would work for her for a while, but would lose its appeal as she lost motivation, whereas a heating system, which took control itself, would potentially have a greater effect in the long term.

The cards and the discussion they provoked provided interest for the subsequent interviews around other aspects of energy use in the home, and integrating qualitative self-reporting and quantitative sensor data (see Chap. 12 'In situ and mixed design interventions'), and by including some more 'controversial' cards, it was possible to elicit opinions around wider issues of control and agency, topical issues around renewable energy, and householders' perceptions of and worldviews around sustainability. The method will be developed further in future projects, but from a practical design for sustainable behaviour perspective, the diversity of responses suggests that designing tailored interventions, to match the realities

Fig. 7.3 Cards developed from Design with Intent patterns, applied to household energy questions, used to prompt discussion with householders. *Top row* applies patterns from Architectural, Errorproofing and Interaction lenses; *middle row* Ludic, Perceptual and Cognitive lenses; *bottom row* Machiavellian and Security lenses. The examples include the Equarium (Keyson et al. 2013), Giacomin and Bertola's (2012) thermal energy visualisation work, and Brighton's Tidy Street (Bird and Rogers 2010)

of people's everyday experience and decision-making—and which perhaps can evolve over time—may be a sensible strategy.

Fig. 7.4 Householders in Dartford, UK, sorting and discussing the cards

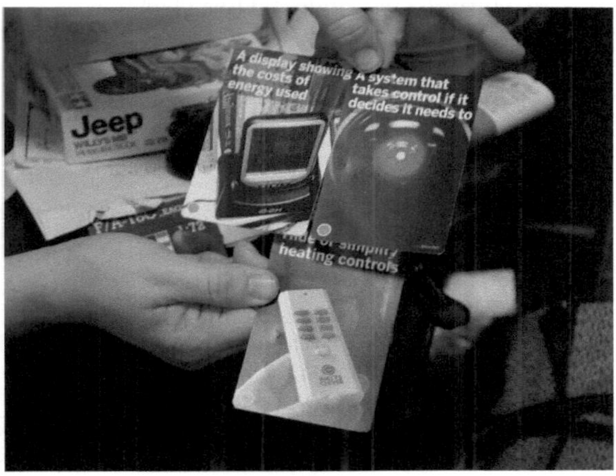

Fig. 7.5 Householders in Dartford, UK, sorting and discussing the cards

7.5 Discussion

The huge scope of sustainability is such that large infrastructural changes are needed, with large changes in not just interaction behaviour with products, but as noted earlier, adoption of new everyday practices at home and at work, decisions about purchases, and fundamental shifts in assumptions, attitudes and political norms. This is where small interventions such as redesigning a heating system interface can seem insignificant, and the value of concepts inspired by something like *Design with Intent* can seem irrelevant in the larger scheme of things. However, it is important to remember that everything around us that has been designed, from the layout of our cities to the structure of our governments, in some way influences how we live, how we make decisions, what resources are used, what is easy and what is difficult. It also, over time, affects how we think, and how we understand the world that we are part of, both individually and together as a society, historically, at present, and looking forward to the future.

So, design which focuses on behaviour is perhaps most usefully understood as design which *reflectively considers* its effects on human action, although there is an inherent presumption towards change. If this reflection can be incorporated into the design process, through questioning assumptions and working with people to understand their lives, and the contexts in which everyday decisions are made, we can be more effective in reaching sustainability goals and in improving quality of living.

It is clear that sustainability needs to be about actions more than just awareness: as Tonkinwise (2004) puts it, "sustainability is a strangely hypocritical politics: even when issues are well understood, actions fail to result; strong and comprehensive awareness of sustainability fails to translate into sustainable behaviour". Design which focuses on people's actions, whether at the level of products, services, environments or larger systems will, inevitably, play an important role in the way we construct our future on this planet.

References

Akrich, M. (1992). The description of technical objects. In W. Bijker & J. Law (Eds.), *Shaping technology, building society: studies in sociotechnical change* (pp. 205–224). Cambridge, MA: MIT Press.

Anderson, S. (1966). Problem-solving and problem-worrying. talk delivered at architectural association. London.

Bernays, E. L. (1928). *Propaganda*. London: Routledge.

Bird, J., & Rogers, Y. (2010). The pulse of tidy street: Measuring and publicly displaying domestic electricity consumption. In *Workshop on Energy Awareness and Conservation through Pervasive Applications, Pervasive 2010*.

Brand, S. (1994). *How buildings learn: What happens after they're built*. London: Viking.

Broady, M. (1966). Social theory in architectural design. In R. Gutman (Ed.), *People and buildings*. New York: Basic Books.

Brynjarsdóttir, H., Håkansson, M., Pierce, J., Baumer, E. P. S., DiSalvo, C., & Sengers, P. (2012). Sustainably unpersuaded: How persuasion narrows our vision of sustainability. In *Proceedings of CHI 2012, Austin, Texas*.

Daae, J. Z., & Boks, C. (2014). Dimensions of behaviour change. *Journal of Design Research, 12*(3).

Desmet, P. M. A., & Hekkert, P. (2007). Framework of product experience. *International Journal of Design, 1*(1), 57–66.

Dilnot, C. (2015). History, design, futures: Contending with what we have made. In: T. Fry, C. Dilnot, & S.C. Stewart (Eds.), *Design and the question of history*. London: Bloomsbury.

Dorst, K. (2015). *Frame innovation*. Cambridge, MA: MIT Press.

Dubberly, H., & Pangaro, P. (2007). Cybernetics and service-craft: Language for behavior-focused design. *Kybernetes, 36*(9/10), 1301–1317.

Ehrenfeld, J. R. (2008). *Sustainability by design*. New Haven, CT: Yale University Press.

Eikhaug, O., & Gheerawo, R. (Eds.). (2010). *Innovating with people: The business of inclusive design*. Oslo: Norwegian Design Council.

Fantini van Ditmar, D., & Lockton, D. (2016). Taking the code for a walk. *Interactions, 23*(1), 68–71.

Fogg, B. J. (2009). A behavior model for persuasive design. In *Persuasive 2009: Proceedings of the 4th International Conference on Persuasive Technology*. New York: ACM.

Foucault, M. (1977). *Discipline and punish: The birth of the prison*. New York: Pantheon Books.

Fry, T. (2015). Whither design/whether history. In: T. Fry, C. Dilnot, & S.C. Stewart (Eds), *Design and the question of history*. London: Bloomsbury.

Giacomin, J., & Bertola, D. (2012). Human emotional response to energy visualisations. *International Journal of Industrial Ergonomics, 42*(6), 542–552.

Gintis, H. (2007). A framework for the unification of the behavioral sciences. *Behavioral and Brain Sciences, 30*, 1–61.

Irwin, T., Kossoff, G., Tonkinwise, C., & Scupelli, P. (2015). *Transition design 2015: A new area of design research, practice and study that proposes design-led societal transition toward more sustainable futures*. Pittsburgh, PA: Carnegie Mellon University.

Jelsma, J., & Knot, M. (2002). Designing environmentally efficient services; a 'script' approach. *The Journal of Sustainable Product Design, 2*(3), 119–130.

Kanis, H. (1998). Usage centred research for everyday product design. *Applied Ergonomics, 29*(1), 75–82.

Keyson, D. V., Al Mahmud, A., & Herrera, R. N. (2013). Living lab and research on sustainability: Practical approaches on sustainable interaction design. In Augusto J.C. et al. (Eds.), *Proceedings of AmI 2013* (Vol. 8309, pp. 229–234). Springer LNCS.

Kuijer, L., de Jong, A., & van Eijk, D. (2013). Practices as a unit of design: An exploration of theoretical guidelines in a study on bathing. *ACM Transactions on Computer-Human Interaction, 20*(4), article 21.

Langdon, P., & Thimbleby, H. (2010). Editorial: Inclusion and interaction: Designing interaction for inclusive populations. *Interacting with Computers, 22*(6), 439–448.

Laschke, M., Hassenzahl, M., & Diefenbach, S. (2011). Things with attitude: Transformational products. In *Proceedings of Create 11*.

Lawson, B. R. (2004). Schemata, gambits and precedent: Some factors in design expertise. *Design Studies, 25*(5), 443–457.

Liedtke, C., Ameli, N., Buhl, J., Oettershagen, P., Pears, T., & Abbis, P. (2013). Wuppertal institute designguide. *Wuppertal Spezial, 46*.

Lilley, D. (2009). Design for sustainable behaviour: Strategies and perceptions. *Design Studies, 30*(6), 704–720.

Lockton, D. (2012). POSIWID and determinism in design for behaviour change. Working paper, Social Science Research Network.

Lockton, D. (2017). *Design with intent: Insights, methods and patterns for designing with people, behaviour and understanding*. Sebastopol, CA: O'Reilly.

Lockton, D., Harrison, D., & Stanton, N. A. (2008). Making the user more efficient: Design for sustainable behaviour. *International Journal of Sustainable Engineering, 1*(1), 3–8.

Lockton, D., Harrison, D., & Stanton, N. A. (2010). The design with intent method: A design tool for influencing user behaviour. *Applied Ergonomics, 41*(3), 382–392.

Lockton, D., Harrison, D., & Stanton, N. A. (2012). Models of the user: Designers' perspectives on influencing sustainable behaviour. *Journal of Design Research, 10*(1/2), 7–27.

Lockton, D., Harrison, D., & Stanton, N. A. (2013a). Exploring design patterns for sustainable behaviour. *The Design Journal, 16*(4), 431–459.

Lockton, D., Bowden, F., Greene, C., Brass, C., & Gheerawo, R. (2013b). People and energy: A design-led approach to understanding everyday energy use behaviour. In *Proceedings of EPIC 2013: Ethnographic Praxis in Industry Conference*.

Maher, L., Poon, J., & Boulanger, S. (1996). Formalizing design exploration as co-evolution: A combined gene approach. In J. Gero & F. Sudweeks (Eds.), *Advances in formal design methods for CAD*. London: Chapman and Hall.

Niedderer, K., Cain, R., Clune, S., Lockton, D., Ludden, G., Mackrill, J., Morris, A., Evans, M., Gardiner, E., Gutteridge, R., & Hekkert, P. (2014.) *Creating sustainable innovation through design for behaviour change: Full report*. University of Wolverhampton.

Packard, V. (1957). *The hidden persuaders*. New York: D. McKay.

Papanek, V. (1971). *Design for the real world: Human ecology and social change*. New York: Pantheon Books.

Payne, O. (2012). *Inspiring sustainable behaviour: 19 ways to ask for change*. Earthscan.

Pettersen, I. N. (2015). Towards practice-oriented design for sustainability: The compatibility with selected design fields. *International Journal of Sustainable Engineering, 8*(3), 206–218.

Reckwitz, A. (2002). Toward a theory of social practices. *European Journal of Sociology, 5*(2), 243–263.

Schmidt-Bleek, F., & Tischner, U. (1993). Designing goods with MIPS. *Fresenius Environmental Bulletin, 2*, 479–484.

Schwartz, E. (1966). *Breakthrough advertising*. Englewood Cliffs, NJ: Prentice-Hall.

Scott, J. C. (1999). *Seeing like a state: How certain schemes to improve the human condition have failed*. New Haven, CT: Yale University Press.

Scott, K., Bakker, C., & Quist, J. (2011). Designing change by living change. *Design Studies, 30*, 279–297.

Selvefors, A., Renström, S., & Strömberg, H. (2014). Design for sustainable behaviour: A toolbox for targeting the use phase. In *Ecodesign Tool Conference, Gothenburg*.

Shove, E., Pantzar, M., & Watson, M. (2012). *The dynamics of social practice*. London: Sage.

Stanton, N. A., & Baber, C. (2002). Error by design: Methods for predicting device usability. *Design Studies, 23*(4), 363–384.

Sterling, B. (2005). *Shaping things*. Cambridge, MA: MIT Press.

Strengers, Y. (2011). Designing eco-feedback systems for everyday life. In *Proceedings of CHI 2011*.

Strömberg, H., Selvefors, A., & Renström, S. (2015). *International Journal of Sustainable Engineering, 8*(3), 163–172.

Thaler, R. H., & Sunstein, C. R. (2008). *Nudge: Improving decisions about health, wealth, and happiness*. New Haven, CT: Yale University Press.

Tonkinwise, C. (2004). Ethics by design, or the ethos of things. *Design Philosophy Papers 2*.

Tromp, N., Hekkert, P., & Verbeek, P.-P. (2011). Design for socially responsible behavior: A classification of influence based on intended user experience. *Design Issues, 27*(3), 3–19.

Watson, J., Clegg, C., Cowell, C., Davies, F., Hughes, C., McCarthy, N., & Westbury, P. (Eds.) (2015). *Built for living: Understanding behaviour and the built environment through engineering and design*. London: Royal Academy of EngineeringLondon.

Wever, R. (2012). Editorial: Design research for sustainable behaviour. *Journal of Design Research, 10*(1/2), 1–6.

Wever, R., van Kuijk, J., & Boks, C. (2008). User-centred design for sustainable behaviour. *International Journal of Sustainable Engineering, 1*(1), 9–20.

Whitson, J. (2014). Foucault's fitbit: Governance and gamification. In S. P. Walz & S. Deterding (Eds.), *The gameful world: Approaches, issues, applications*. Cambridge, MA: MIT Press.

Wiek, A., Talwar, S., O'Shea, M., & Robinson, J. (2014). Toward a methodological scheme for capturing societal effects of participatory sustainability research. *Research Evaluation, 23*(2), 117–132.

Wilhite, H. (2013). Energy consumption as cultural practice: implications for the theory and policy of sustainable energy use. In S. Strauss, S. Rupp, & T. Love (Eds.), *Cultures of energy: Power, practices, technologies*. CA: Left Coast Press, Walnut Creek.

Chapter 8
Architectural Research in Living Labs: Exploring Occupier Driven Changes in Homes

Paula Femenias, Liane Thuvander, Cecilia Holmström, Lina Jonsdotter and Madeleine Larsson

Abstract The chapter reports from a project developing architectural research in connection to a Living Lab. The aim is to create innovative design solutions in order to decrease the environmental loads from material flows over time focusing on occupier driven renovations and alterations to layout, materials and installations of apartments in multi-residential buildings. In a first step, empirical insights from over 300 owner-occupied apartments answers the questions: what changes are made by occupiers, what motivates these changes, and can these changes be linked to different architectural designs? In the continued research the material flows and the environmental impact attributed to these occupier driven renovations and alterations will be estimated giving further indications for more sustainable design of homes.

Keywords Housing design · Renovation · Alterations · Material flow · Owner-occupier

P. Femenias (✉) · L. Thuvander · L. Jonsdotter · M. Larsson
Department of Architecture, Chalmers University of Technology, Gothenburg, Sweden
e-mail: femenias@chalmers.se

L. Thuvander
e-mail: thuvander@chalmers.se

L. Jonsdotter
e-mail: jlina@chalmers.se

M. Larsson
e-mail: madeleine.lrssn@gmail.com

C. Holmström
Tengbom Architects, Stockholm, Sweden
e-mail: cecilia.holmstrom@tengbom.se

D.V. Keyson et al. (eds.), *Living Labs*, DOI 10.1007/978-3-319-33527-8_8

8.1 Introduction

In recent years, low-energy construction for new homes has increasingly become standard. However, focus has been on delivering technical infrastructures and less has been done to understand user-centred and behaviour-based perspectives for sustainable living (Scott et al. 2012). Several studies point to the necessity to address behaviour and home-related practices in the quest to reduce resource use and reach more sustainable homes (Gram-Hanssen 2011; Janda 2011). Improved knowledge about user habits is paramount but also the role of technologies and design in these processes, as they will shape habits, routines and practices (Shove and Walker 2010; Scott et al. 2012).

Presently, housing design is driven by market forces and their definition of sustainable homes, not on knowledge of homes-related practices and the role of design in these (Hagbert and Femenías 2015). Little attention is attributed to actually providing structures for sustainable living that can lead to decreased resource use (Hagbert 2014). There is also a misfit between an accelerating diversity in consumer preferences and the housing provision with few possibilities for occupiers to change or alter their home to fit their (changing) needs (Braide 2015). Schneider and Till (2007) argues for a more flexible housing design to enable social sustainability by responding to demographic and cultural changes as well as changed residents' need.

This paper reports on research focusing on occupier driven alterations and refurbishment of homes with respect to the material flows these activities will create and their related environmental impact. Studies of the environmental load from residential buildings have so far mainly focused on the operation and use phase. A recent study shows that the embodied environmental impact and transports related to production equals the environmental impact from the use phase of new multi-residential energy efficient housing (Liljenström et al. 2014). This indicates a need to take material use into consideration to reach more sustainable housing.

One limitation of these studies is that they do not consider internal changes over time in the dwellings, an area that represents a knowledge gap. Recent increase in material waste from home make-over supports initiatives for research in the field. According to the municipal newsletter Kretslopp in Göteborg, the waste from what is called 'large' waste and which can be connected to renovation increased with 5 % during 2013 and now exceeds the top notation since the last economic downturn in 2008.

8.2 Aim

The overall aim of the research is to create innovative design solutions in order to decrease the environmental loads from material flows in housing focusing on occupier driven renovations and alterations to layout, materials and installations of apartments in multi-residential buildings. This chapter reports from an empirical insight study representing the first step in defining further experimentation in connection to the HSB Living Lab. One of the main objectives for the insight study is to reveal links between architectural design of homes and interior changes made over time with the question if different layouts and design of homes are more or less prone to changes and thus environmental impact from material flows. We have limited the study to apartments in multi-residential owner occupied dwellings, which represents 40 % of all housing in Sweden (SCB 2012). Further we focus on recently built housing thus casting light upon the effect of material flows from contemporary architecture. The study will provide insight in what changes and renovations are made in recent owner-occupied and what motivates the occupiers to do these changes. It has been stated that the replacement of building components and materials are seldom driven by technical or functional end of service-life (Bradley and Kohler 2007, Feilden 2007). Economy, fashion and other non-technical factors are presumed to motivate change.

The research is done in collaboration between Chalmers University of Technology, Department of Architecture, Tengbom Architects, and Bengt Dahlgrens technical consultants with funding from the Swedish Energy Agency. All participating organisations are partners in the HSB Living Lab. HSB Living Lab is developed as a purpose-built facility for research and development on sustainable living in collaboration with partners from academy and industry at the campus of Chalmers University of Technology in Gothenburg, Sweden. The purpose-built facility will provide accommodation for students consisting of smaller individual units for sleeping and working together with a larger area with common shared facilities, with a few larger apartments for guest researchers. The Lab is under construction and tenants will move in 2016. HSB, the client for the HSB Living Lab is one of the largest Swedish developers of owner-occupied housing. Their business model is to sell housing property to owner co-operations. HSB also provides maintenance for these owner co-operations. HSB Living Lab could potentially include testing in their other production of housing beyond the purpose-built facility.

The research has been planned using the Three Tier Model promoted by the SusLabNWE (Baedeker et al. 2014) distinguishing the phases (a) insight studies, (b) explorations, test and evaluations in prototype homes and (c) the introduction and up-scaling on the market. The here reported insight study prepares for continued research in which experiments for more resource efficient and less material flow driving designs of homes will be tested in connection to HSB Living Lab.

Results address clients, architects and housing policy makers, but also occupiers and users of home in a bid to induce changed habits by making material flows and environmental loads from home makeover tangible.

8.3 The Need of Feed-Back from Use

The learning loops for building related knowledge are long. Many years will pass between planning and design to actual feed-back from operation and use, and many and diverse contextual factors will influence what happens after a building is constructed (Brand 1994). In addition, most building projects, lacks of post-occupancy evaluations and feed-back from use(rs).

Contemporary development and innovation of housing lay in the hand of the market and is slow, risk averse, and has stagnated in recent decades. In the early post-war era, the Swedish government financed home-related research. Based on empirical studies universal layouts for good homes were designed and implemented with regulatory instruments. Since the abolishment of state involvement in housing development in the 1990s few empirical studies of home-based practices have been conducted.

During later years, housing design has gone from homogenous and universal standards to a more heterogeneous and varied provision (Nylander 2013). However, it is not proved that this diversity truly recognises actual needs. Basically, 20th century housing schemas are reproduced without taking into account recent and drastically changing conditions for modern living with altering demographics, globalisation, digitalisation and climate change. Modern living is changing quickly, with varying family configurations, and new working and moving patters (Braide 2015). Families with children are for example to a higher degree seeking near city living rather than sub-urban settlements, an example of a trend that has not been anticipated by the market. Regarding sustainability, research has also detected an 'efficiency gap' indicating a broader willingness to live smaller and to use less resources than what is recognised by the market (Hagbert 2016).

With respect to material flows driven by refurbishment and changes to buildings, flexible building concepts where users can adapt their living environment, has been proposed as a path to achieve more sustainable architectures (Kendall 1999). Open building concepts or skeleton buildings are since long used in the commercial sector allowing for user-customised solutions fitted for individual needs regarding unique interior spaces, equipment and systems. However, such solutions are still scares in housing design. Our 19th century modernistic housing ideals are based on 'rigid' single-use expert-made designs that are neither prepared for changes, renovation nor repair (Scott 2008).

8.4 Living Labs to Support Design of Sustainable Homes

Living Labs have been brought forward as a means to support sustainable innovation and transitions to improve knowledge about user habits and the test of practice-oriented design experimentations. They represent platforms through

Fig. 8.1 Next 21 open building housing experiment in Osaka. Photo Paula Femenías

which housing development and innovation can benefit from tests that are more focused on influencing external factors than normal production lines. They will have limited economic risk, in comparison to direct implementation without test and can shorten the time span from design to evaluation (Femenías and Hagbert 2013). Living labs are based on co-design and user-centred processes (Liedtke et al. 2015) with the object of also inducing changed habits by 'living change' (Scott et al. 2012).

While a larger body of research papers reports on developing methods for monitoring and inducing change in user habits in Living Labs (Baedeker et al. 2014) and on the development of sustainable products and services (Scott et al. 2012) there are a few that report on architectural design. Thomsen and Tjora (2006) report on an experiment with a design-build-live project for compact collective student housing in which students would experience design made to improve the sharing of living space. Other architectural research on the design-build-live concept has been carried out in relation to Solar Decathlon projects (Masseck 2014). Full-scale open housing experiments have been carried out in the Netherlands, the UK, Finland and in Japan (Kendall 1999). One example is the Next 21 project in Osaka experimenting (Fig. 8.1) where the exterior cladding, the mechanical systems and the interior design has been constructed as separate systems in order to facilitate renovation and adaptations.

8.5 Method and Approach

In order to get empirical insights in renovation cycles and changes made in contemporary owner-occupied apartments, a questionnaire was sent out to five selected housing co-operations. All projects have been developed by HSB and then sold to the housing co-operations. The co-operations consist of a group of owner-occupier with individual apartments and a collectively owned building infrastructure. Guidelines for the selection of cases were approximately 10 years in use, 100 apartments and a diversity regarding construction and design. Housing older than 15 years (before 2000) were not considered due to data availability but also deliberately, as we wanted to focus on contemporary housing design. Further, we choose projects with different solutions for facades as we did not want other problems with the building to overshadow our results, for example the EIFS (Exterior Insulation and Finishing System), commonly used in the early 21st century, is often connected to mould problems.

Four housing co-operations were chosen in Göteborg Cases B–E, the second largest city in Sweden with a comparative case, Case A, in Stockholm, the capital (Table 8.1). All five housing projects are located in new waterfront developments where energy efficient and sustainable housing have been guiding the design and development.

We contacted the board of the housing associations to ask for their interest to participate in the study and their cooperation with a questionnaire. A questionnaire was then sent out to the 462 households in the five co-operations. The questionnaires collect data on changes made to different rooms including: floor and wall layers in each room; removal or addition of walls; changes of opening and doors; new layout of kitchen and bathrooms; new or altered equipment in kitchen and bathroom and/or altered location; changes to storage and balconies/terraces; new or altered installations (electric or other). The questionnaire consisted of 64 binary scale questions only allowing yes/no answers with possibilities to add free text. Together with the questionnaire we sent out the original layout of the apartments with an appeal for the occupier to draw the changes made to the apartment.

In a planned second step, in-depth interviews will be carried out with a few households (~5–10) living in apartments that have been subjected to major changes but also representing different kinds of changes driven by the specific design of the apartment. The interviews will study the changes and also the motives for these. When all data is collected, workshops will be organized in the project team (mainly Chalmers and Tengbom architect, and HSB invited at a larger stage) to discuss how changes made in the apartments can be linked to differences in architectural design. Other factors influencing changes to the apartments such as demographics of the user, size of apartment, number of owner changes, and normative practices among owner-occupier of homes will be also discussed.

Finally, in a third step the environmental impact from the owner-driven changes will be estimated, based on typical changes of apartments, their related material flows and environmental impact using appropriate software. Usually internal

Table 8.1 Presentation of the five co-operative housing associations addressed with a questionnaire

	Case A	Case B	Case C	Case D	Case E
Built	2001	2002	2004	2006	2008
No of apartments	110	95	70	135	55
Response rate (%)	63	84	70	46	51
Turn over per year (%)	7	6	9	11	12
Building form	Three buildings, two in split level, up to six floors. Includes commercial premises	Two free-standing towers and two detached block houses	One floor slab house connected to a tower with seven floors	Five free-standing towers blocks with seven floors	One nine floor tower
Apartment type	Very spacious apartments with possibilities to change and separate rooms	Large variety of spacious apartments with open plan and possibility to close kitchen and create additional rooms	Compact apartments with very open layout, cupboards to delimit kitchen from living room	Open plan floors and compact design with economic small apartments	Award winning designed for social interaction and flexibility, open but compact plans with possibility to close kitchen

user-driven renovation and material flows are allocated to the household, and not to the building. This study will thus provide original insights into the relative environmental impact, estimated for a period of 50 years, of internal renovations in comparison to the environmental impact of the whole building when constructed.

8.6 Preliminary Results and Discussion

The insight study is still on-going and data is under collection and analysis. There has been a large interest among housing co-operations to participate in the insight study, and all housing co-operations that we initially approached were interested in participating and five were selected. There has been a large response rate for the questionnaires 315 answers and an average response rate of 68 %. Many respondents have also delivered drawings in which their changes are marked.

Two main categories or renovations have been found. Not surprising, the most common changes are repainting, new wall-papers and new flooring but also new inner doors, changed white goods and sanitary equipment. Some of these changes are driven by preference and wish to set a personal touch to the home but many are driven by technical damage driven by low quality original materials. Considering that the apartments in the study only are 8–15 years old, these changes are remarkable as sanitary components such as WC could last for 30 years and wooden parquet for 40 (Willis-Insurance-Pool 2015).

Second, and most important for this study, are changes driven by the layout and the architectural design of the apartments. Based on the questionnaires and the drawings the respondents handed in, we have categorized different changes made to the original design. The most prominent alterations to the original layout of the apartment regards the design of the kitchen, the contact between the kitchen and the living room, addition or removal of walls to create more or fewer bedrooms, and measures to limit excessive space only used for communication between rooms. The largest number of changes is found in apartments in Case A in Stockholm where many owners have added walls and created more rooms. Apartment with originally two or three bedrooms have been altered into four or even five bedroom apartments to house families with children. Case A and Case B in Göteborg, are the oldest in the study and have been designed with more generous apartments and the possibility to separate more rooms than in the original configuration. Case C, D and E are more recent and represent more compact apartment designs. The compact apartment designs seem to have less possibility for alterations and can be seen as more locked-in designs in terms of adaptation. With respect to an observed lower turn-over in the more generous apartment designs (as in Case A and B) this could indicate that when apartments have possibilities for alteration, thus having a built-in flexibility, this could lead to lower turn-over. That is, if the occupiers have the possibility to alter the apartment design according to their changed needs, they are more prone to stay, something that could contribute to social sustainability of the area (Schneider and Till 2007; Braide 2015).

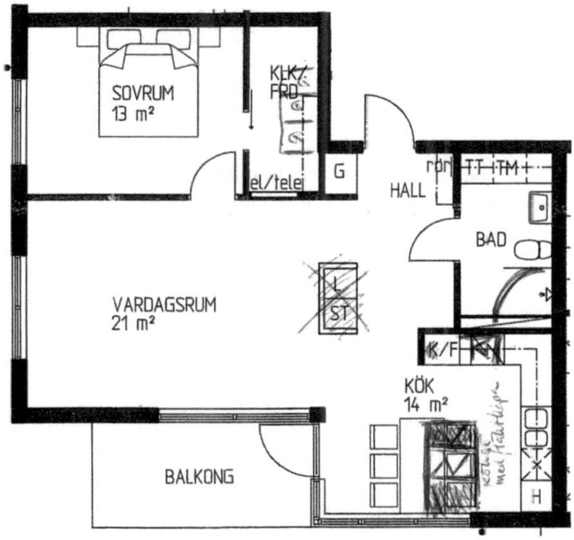

Fig. 8.2 Example of layout of apartment in Case D

Figure 8.2 shows an example of layout of apartment in Case D, where the owner-occupiers have drawn their changes to the original layout. The owners have removed a cupboard unit, and thus also changed the whole floor cladding, originally designed to create a separation between kitchen and living room for better sight between the two spaces.

8.7 Conclusions and Implications for Continued Research

The insight study provides us with empirical insights from owner-occupier changes to apartments in a few new developments of multi-residential water-front developments in Swedish growth areas. The results have valuable input for architects and developers of housing but also for developing planned experimentation with more flexible housing design within the framework of the HSB Living Lab.

There has been a large interest to participate in the insight study. This could indicate that there is an interest among owner-occupiers in HSB housing associations also for continued research and experimentation in the field, where we will define design experiments in relation to more open house concepts with a broader possibility for user adaptations. The research is now continued with in-depth interviews to better understand motives for changes and renovations. In parallel, the environmental impact of the changes and renovations is estimated.

The empirical material also point to complementary perspectives that might be of interest for continued analysis and studies. One of these perspectives is to highlight the process from design to production, where initial intention of the architect designer has been altered during production on initiative of the contractor and builder, and in what sense this affect the use of the building. Another perspective is that of risk. If the owner-occupier of the individual apartment renews or changes installations, piping or materials, this could affect the building's systems. For example, if the kitchen fan is removed or is in another way altered this could affect the balance in the ventilation system. If the owner-occupier installs additional heating sources or adds insulation in non-heated areas this could challenge the buildings overall energy performance and maybe even ratings for sustainability assessments. New claddings on terraces and balconies also present a risk regarding mould in nearby façade constructions.

Finally, our study presents pioneering research towards a broader understanding on how our built environment changes over time, what is the role of design and architecture in these processes and in what way is design linked to material flows and related environmental impact. Our intention is to improve knowledge of more open building concepts but also to increase awareness of environmental load, and eventually risks, from interior changes driven by the user. We also hope to provide input to policy makers in the housing area and for future building regulations by providing empirical studies of use of our homes.

References

Baedeker, C., Greiff, K., Grinewitschus, V., Hasselkuß, M., Keyson, D., & Knutsson, J., et al. (2014). Transition through sustainable product and service innovations in sustainable living labs: Application of user-centred research methodology within four living labs in Northern Europe. In *IST 2014: 5th International Conference on Sustainability Transitions, 2014*.

Bradley, P. E., & Kohler, N. (2007). Methodology for the survival analysis of urban building stocks. *Building Research and Information, 35*(5), 529–542. doi:10.1080/09613210701266939.

Braide, E. A. (2015). Residential design implementing social sustainability: Towards a paradigm shift within design thinking? In *Paper Presented at the The Future of Open Building Conference, ETH Zürich*.

Brand, S. (1994). *How buildings learn: What happen after they are built*. London: Viking Press.

Feilden, C. B. (2007). The environmental handbook. London: Right Angle Publishing.

Femenías, P., Hagbert, P. (2013). The Habitation Lab. Using a design approach to foster innovation for sustainable living. Technology innovation management review (November), pp. 15–21.

Gram-Hanssen, K. (2011). Understanding change and continuity in residential energy consumption. *Journal of Consumer Culture, 11*(1), 61–78.

Hagbert, P. (2014). *Interpreting the sustainable home*. Göteborg: Chalmers University of Technology.

Hagbert, P. (2016). *A sustainable home? Reconceptualizing home in a low-impact society*. Department of Architecture. Diss. Chalmers University of Technology.

Hagbert, P., Femenías, P. (2015). Sustainable homes, or simply energy-efficient buildings? *Journal of Housing and the Built Environment* 1–17. doi:10.1007/s10901-015-9440-y.

Janda, K. B. (2011). Buildings don't use energy: people do. *Architectural science review, 54*(1), 15–22.

Kendall, S. (1999). Open building: an approach to sustainable architecture. *Journal of Urban Technology, 6*(3), 1–16.

Liedtke, C., Baedeker, C., Hasselkuß, M., Rohn, H., Grinewitschus, V. (2015). User-integrated innovation in Sustainable LivingLabs: An experimental infrastructure for researching and developing sustainable product service systems. *Journal of Cleaner Production, 97,* 106–116. doi:http://dx.doi.org/10.1016/j.jclepro.2014.04.070.

Liljenström, C., Malmqvist, T., Erlandsson, M. (2014). Byggproduktionens miljöpåverkan i förhållande till driften.

Masseck, T. (2014). Living labs in architecture: open innovation and co-creation towards a more sustainable architecture and lifestyle.

Nylander, O. (2013). *Svensk bostad 1850-2000.* Lund: Studentlitteratur.

SCB. (2012). Yearbook of *Housing and Building Statistics 2012*. Stockholm.

Schneider, T., & Till, J. (2007). *Flexible housing*. Elsevier, Oxford: Architectural Press.

Scott, F. (2008). *On altering architecture*. New York: Routledge.

Scott, K., Bakker, C., & Quist, J. (2012). Designing change by living change. *Design studies, 33*(3), 279–297. doi:10.1016/j.destud.2011.08.002.

Shove, E., & Walker, G. (2010). Governing transitions in the sustainability of everyday life. *Research Policy, 39*(4), 471–476. doi:10.1016/j.respol.2010.01.019.

Thomsen, J., & Tjora, A. (2006). Changeable space as temporary home: A qualitative exploration of life in an experimental student house. *Nordic Journal of Architectural Research, 19*(3), 13–22.

Willis-Insurance-Pool. (2015). Teknisk livslängd för byggnadskonstruktioner [Technical life-time for building constructions].

Part III
Sustainable Living Labs

Chapter 9
DARE2Build

Shea Hagy, Peter Selberg, Larry Toups and Paula Femenias

Abstract Education efforts, both at Chalmers and elsewhere, could be developed to better suit the needs of competences that would help with solving the societal challenges of coming and current generations. The educational framework Dare2Build aims to be a part of this development by immersing its students in cross-disciplinary teams, working with co-creational methods, to solve practice-based tasks. The proximity to the explorative research environment surrounding a living lab, and the prototyping infrastructure in particular, the HSB Living Lab, is expected to incorporate innovation and entrepreneurship into the participants' learning experiences and thereby promote utilization of activities of the university sphere.

Keywords Education · Practice-based learning · Cross-disciplinary · Co-creation

9.1 Introduction

Dare2Build is a proposed educational framework that connects students from various disciplines and includes the use of a third generation Living Lab as a crucial part of that framework. The development of this educational framework began within the European Union funded SusLab Project, which aims to support the

S. Hagy (✉) · P. Selberg
Building Technology, Chalmers University of Technology, Gothenburg, Sweden
e-mail: Shea.hagy@chalmers.se

L. Toups
Exploration Mission Planning Office NASA, Johnson Space Center,
Chalmers Architecture, Houston, USA

P. Femenias
Department of Architecture, Chalmers University of Technology, Gothenburg, Sweden

D.V. Keyson et al. (eds.), *Living Labs*, DOI 10.1007/978-3-319-33527-8_9

Fig. 9.1 Render of HSB Living Lab on Chalmers Johanneberg Campus, Gothenburg, Sweden (*Photo* Tengbom Architecture)

development of an international Living lab infrastructure network for testing and promoting innovation within a real-life context to create new ways of improving sustainability within the home environment ([suslab.eu/about/suslab-mission]). From work at Chalmers within the SusLab Project emerged the concept of a Hab Lab, as described by Femenias and Hagbert (2013). At the same time the Climate-KIC flagship project, the building Technology Accelerator (BTA) was being developed to continue and extend the work done within the SusLab Project with the BTA supporting the creation of a living lab network. One of the living labs within the network will be located at Chalmers University of Technology in Gothenburg Sweden, the HSB Living Lab (HLL). The HLL is a living lab infrastructure in the form of student housing for the real world testing and prototyping of sustainable innovation in the home (Fig. 9.1). A prototyping lab has been Included within the HLL infrastructure to foster the innovation and prototyping process.

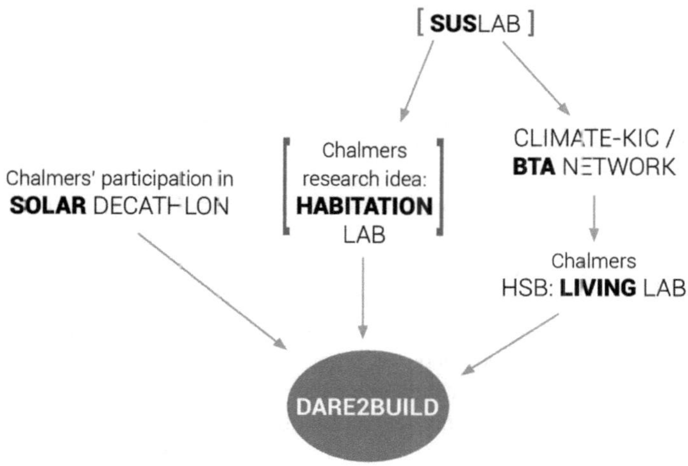

Fig. 9.2 The idea for the DARE2Build concept has grown from several different sources

A Dare2Build pilot curriculum is currently being planned to run at Chalmers University in Gothenburg, Sweden in cooperation with Rice University in Houston, Texas, providing the tools and support to create an innovative living unit founded in an advanced understanding of what sustainability means in the building sector. It intends to take advantage of the Chalmers HLL as a crucial piece of Dare2Build in the conceptualization, building, prototyping and evaluation of technologies and user behaviors that promote those sustainable innovations for future homes.

Also, in 2013, Chalmers entered the first Swedish entry into the Solar Decathlon competition in China, Halo, a plus energy solar home designed and constructed by an interdisciplinary team of students (http://www.halosweden.com/). Since this was a rewarding practice-based learning experience for both students and staff, the intention is for Dare2Build to keep developing entries to future Solar Decathlon competitions (see Fig. 9.2).

9.2 Call for Innovative Education: The Why

In the building sector, collaboration between researchers and non-academic stakeholders will influence the legitimacy, ownership and accountability of solutions that potentially can speed up innovation. The Dare2Build-course aims to facilitate this collaboration, and its structure draws on experience and knowledge at Chalmers University as well as international research within the field. There is an increasing agreement that the complex issues associated with sustainability and the built environment are in need for integrated and transformative solutions which can be offered by transdisciplinary research and triple-helix formations (Lang et al. 2012). Constructive input is needed for sustainability from various communities of knowledge to reduce short-sighted, non-holistic perspectives and that are sometimes present within individual competences.

Education in the built environment needs to adapt to the new demand for competence and capability regarding sustainability in urban planning and design. Referring to the 'wicked' or 'non-tame' character of urban and architectural challenges (Rittel and Webber 1973) education of architects has since long been centered around near practice and reality based studios in collaboration with stakeholders. However, sustainability will set emphasis on social learning and transformation whereas education of built environment professionals has mainly been technical. These preconditions place some demands on what skills future professionals in the built environment should have when they enter the work force, including effective communication across disciplinary borders and with non-academic stakeholders, putting one's knowledge and abilities into context, and handling complex problems throughout all levels of the design process to drive sustainable development.

Recently there has been an increase in reality-based design studios to improve knowledge in sustainable solutions and processes of the type design-and-build in inter and transdisciplinary settings (Femenías and Hagbert 2013). Examples of this are found in the Rural Studios (Oppenheimer Dean and Hursley 1998) and the Solar Decathlon competitions. Some Universities have experimented with labs for housing experiments like the SpaceLab at MIT. TreStykke in Norway, on the concept design-build-live (Thomsen and Tjora 2006), and a Living Lab in the Netherlands (Scott et al. 2012) show examples of how students gain personal experiences in the design and use of space and facilities. This is beneficial as it broadens the use of co-creation towards sustainability, advancing knowledge into teaching, and thereby challenges the unsatisfactory trend of how it is increasingly normalized in sustainability science but teaching efforts have so far been mainly faculty research. (Trencher et al. 2015), and builds upon how the United Nations Decade of Education for Sustainable Development has highlighted the need for real-world situations for education involving societal stakeholders (Wals 2014).

In order to achieve these skills students should be given the opportunity to engage in an ambitious, reality-based, cross-disciplinary and co-creative learning environment. Contemporary sustainable challenges demand inter-disciplinary collaborations and more innovative solutions to address the quickening pace of both environmental and societal changes. The building industry, however, is slow at changing and very risk averse, making it more essential that students before entering the industry are given the opportunity and experience to work within this kind of innovative environment allowing them to develop the skills and knowledge needed to speed-up innovation processes and more effectively address complex problems and challenges (Femenías and Hagbert 2013).

To further understand how to integrate an innovative educational cross-disciplinary practice-based learning platform into the Chalmers' campus, a 2-day workshop was held with representatives and students from Chalmers' architecture and engineering departments as well as representatives from Rice University

Fig. 9.3 Photos from the cross-disciplinary and practice-based learning co-creation workshop #2. held at Chalmers University of Technology, Gothenburg, Sweden, November 20th and 21st, 2014 (*Photo* JIG EF, teamjig.com)

and Chalmers' Areas of Advance: Energy, and Built Environment (see Fig. 9.3). Through this workshop many ideas were generated together with a series of values for the various stakeholders (see Table 9.1), to be used later in the upcoming curriculum development.

Table 9.1 Value of practice-based and cross-disciplinary learning

Value for students	• Getting prepared for the profession while studying (can replace some time of professional experience) • To get skills and knowledge for planning and facing work environment challenges • To get improved communication and teamwork skills • To understand the complexity surrounding the realization of an idea (concept or design) • To apply/test theoretical knowledge in practical work • To understand the necessity of getting multi-disciplinary perspectives on problems in order to develop thorough solutions • To learn how to express one's idea through an adapted terminology
Value for teachers	• To have insight into the practice world • To develop teaching skills by facing practice challenges • To make courses more relevant for the education of engineers and architects • To design courses that are more relevant for the education of future professionals • To learn from other disciplines • To be aware of future research needs
Value for universities	• To gain a higher quality of education by providing students with practical experiences • To potentially obtain better international ranking by creating a practical profile • To improve interactions between different departments and active internal organizations • Achieving a better profile by sending a better-prepared working force to the industry • To strengthen the collaboration between different departments through common courses
Value for industry	• To create networks between students and companies (to serve good ground for future employment) • To receive graduates with good understanding of real-world challenges • Opportunity for companies to develop innovative solutions by collaborating with students and academia • Newly-graduated students that are better prepared and more efficient at work

At Chalmers there exists a unique opportunity to connect education to a 3rd generation living lab, the HSB Living Lab. This will allow students to not only participate in design-build processes but also the operation and evaluation of the built structure which they themselves have designed and built, and the insights that come with this. This in turn will connect industry and academia creating new pathways for innovation penetration into the building industry, where ideas move from academia to companies both through students having more fully developed innovative ideas and knowledge when they enter the workforce and through the collaboration with industry partners in the Dare2Build framework and educational process. This can provide a catalyst to break the business-as-usual mindset of many graduated students stemming from tradition and culture at both professional and university levels.

Thuvander et al. looked at 25 years of the sustainable building curriculum at Chalmers architecture department and found support for courses like

Dare2Build; "In order to keep up with knowledge development, integration of different disciplines and stakeholder groups from the construction sector is necessary [...]. Important factors for success are integration, collaboration, continuity, progression, and enthusiasm". This supports the idea that Dare2Build is a next step in the evolution of sustainable building education.

9.3 Course Creation: The How and the What

Dare2Build uses co-creation design process, in cross-disciplinary teams, working on real-world problems. The main difference between Dare2Build and similar courses at Chalmers is that it utilizes the entire CDIO-framework, an "*innovative educational framework for producing the next generation of engineers.*" ([cdio.org/about])—and incorporates other disciplines into this framework, similar to Building Information Modelling (BIM) collaborations that are becoming more common on the professional level.

The structure of Dare2Build follows that of the CDIO-initiative, which is applicable also for architects. The four stages of a CDIO project are: Conceive, where the customer needs, technologies and concepts are defined; Design, where the product is defined through drawings, models or algorithms; Implement, where the design is transformed into a product through manufacturing or coding as well as testing and validation; and finally Operation, where the product is applied in its intended context, used and maintained. Regardless of the type of product, there would need to be a mix of competences present to be able to reach a good result, since the full complexity of the task needs to be handled.

Each year of Dare2Build will be anchored in a special theme surrounding sustainability in the built environment. These would be gathered from global trends, societal conversation and ongoing research, to reach an adequately innovative, efficient and relevant way of driving sustainability and bridge education, research and utilization. The participation in the international sustainable building student competition Solar Decathlon will force the project to new heights (if applicable for the given year), as will the cooperation with Rice University in Houston (See Fig. 9.4).

This enterprise needs a well-defined organizational structure. The Area(s) of Advance, an organizational element within Chalmers that connects researchers, companies and other stakeholders within a certain theme, e.g. Built Environment, appear as a natural point of departure to serve as a platform for meeting, contact agency and supportive patron. The proximity to non-academic actors within the Areas of Advance can also contribute to quality assurance and market needs.

The end result of the curriculum aims at creating a fully functioning housing unit (building system) that is net positive on electricity, i.e. produces more electricity than it consumes. This demands cooperation between students with varying competences to realize the level of complexity needed to construct such a building. The main challenge inherent in the task, the students themselves are the

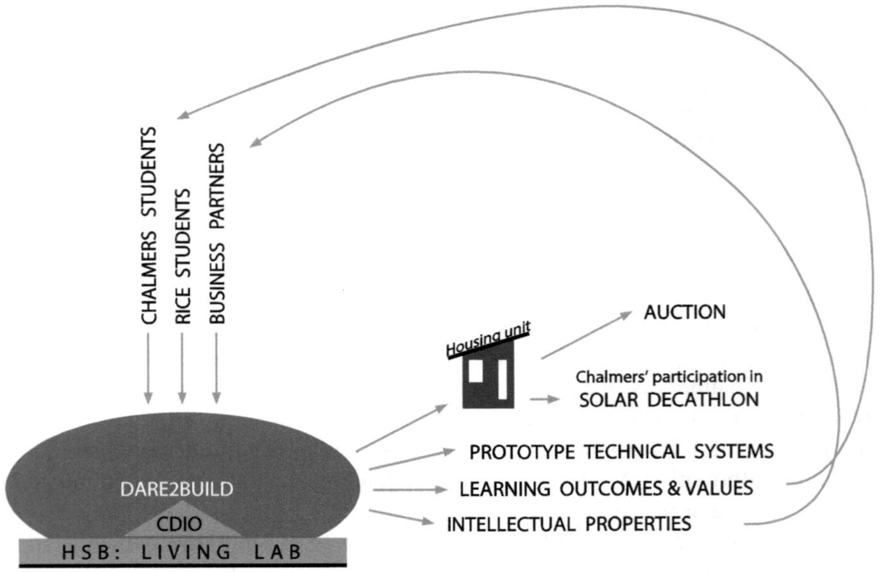

Fig. 9.4 The circulation of identified inputs and outputs from the DARE2Build curriculum

craftspeople in charge of executing their own design, will by necessity bring about a heightened level of cohesion between the participants increasing the understanding of cross-disciplinary collaborations.

9.4 Next Steps

Dare2Build will open up new possibilities for everyone involved. We believe that it, and its accompanying processes, will contribute to an educational platform that operates as a living laboratory and open arena which challenges normative and cemented paradigms, rallies competences and individuals to joint journeys of evolution, and that opens up the solutions to industry and society. Aiming to encourage more effective collaborations across industries and disciplines through such a course, is crucial within the built environment if we are to meet the challenges we face today. Students, teachers, and professionals across the building industry need to rethink and restructure learning and innovation processes and methods in order to reduce impact on global climate, resource use and to progress towards a more sustainable built environment. Dare2Build is an evolution towards this goal. Currently discussions with department heads on how to implement DARE2build are taking place at Chalmers, as well as development of a clear model for collaboration with Rice University. A part of this will include improving the initial set of learning outcomes that have been developed based on ideas, input from stakeholders and workshops associated with Suslab, BTA, Solar decathlon and HLL projects.

References

CDIO Initiative—Collaborating Institutions [Internet]. (2015, July 27). Retrieved from http://www.cdio.org/

Femenías, P., & Hagbert, P. (2013). The habitation lab. using a design approach to foster innovation for sustainable living. *Technology Innovation Management Review* (November), 15–21.

Lang, D. J., Wiek, A., Bergmann, M., Stauffacher, M., Martens, P., Moll, P., & Thomas, C. J. (2012). Transdisciplinary research in sustainability science: Practice, principles, and challenges. *Sustainability Science, 7*(1), 25–43.

Oppenheimer Dean, A., & Hursley, T. (1998). *Proceed and be bold: Rural Studios after Samuel Mockbee*.

Rittel, H. W., & Webber, M. M. (1973). Dilemmas in a general theory of planning. *Policy Sciences, 4*(2), 155–169.

Scott, K., Bakker, C., & Quist, J. (2012). Designing change by living change. *Design Studies, 33*(3), 279–297. doi:10.1016/j.destud.2011.08.002.

SusLabNWE [Internet]. (2015, November 15). Retrieved from http://suslab.eu/about/suslab-mission/

Thomsen, J., & Tjora, A. (2006). Changeable space as temporary home: A qualitative exploration of life in an experimental student house. *Nordic Journal of Architectural Research, 19*(3), 13–22.

Trencher, G., Terada, T., & Yarime, M. (2015). Student participation in the co-creation of knowledge and social experiments for advancing sustainability: Experiences from the University of Tokyo. *Current Opinion in Environmental Sustainability, 16*, 56–63.

Wals, A. E. (2014). Sustainability in higher education in the context of the UN DESD: A review of learning and institutionalization processes. *Journal of Cleaner Production, 62*, 8–15.

Chapter 10
The Storyline for the Design Process that Shaped the HSB Living Lab

Peter Elfstrand, Gregory M. Morrison, Larry Toups and Shea Hagy

Abstract The storyline for the design of a Living Lab is presented from the idea stage through to building construction. The design process was a co-production of social and technical knowledge within the emerging partnership, which was structured around six defined focus fields. It also describes how a unique participation within Chalmers, focused on linking research in space, added to the inspiration. The outcome is that once built, HSB Living Lab is both an innovative building and a real-life user-centered facility with the flexibility to enhance and enable innovation through the lifetime of the building.

Keywords Living lab design · Private and shared space · Space living lab · Exchangeable facades · Social space · Architecture · Innovation

---

P. Elfstrand
Tengbom Architecture, Gothenburg, Sweden
e-mail: peter.elfstrand@tengbom.se

G.M. Morrison (✉)
Curtin University Sustainability Policy Institute, Curtin University, Perth, Australia
e-mail: greg.morrison@custin.edu.au

L. Toups
Johnson Space Centre, NASA, Houston, Texas, USA

L. Toups
Chalmers University, Gothenburg, Sweden

S. Hagy
Building Technology, Civil and Environmental Engineering,
Chalmers University of Technology, Gothenburg, Sweden

D.V. Keyson et al. (eds.), *Living Labs*, DOI 10.1007/978-3-319-33527-8_10

Fig. 10.1 Exterior of HSB Living Lab Version 1.0. The tenants has moved in 1st June 2016

10.1 Introduction

This article considers the design process for shaping a Living Lab, the HSB Living Lab (in Gothenburg Sweden) Fig. 10.1 from an idea stage to the construction of the building itself. The lab design process is both an early preparation for the co-production of knowledge (Evans et al. 2015) and the place for the co-creation of partner value propositions (Lai et al. 2011). The storyline that is presented here therefore provides a critical insight into the Living Lab design process that is underpinned by a transdisciplinary innovation arena of industry, society and academics. The lessons learned are important as a Living Lab provides a heterogeneous spectrum of social-centered to technical-centered innovation manifests (Franz 2015) and done properly, the design of space and place should facilitate, but if done wrong might hinder, innovation (Bergvall-Kareborn et al. 2015). Further, the lessons learned should also aid others in providing a Living Lab design that caters for the long-term viability of future Living Labs (Mastelic et al. 2015) and support business models for that purpose (Rits et al. 2015).

10.2 Storyline

It might seem unusual for a number of reasons, including business economics and reputation, for an architect to be involved in the early stages of a project that is simply a vision for a place, a space, for business and academics to work together and in the absence of a clear idea of a program for either use or function. Nevertheless, as an architecture company (Tengbom Architecture of Sweden) we saw this as a unique opportunity, a challenge, a market niche and a value proposition. We led a design by research process that was close to an artistic approach which brought together the collective creativity of the business and academic partners (Lehmanm et al. 2015) with the free research methods setting the tone for the early stages of Living Lab design. While the main effort was directed towards the design process the project also became a communication and knowledge co-production process for the partners within the project.

The origin of the Living Lab can be traced back to a first meeting between an employee of HSB (see acknowledgements) and a professor at Chalmers University of Technology (the second author). Both identified the need for a real-life place (Femenias and Hagbert 2013) for studies on how we actually live (social-centered, Franz 2015), as well as a place to test new materials and construction techniques (technical-centered, Franz 2015). The fact that the Living Lab founding partners together represented the residential business sector and academia, respectively, was the start of the co-production of knowledge that is the signature for, and has driven the design evolution of, HSB Living Lab.

Tengbom architects joined the partnership with the stated intention to enable and create good and sustainable living solutions through architectural endeavor, this being the value proposition for their business. For this purpose Tengbom co-assisted in several of the early co-creation workshops, to allow the definition of value and form, the latter being early knowledge and ideas. Six thematic themes emerged out of the workshops which were established as focus fields for the partnership and the design process viz: Architecture, plans and accessibility; Architecture and movability; Ecological sustainability; Energy systems; Laundry room; Research as a function.

10.2.1 Design Pilot

In parallel with establishing the focus fields, Tengbom initiated a design pilot to collate global knowledge within the focus fields and for current and proposed Living Labs. For the latter, the design pilot was partly influenced by the Living Labs for sustainability of Suslab and Climate-KIC BTA (see acknowledgements) and partly by other established Living Labs such as those in pervasive healthcare (Favela et al. 2015) and Green Dormitories (Watson et al. 2015). The team for the design pilot also considered, filtered and collated design solutions at interior and building scales. This provided inspiration and associative methods and an important library of knowledge.

A design process for architecture balances between free mind and structured thinking,

Tengbom assisted with systematically organizing this information within the focus fields. Ideas were arranged under the fields and colour coded where green meant go, yellow meant parked for possible future use and red meant outside the scope of the project. The colour coding allowed the architects, in close cooperation with researchers from Chalmers, to shape the focus fields for a final round, as well as to provide a database for the Living Lab design (version 1.0), for possibly future work.

Stakeholders within the housing association's service ecosystem who were considered key stakeholders for the Living Lab open innovation (Lapointe and Guimont 2015), but outside the initial partnership, were engaged to enrich the reviews of the state-of-the-art and to provide proposals for further studies within the focus fields. The design pilot was finalized in two days synthesis workshops for each focus field, which Tengbom used to form the starting point and foundation for the architectural program, which at this point was excessive.

10.2.2 Concepts and Building Program

In most contracted situations, the architects' client has a clear idea of the program for the building or construction. The situation with HSB Living Lab was different insofar as although the general guidelines were to build a Living Lab for knowledge-innovation as student apartments, a wide degree of freedom from a large knowledge base confounded the attempt to simplify the program. This situation challenged the architects to be exceptional in communication, to interpret the response from the main client as well as other partners, and to slowly iterate the qualitative and quantitative input into a focused conceptual program. Placemaking was one of the models to communicate and define the project (Fig. 10.2).

To add structure to this complexity, the architect employed methods of analysis and process, as well as benefiting from close collaboration with dedicated experts at Chalmers, to figure out their initial needs for the building, as well as to discover important features of the building design that would add value to the research process (Fig. 10.3). Fortunately, the Chalmers researchers brought an experience from the European Suslab project and needs for the Climate-KIC BTA flagship (see acknowledgements) and a relatively precise definition of a Living Lab with sustainability and innovation as key features (Brynarsdottir 2012; Dell'Era and Landoni 2014; Leminen and Westerlund 2012).

10.2.3 Program

At the same time as the initial design process was occurring, a unique source of inspiration emerged. While looking at examples of previous terrestrial living lab examples, a space living lab, the International Space Station (ISS), was also brought into the conversation. The modularity, flexibility, and confined spaces of the ISS were all similar to design features needed for the HSB Living Lab. The

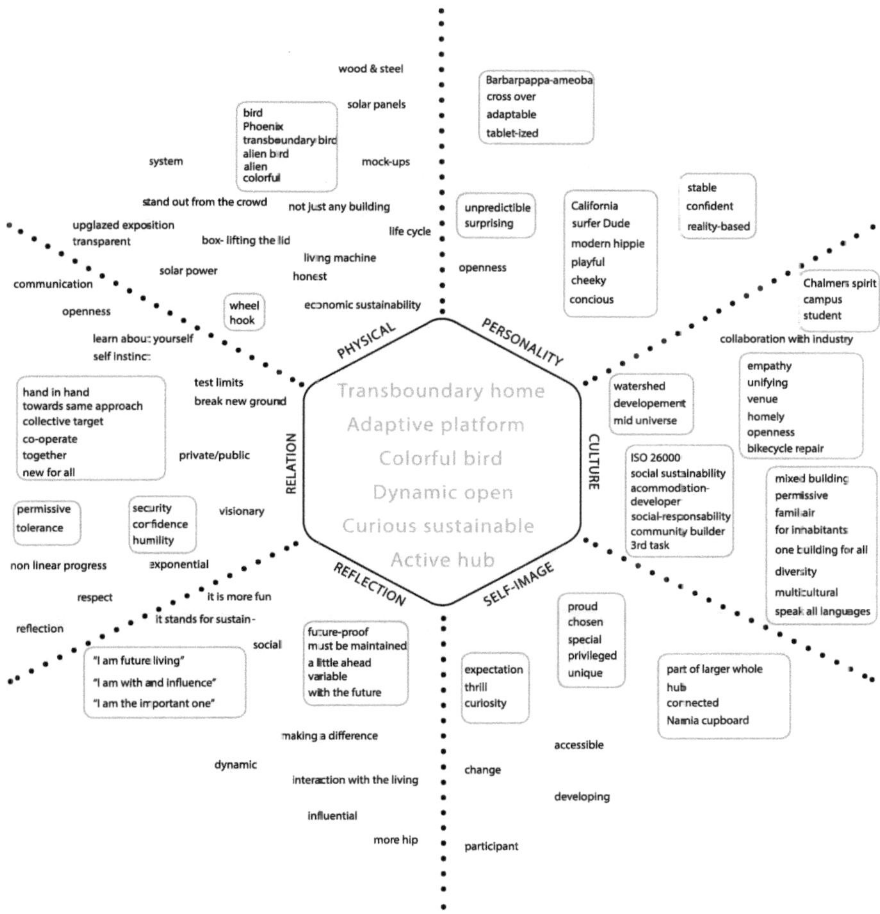

Fig. 10.2 The design process included the collation of thoughts and data followed by arranging into words and symbols and finally the values behind the focus fields

ISS adds structure to complexity that enables science, technical and behavioral research to be conducted, but in a low Earth orbiting platform.

One research need was the provision of equal living unit size and layout, as well as quantified numbers of equal situations. This challenged aspects of surface, economy and planning, providing a key question—how do we design at least 20 equal living units on a heavily limited footprint? The design result was the organization of social living units with relatively small private surface space, but fairly generous shared space that could be evaluated against each other for a wide range of research activities including different material choices or difference in social behavior (Fig. 10.4).

Fig. 10.3 Space syntax for the Living Lab, the basic concept is presented for space for each inhabitant

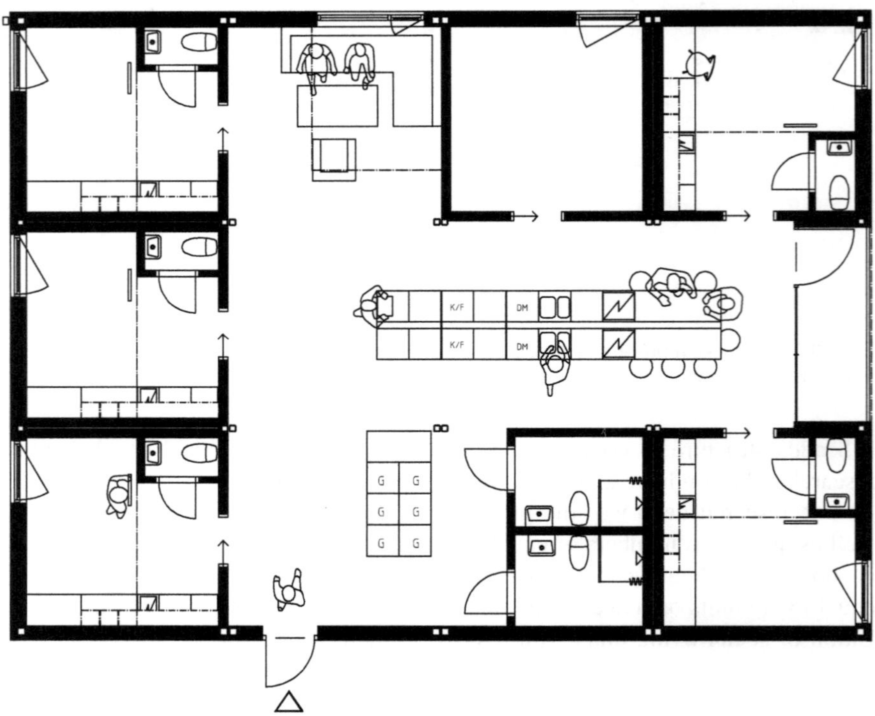

Fig. 10.4 Layout of 6 equally sized student apartments, mirrored to provide 12 apartments on the first floor. The equal size provides the basis for comparative research on a home energy management system and indoor climate, with extensive sensor systems in place. The shared space for the kitchen and living enables the introduction of compact private space

With slight adjustments, this organization of living space is repeated four times in the building to further enhance quantitative replication between groups. One floor, situated in mid-level of the building, is standardized, close to present market solutions, which provides the potential to use this as a test platform for small configuration with good economy that can be adapted for contemporary building form and practice.

The structure and order of the building concept, was also designed for feasible layout with the intention to allow the stakeholder and research communities to engage and communicate. This is important as the Living Lab platform will attract partners that may have less knowledge about built environments and/or research, although they nonetheless have a valuable contribution from their specific professional field. The architecture was therefore ordered to make it easy to understand and communicate around with three attributes of being generic, democratic, and finally, a built research and living platform.

10.3 Integration of the Six Focus Fields into the Design

10.3.1 Architecture, Plans and Accessibility

The design parameters are thoroughly pragmatic and the building is designed following simple principles of measurement, structures and surfaces and all with the aim to provide direct use of the building and to provide the flexibility for future changes and alterations in plan layouts. As an example of this, one of the main inspirations for a conceptual idea as well as built structure has been the Japanese tatami carpet (Daniels 2012), which as a single artifact may be used to inspire and organize spatial units. To this has been added an interest in spatial units and the use of volume (Fig. 10.5)—mainly in communal apartment groups—in the form of a higher ceiling height, small housing units with loft and common spaces with loft and contact with two different facades. The design incorporated simplicity and repetition, with small displacements, to create interesting differences between the floor levels but still offer equality for the quantitative research.

10.3.2 Architecture and Movability

A high priority from the start of the project has been to be able to reuse parts or move all of the building at a later date. The architect employed simple measurement principles that would work in conjunction with the current national building standards for residential volumes. The idea is that this building or a similar design can be moved and adjusted/programmed with small adjustments, both to the unique location of the Living Lab, but also for use as an infill project along a city street or solitary in an open area. For this purpose the building is divided into

Fig. 10.5 The compact private units feature a shift of paradigm in interior design from surface (m2) to volume (m3) through raised sleeping and storage space

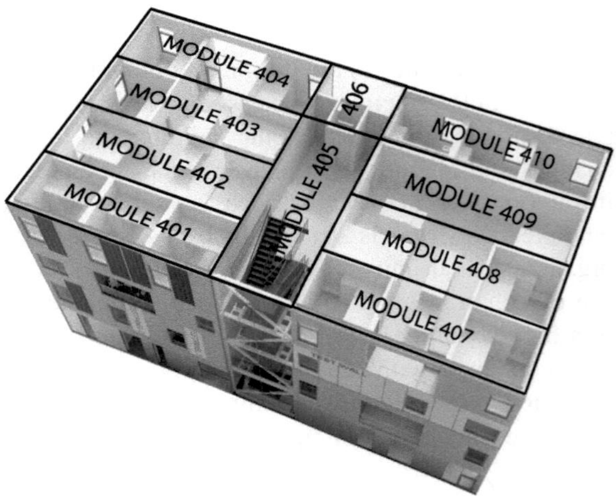

Fig. 10.6 The orientation of the 44 factory built modules and structures. This design concept has other potential applications other than an integrated student apartment/ Living Lab building, including infill of valuable urban space

44 modules that were prefabricated and then moved to the site and this enables the possibility for disassembly. A lot of effort and thinking went into making this field of design marry with the other required functions (Fig. 10.6). The design also allows for housing units to be lifted up on the roof at a later date, for 3D property development.

10.3.3 Ecological Sustainability

Ecological sustainability includes an overall thinking about the life-cycle perspective and the materials and products to use during construction, but also the flows of energy, water and resources through the building during use. Revisiting the analogous situation with space travel, this time looking beyond low Earth orbit, beyond the ISS, to future missions into deep space. These future missions, for example to Mars, will require sustainability not available to present day astronauts. Life support systems that recycle the water and revitalize the air and the growing of plants to produce fruits, herbs and vegetables locally are just two examples of the ecological sustainability that will be a required part of space exploration.

We have deliberately posited open questions (that do not lock into predetermining values, ideas and trends) have found in the design process that ecological sustainability is all about developing and changing our mindset as users and for the building to create social environments that can deliver experimental studies during the operation stage of the building.

One example of ecological sustainability was designing the space for private or common gardening has been planned through shared balconies for the communal apartment floors, as well as on the balconies of the more private floor level. This gardening activity is designed for a seamless transition into the common kitchen which enables research into reducing food waste (Fig. 10.7), the latter being based on our preliminary household metabolism studies (Harder et al. 2014).

Fig. 10.7 The common kitchen provides the shared space for the efficient use of water, energy and resources while contributing to a compact building footprint. Each common kitchen links the indoor and outdoor space through a covered balcony where the users' relation to gardens and urban ecology can be studied

The water engineering has been designed with the flexibility to allow rainwater harvesting for the gardens, a rain garden to demonstrate climate adaptation through stormwater infiltration and the selective recycling of wastewater. A further research facility is also being planned by Chalmers researchers, with free standing greenhouses, either on the roof of the building or on surrounding land.

10.3.4 Energy Systems

A Living Lab provides the possibility to study user interaction with heating/cooling and other energy related services. Early studies in the Suslab project (see acknowledgements) focused on the use of indoor climate sensors to support practice based design (Gram-Hansen 2009; Kuijer and de Jong 2012) for improved home energy systems. Meanwhile, Climate-KIC BTA (see acknowledgements) as a project was designed in parallel with HSB Living Lab to study the development of a home energy management system. As a consequence of these needs, 20 of the living units will be filled with sensors from start, close to 2000. The use of a central stairway space (Fig. 10.7) with direct vertical access through an open vertical shaft (with open and accessible systems) allows further testing and improvement of the system when required.

The fully visible systems design simulated the trunk of a tree that branches out on each level to support each living unit separately. The system has been termed Plug n'play by the architect as it is a fully flexible system that may be built at the preconstruction site and assembled to fully function when each module is mounted on site, as well as disassembled when parts of or the whole building is moved (Fig. 10.8).

10.3.5 Social Laundry Room

During the early design phase, parallel inspiration workshops were carried out to provide co-creation between the partners and researchers with the aim to provide the basis for both user and technically relevant prototypes. The first co-creation workshop was given the title "A Hackathon for Next Generation Clothing and Laundry Systems" and arranged by Chalmers, NASA and Rice University and run by the Royal College of Art in London. This methodology proved very effective for bringing the partners together and has now become an integral part of a multifaceted and creative design process.

The co-creation workshop was carried out in 2014 with students at Chalmers and Rice University (Houston, Texas USA), with engineers at NASA's Johnson Space Center and with representatives of the now burgeoning HSB Living Lab partnership (including, and particularly relevant, the Electrolux Group). The workshop unraveled innovative concepts for clothes washing that will be tested within

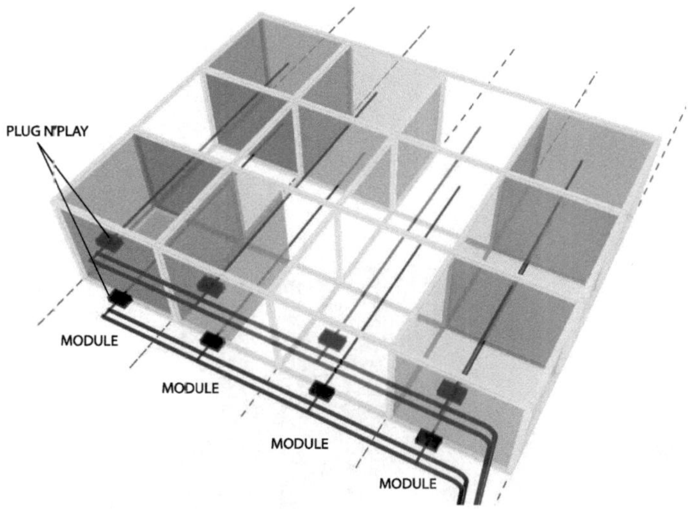

Fig. 10.8 The illustration represents the group of modules, arranging living units/apartments and their connection to one centralized energy system merging from one central shaft, thus making the systems visible and transparent

the HSB Living Lab, as well as informing the experts at NASA dealing with laundry in Space (Fig. 10.9). Two concepts emerged, the social washing space (see Fig. 10.10) and the refreshment cabinet for unsoiled clothing.

The results of the workshop were translated into the design of the building infrastructure through version 1.0 of the social washing room (Fig. 10.10) where a mock up version of laundry furniture is positioned and the washing machines connected in the very centre of the main entrance level space. This position is designed to differ from standard common laundry solutions where all the functions are positioned up against one free wall. Here the architect has designed a configuration to form furniture around which the users may interact and do very much more than just laundry, a dynamic meeting place (Fig. 10.10). This solution demonstrates how a Living Lab may communicate and interact ideas with society and accelerate innovation as this solution has already gained considerable interest in the Swedish media.

The social laundry room is the first example of how innovative concepts can be tested in the HSB Living Lab to uncover new ways to recycle and reuse natural resources in a small group environment of 4–6 students in a communal setting (Fig. 10.4) or in the Lab as a whole. The design process does not end with the building phase, as the innovation process will continue in this integrated environment of daily life which includes washing clothes, preparing meals and conducting personal hygiene. A second co-creation workshop held in 2015 focused on a next generation kitchen, both on Earth and in Space. Concepts for food preparation, storage, water use and reclaiming waste were investigated during the workshop.

Fig. 10.9 Astronaut Chris Hadfield demonstrates washing clothes on the International Space Station

10.3.6 Research

A Living Lab is a real-life place for knowledge innovation and as such should be supported by cutting-edge ideas in research. Contemporary thinking is that involvement of citizens in the co-creation process provides a greater user-centered value and thereby shifts focus from a rational organizational basis to a more novel and creative process (Ind and Coates 2013; Leminen and Westerlund 2012) which is both iterative and reflective (Vicini et al. 2013). So research supports the Living Lab user-based innovation process and was therefore an important aspect in the design.

Fig. 10.10 Version 1.0 of the laundry module test configuration has a spatial positioning in the room, rather than along a wall, with surrounding furniture and work stations that can be drawn apart and reassembled for the purpose of the student users and/or research

There has been extensive exchange of ideas and co-operation between Tengbom and Chalmers to analyze and identify the research needs from the building. A key research project was developed in the Climate-KIC BTA (see acknowledgements) to prototype new façade combinations and solutions. An exchangeable test panel system was designed as an integral part of the internal and external building walls (Fig. 10.11) and is sufficiently flexible to permit the prototyping of other indoor/outdoor systems such as thermal storage or adaptive cooling systems.

10.4 Insights

Designing a Living Lab requires an extra awareness of an open mindset and out of the box thinking by designers. But one should not forget that behind this is partners working together in a true spirit of co-operation with each and every interest emerging with time and often enriching the design process. The consequence is that in this project the architects has become much more than designers alone, but also informal project leaders, directing the dialogue and communication of the process, the interaction of partners and their input into the design of the building.

The design focus has been to create a platform that can be adjusted and tailored as the research reveals user practice and behavior related to layout. This then is the very essence of the Living Lab; people living in the Living Lab environment and thereby affecting the outcome of studies which can then influence ideas and new studies, both physically and socially.

Fig. 10.11 The building includes 12 interchangeable wall/facade test panels that can be replaced through a cassette system, as well as 2 full sections of the façade on the third floor

The design/building project has become the very generator and the vital symbol that, with its conceptual ideas has interacted and brought the partners together. The design team, the researchers from Chalmers and the co-author from NASA have continually challenged the partners to take important steps out of their professional comfort zones. The result has been a quality Living Lab within reasonable economic constraints and, importantly, meeting partner expectations.

The HSB Living Lab at Chalmers has been built to test and demonstrate sustainable technologies and practices for future homes. However, there are other communities that might learn from the research conducted, and the findings that will result from the use of the Living Lab. One of those communities is the international group of Space researchers looking at more sustainable habitats for future human missions beyond low Earth orbit. The long duration of such missions, to destinations such as Mars, actually parallel the conditions and constraints that will be present in the HSB Living Lab. The extended duration of testing of

Fig. 10.12 Stairway space of the HSB Living Lab, displaying the unique and transparent, easy accessibly, systems

technologies and monitoring of user behaviors within the Living Lab are relevant analogs to space habitats, which will need to be more sustainable and make wise use of finite resources such as water, air and food.

These are excellent examples of a required activity needed, whether you are living on Earth or traveling to Mars. The HSB Living Lab provides the platform, the analog, to demonstrate and test these concepts, in an "integrated environment", over an extended time, with people. While this is being conducted within the next few years at the HSB Living Lab, it could actually benefit those looking 20–30 years into the future of human space exploration. Space agencies such as NASA have a unique opportunity to change the way it thinks about conventional concepts like kitchens and laundry. Combine that with the necessity of even more innovation for the journey to Mars and the natural inclination of students to take a fresh look at things and the possibilities multiply exponentially. The HSB Living Labs has been successfully built and will open in the summer of 2016 (Fig. 10.12).

Acknowledgments Anna Olofsson of HSB was a pioneer of the Living Lab. Sanna Edling of HSB demonstrated leadership in bringing the partnership together through regular meetings and co-creation workshops.

The Living Labs in the Interreg NWE Suslab program and the EIT Climate-KIC BTA (Building Technology Accelerator) flagship program were a source of inspiration for this publication.

All illustrations are reproduced with permission of Tengbom Architecture and NASA.

Photographs by architect Peter Elfstrand reproduced with full permission.

References

Bergvall-Kareborn, B., Ihlstrom-Eriksson, C., & Stahlbrost A. (2015). Places and spaces within Living Labs, *Technology Innovation Management Review, 5*(12), 37–47.

Brynarsdottir, H. (2012). Sustainability unpersuaded: How persuasion narrows our vision of sustainability, CHI'12 Texas, pp. 947–955.

Daniels, I. (2012). Material cultures of domestic interiors: Japan, Home/Homelessness, pp. 211–216, Elsevier.

Dell'Era, C., & Landoni, P. (2014). Living Lab: A methodology between user-centred design and participatory design, *Creativity and Innovation Management, 23*, 137–155.

Evans, J., Jones, R., Karvonen, A., Millard, L., & Wendler, J. (2015). Living Labs and co-production: University campuses as platform for sustainability science. *Current Opinions in Environmental Sustainability, 16*, 1–6.

Favela, J., Kaye, J., Skubic, M., Rantz, M., & Tentori, M. (2015). Living Labs for pervasive healthcare research. *Pervasive Computing*, April–June, 86–89.

Femenias, P., & Hagbert, P. (2013). The Habitation Lab: Using a design approach to foster innovation for sustainable living. *Technology Innovation Management Review, 3*, 15–21.

Franz, Y. (2015). Designing social living labs in urban research. *Info, 17*(4), 53–66.

Gram-Hansen, K. (2009). Standby consumption in households analysed with a practice theory approach. *Journal of Industrial Ecology, 14*(1), 150–165.

Harder, R., Kalmykova, Y., Morrison, G., Mangold, M., Feng, F., & Dahlen, L. (2014). Quantification of goods purchases and waste generation at the level of individual households. *Journal of Industrial Ecology, 18*(2), 227–241.

Ind, N., & Coates, N. (2013). The meanings of co-creation. *European Business Review, 25*(1), 86–95.

Kuijer, L., & de Jong, A. (2012). Identifying design opportunities for reduced household resource consumption: Exploring practices of thermal comfort. *Journal of Industrial Design, 10*(1/2), 67–85.

Lai, H. C., Hu, K. K., & Chen, L. W. (2011). The activities and typologies in service innovation design and deployment: A socio-technical perspective on University based Living Lab, 2011, *Proceedings of the 2011 IEEE IEEM*, 1155–1159.

Lapointe, D., & Guimont, D. (2015). Open innovation practices adopted by private stakeholders: Perspectives for living labs. *Info, 17*(4), 67–80.

Leminen, S., & Westerlund, M. (2012). Towards innovation in Living Labs networks. *International Journal of Product Development, 17*, 43–59.

Lehmanm, V., Frangioni, M., & Dube, P. (2015). Living Lab as a knowledge system: An actual approach for managing urban service projects. *Journal of Knowledge Management, 19*(5), 1087–1107.

Mastelic, J., Sahakian, M., & Bonazzi, R. (2015). How to keep a Living Lab alive? *Info, 17*(4), 12–25.

Rits, O., Schuurmann, D., & Ballon, P. (2015). Exploring the benefits of integrating business model research with Living Lab projects. *Technology Innovation Management Review, 5*(12), 19–27.

Vicini, S., Bellini, S., & Sanna, A. (2013). User-driven service innovation in a smarter city Living Lab. *International Conference on Service Science*, pp. 254–259.

Watson, L., Johnson, C., Hegtvedt, K. A., & Parris, C. L. (2015). Living green: Examining sustainable dorms and identities. *International Journal of Sustainability in Higher Education, 16*(3), 310–326.

Chapter 11
Exploring the German Living Lab Research Infrastructure: Opportunities for Sustainable Products and Services

Justus von Geibler, Carolin Baedeker, Christa Liedtke, Holger Rohn and Lorenz Erdmann

Abstract Living Labs for Sustainable Development aim to generate low-resource innovations in production-consumption systems by integrating users and actors. This chapter presents the results of a German study investigating potentials of and measures towards the realization of a German Living Lab infrastructure to support actor-integrated sustainability research and innovations in Germany Geibler et al. (2014). Generally, as the status quo analysis revealed, the sustainability and Living Lab communities in Germany are hardly intertwined. However, twelve Living Labs that explicitly consider sustainability aspects could be identified. The analyses of drivers and barriers as well as SWOT (Strengths and Weaknesses, Opportunities and Threats) provided the foundation to identify options for the promotion of a user-integrating research infrastructure supporting sustainable products and services. The measures suggested for Germany include a funding program for actor-integrated, socio-technical research based on a Living Lab network, a communication campaign, and programs to foster networking and the inclusion of SMEs. Some of the suggested measures have already been taken up.

Keywords Living lab · User integration · Innovation · Sustainability · Resources · Resource efficiency · Innovation system · Research infrastructure

J. von Geibler (✉) · C. Baedeker · C. Liedtke
Wuppertal Institute for Climate Environment and Energy,
Döppersberg 19, 42103 Wuppertal, Germany
e-mail: justus.geibler@wupperinst.org

H. Rohn
Faktor 10—Institut für nachhaltiges Wirtschaften gemeinnützige GmbH,
Alte Bahnhofstraße 13, 61169 Friedberg, Germany

L. Erdmann
Fraunhofer Institute for Systems and Innovation Research ISI,
Breslauer Straße 48, 76139 Karlsruhe, Germany

D.V. Keyson et al. (eds.), *Living Labs*, DOI 10.1007/978-3-319-33527-8_11

11.1 Introduction

Today's challenges such as climate change, migration, and loss of productive land (Rockström et al. 2009; Münz and Reiterer 2009; Bringezu et al. 2014) are negative effects triggered by the ever-increasing worldwide consumption of resources (Fischer-Kowalski et al. 2011). Although more and more efficient production methods are developed, a major turnaround is out of sight. Key drivers of the over-consumption of natural resources by the human technological system are basically our current lifestyles in industrialized countries and corresponding household consumptions (Baedeker et al. 2008; Welfens et al. 2010). Based on the material footprint, which includes resources used for energy, a study sketched out a reference frame of the magnitude of reduction needed in terms of resource consumption (Lettenmeier et al. 2009). As a long-term goal, an annual material footprint of 8 tonnes/person should be targeted for sustainable household consumption by 2050 (Lettenmeier et al. 2014). To achieve this target, the present material footprint of household consumption must be reduced by 80 % (Lettenmeier et al. 2014). It is clear that a fundamental transition of the current production-consumption system is necessarily required to reach a reduction of this magnitude. Its drivers are not only the development of innovative, low-resource technologies, products and services, but also lifestyle changes. Research and innovation systems that can effectively generate these low-resource product and service innovations are an essential component of such a transition (BMBF 2012).

Over the last decades, a large number of potential innovations towards a sustainable development, including efficient technologies as well as products and services, were developed without considerable integration of consumers—for example car sharing, new heating systems, alternative light bulbs or detergents and electric kettles. Unfortunately, most of these innovations did not lead to the desired effects due to unexpected user behaviour and changing life styles. In fact, the technologically possible efficiency gains were even overcompensated by changing behavioural patterns (for an overview see Buhl et al. 2015). Many product service innovations with a high sustainability potential fail because they are rejected by consumers or create negative rebound effects (Sorrell 2007; Druckman et al. 2011).

In both research and product development, the interfaces of socio-ecological transformation of consumption patterns have obviously received little attention so far. Amongst other factors the indifference is caused by the fact that facilities for explorative and experimental research in real-world surroundings (e.g., household laboratories) are still lacking (Jackson 2005; Talwar et al. 2011). To counteract this lack of attention Living Labs for Sustainable Development present a promising network and actor-interaction approach (Liedtke et al. 2012, 2015). Based on an earlier article this chapter points out "*the potential for and the keystones of a user- and actor oriented future research- and innovation infrastructure in Germany, which will use the Living Lab approach to foster energy- and resource efficiency innovations to contribute to long-term sustainable development*" (Geibler et al. 2014, p. 577). Furthermore, first steps of implementation of suggested measures are described.

11.2 Methodology for Assessing the Potential of a German Living Lab Research Infrastructure

11.2.1 The Living Lab Concept

Originally developed at the Massachusetts Institute of Technology (MIT) (Pierson and Lievens 2005) the Living Lab concept with focus on sustainable innovation has been defined for the study as follows:

"*A Living Lab for Sustainable Development (or Sustainability Living Lab) is a research approach aimed at open socio-technical innovation processes, in which users as well as relevant actors of the value chain and the utilization environment participate in the development and application of new products, services and system solutions. The interactive innovation process takes place in the real environment of the users (e.g., user observation, field tests) and/or in laboratories that are configured for user interactions (e.g., for the development of prototypes). The innovation process is guided by sustainability criteria and aims to contribute to production and consumption patterns that can be applied on the global and long-term scale and are inter- and intragenerationally viable.*" (Geibler et al. 2014, p. 578)

For the German sub-project of SusLabNWE we adopted this definition in a slightly modified version (see Liedtke et al. 2015, p. 107) as a guideline for the research and development process.

11.2.2 Research Steps and Methods

Based on the concept of innovation and technology analysis (ITA), the analysis of the potential for a German Sustainability Living Lab research infrastructure combines research and practical considerations (Zweck et al. 2008; Geibler et al. 2011). Therefore discussions of experts and stakeholders concerning innovations and new technologies are of particular relevance. In total, 41 experts and stakeholders participated. The research was conducted in a five-step approach:

Research Step 1: In order to characterize the Living Lab landscape in Germany, the German research and development (R&D) landscape was screened on the basis of an Internet and literature research. Also Living Labs in neighboring countries and temporary projects were taken into account for benchmarking purposes. To do so, differences were made between laboratories that explicitly considered sustainability aspects and those who implicitly worked on topics of relevance to sustainability. In addition, the level of institutionalization of the laboratory was investigated, by making another differentiation between temporary single projects on the one hand and permanently established Living Labs on the other hand.

Research Step 2: In a second step, fields of application that promise a high potential for benefiting from user integration in the innovation process were identified via a literature search. Besides efficient technologies, the focus was on products and services, which have the potential to significantly reduce resource consumption, i.e., can boost resource efficiency. Next, the identified relevant products, technologies and services were classified and grouped into specific areas of application known to be significant for resource consumption on the household level. Those listed products, technologies and services with the highest potential for resource efficiency and applicability were identified by means of a set of criteria developed by Rohn et al. (2009) and adapted for this study (see Table 11.4 in the Annex). Within the framework of a potential-validation workshop, the results could be supplemented and evaluated through expert interviews and discussions (Workshop "Sustainability potentials of Living Labs" 2012).

Research Step 3: The third step focused on the creation of an innovation-oriented and internationally competitive research and development infrastructure in Germany by analyzing its sufficient conditions and by conducting a further literature analysis and five expert interviews. In two expert workshops an extensive dialog unfurled, in which 350 actors were identified and evaluated with regard to their relevance for this project. 60 of these were rated as highly important based on their potential towards innovation, consumer orientation and sustainability.

Research Step 4: Based on the research results a profile of strengths, weaknesses, opportunities and threats (SWOT) for a German research and innovation system was created. This analysis covered both the general structure and performance of the Living Lab landscape and its orientation towards sustainability. The analysis was evaluated by a series of interviews with Living Lab practitioners and experts.

Research Step 5: The last step leads to a discussion about options for practical measures and strategies for the implementation of a successful research infrastructure of Living Labs for Sustainable Development in Germany. The discussion is based on the SWOT profile and the results from the expert workshops and interviews.

11.3 Points of Departure for the Development of a Sustainability Living Lab Infrastructure in Germany—The Results

11.3.1 Results of Step 1: The Status Quo

Regarding the current practice, the analysis of the existing infrastructure revealed that the term "Living Lab" is used to describe a broad range of laboratories and research approaches. Through desk research and further amendments in the expert workshops, a total of 76 Living Labs were identified in Germany and neighboring countries (see list of Living Labs in the Table 11.5 in the Annex to this section). In Fig. 11.1 their geographical locations, as well as the degree to which sustainability is considered in these Living Labs are depicted. 40 of the 74 Living Labs meet

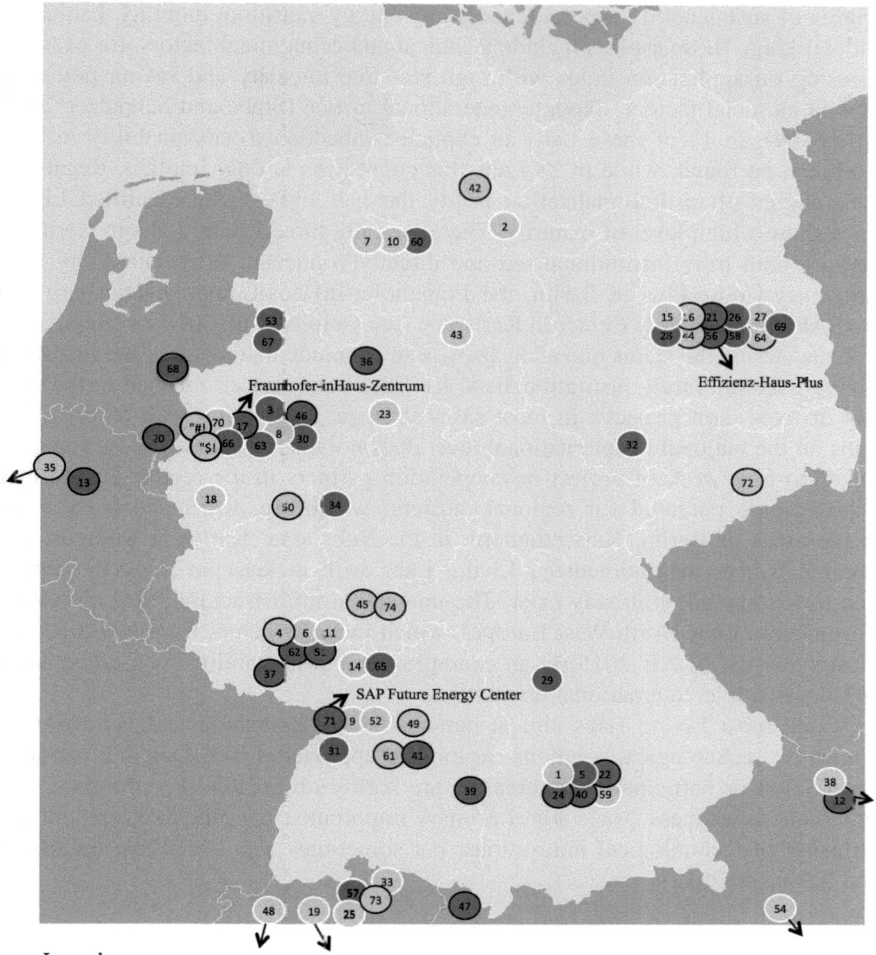

Legend:

Explicit or implicit consideration of sustainability

Connection to sustainability unclear or not present

High degree of institutionalisation

The numbers specify the Living Labs, see Annex (Table 11.5) for their names and research fields.
Arrows indicate names or a location outside the map.

Fig. 11.1 Living Labs identified in Germany and neighboring countries (*Status* April 2012) (*Source* Geibler et al. 2014, p. 581)

criteria of sustainability aspects, concerning energy, nutrition, mobility, habitation and working. Those aspects include technical and economical factors, for example focusing on application fields with high resource intensity and saving potential, as well as social factors, like intergenerational justice (Stabe and Schnalzer 2012, pp. 65–69). In 12 of these Labs an explicit connection to sustainability in their work can be found, while in 28 Labs this connection is only implicit. Regarding their degree of institutionalization nearly the half (34) of the identified Living Labs show a high level of maturity. There are only three Living Labs in Germany that are both fully institutionalized and directly connected to sustainability: The Efficiency-House-Plus in Berlin, the Fraunhofer-inHaus-Center in Duisburg and the SAP Future Energy Center in Karlsruhe (see Geibler et al. 2014 for details).

Considering the status quo analysis we can conclude that many of the identified Living Labs are rarely institutionalized facilities that were established in the context of fixed-term projects. In most cases strategic networking with other Living Labs on the national or international level does not exist. However these Labs are linked to a network of project or cooperation partners in the region. Figure 11.1 reveals some potential for regional clusters, which are distributed particularly in the areas of Berlin, Kaiserslautern, in the Ruhr area, Karlsruhe/Stuttgart and Munich, where institutionalized Living Labs with at least an implicit connection to sustainability already exist. The innovation infrastructure "SusLab NWE" (Sustainable Labs North West Europe), which includes actors from North-Rhine-Westphalia (SUSLAB 2015) is an example for a sustainability focused research consortia with an international orientation.

While most Living Labs aim at demonstration, evaluation and validation of innovative technological solutions explorative approaches like open and unbiased observation or participative co-creation are hardly observable. For the most part economic or business issues are of primary importance, e.g., the development and diffusion of technological innovations, but sometimes also social aspects (Stabe and Schnalzer 2012).

11.3.2 Results of Step 2: What Areas of Application Promise the Greatest Sustainability Potentials?

Table 11.1, which is based on an earlier description by Geibler et al. (2014), lists the identified product groups, technologies and services with high resource efficiency potential and high potential for development in Living Labs. The results of the systematic assessment (see Table 11.4 in the Annex for the evaluation criteria and scoring scales used) are grouped into the three particularly relevant areas of application ((1) Living and working; (2) Town, region and mobility; (3) Retail and gastronomy), as well as the two research perspectives "user behavior" and "product innovation" (Rohn and Leismann 2012).

Table 11.1 Overview of relevant fields of application identified for living labs for sustainable development

Field of application	Research perspective	
	Product/Service innovations	User behavior
Life and work	**Building and infrastructure**, e.g., security, heating and energy supply, insulation, e-energy/energy assistance **Food**, e.g., chilling, storage, preparation, assistants **Health and hygiene**, e.g., medical care, fitness, medical technology **Furnishings of living and working spaces**, e.g., design of electric and electronic equipment, furniture, textiles **Information management**, e.g., communication in the home/out of home, ICT products and their use **Substitution of physical mobility** by "ICT mobility", connection to logistics systems, Smart Grids	**Behavior at home and workplace**, e.g., health and exercise, energy consumption **Nutrition**, e.g., food wastage, shopping, health **Phase of life appropriate design of home/workplace**, e.g., autonomous life at old age, user acceptance of innovations **Integrated design**, e.g., in the area of fields of demand or service design **Furnishings of living and working spaces**, e.g., new workplace concepts, ways of utilization, cascading systems, ICT **Service and time management**, e.g., being mobile, eating healthy, exercise
Town, region and mobility	**Out of home catering**, e.g., delivery services, drive-in restaurants, etc. **Mobility**, e.g., efficient mobility options (logistics), freight, public transport linkages, design of mobility options **Regional networks/"location promotion"**, e.g., health support systems, urban planning, communication systems, regional energy supply, tourism, sharing and renting options	**Mobility**, e.g., use and user acceptance of resource efficient mobility options **Communities/networks**, e.g., urban agriculture, barter systems, neighborhood networks, service concepts and suburb development **Leisure/holiday behavior**, e.g., regional tourism **ICT services**, e.g., integrated ICT, mobility and logistics management
Retail and gastronomy	**Furnishings**, e.g., electric and electronic equipment, lighting, media, online shopping, design **Mobility**, e.g., efficient mobility options **Nutrition**, e.g., food labelling and declaration **Support at old age**, e.g., intelligent appliances	**Intelligent appliances**, e.g., digital product memory **Choice of products**, e.g., influence of advertisement and information campaigns

Source Geibler et al. 2014, pp. 584–585

11.3.3 Results of Step 3: Drivers and Barriers for a Sustainability Living Lab Approach

Based on expert workshops and interviews, the following expertise-related factors can be assessed as particularly beneficial for the development of a Sustainability Living Lab infrastructure:

- Systematic utilization of sustainability indicators such as MIPS, Material Footprint, Carbon Footprint, GRI indicators (Lettenmeier et al. 2009; Bringezu et al. 2009; Spangenberg et al. 1999; Lin et al. 2014; Saurat and Ritthoff 2013; Wiesen et al. 2013) and assessments, which consider long-time horizons and high case numbers; and
- A sensitivity to cultural barriers that could be in the way of considered socio-technical approaches towards more sustainable patterns of production and consumption.

The following expertise-related factors have been stated by interviewed experts to be particularly detrimental to the development of a Sustainability Living Lab infrastructure:

- Time-constrained and reductionist research designs; and
- A lack of competency for inter- and transdisciplinary communication among researchers and between researchers and users.

Regarding the role of power-related factors, experts emphasized that it is necessary to ensure long-term public financing due to the high share of basic research to be conducted.

Experts estimate that sufficient freedom for the designers and prototype developers is required to enjoy creativity and the willingness to take risks, which is conducive for process aspects.

Based on the expert interviews and workshop discussions, barriers are seen in "the high burden of time and effort that need to be invested in ensuring optimal assistance for and interaction with the users (…), risk averse funding principles of public funding bodies that are not ideally tailored to the needs of innovative and sustainability-oriented activities and a lack of the flexibility that is needed in order to support or participate in dynamic innovation processes that include a wide array of different actors" (Geibler et al. 2014, p. 586).

Regarding relational aspects, the workshop focused on user motivation and the importance to test and support the users ability for reflection on their behavior. It is important, that the social relations to users are moved at the center-stage of Living Labs for Sustainable Development.

11.3.4 Results of Step 4: SWOT for the Implementation of the Living Labs for Sustainable Development Approach

Based on the previous results, the Strengths and Weaknesses, Opportunities and Threats (SWOT) for Living Labs for Sustainable Development in Germany are derived. The results are summarized in Table 11.2.

The results of the SWOT analysis show that, like many other innovation projects, research in Living Labs is not entirely free of certain risks. Based on the results of the SWOT analysis, a number of options and strategies for action which are outlined in the next section, can be developed.

Table 11.2 Strengths and weaknesses, opportunities and threats (SWOT) of the German research and innovation system for the development of living labs for sustainable development

Strengths	Weaknesses
• **Acceptance of sustainability**: Sustainability is a broadly accepted goal, which can be referred to • **Existing Living Labs are complementary to the technology-focused research landscape**: Living Labs enable the integration of non-technical aspects in technical innovation processes • **Existing regional clusters**: A "place" is important to connect actors	• **Lack of capacity**: Systems design and mediation capacities should be developed, e.g., to bridge between "soft" approaches like user-integration and "hard" technology-centered approaches • **Opinion of limited commercially usable outcomes**: Results of experimental research do not necessarily yield commercially viable products in the short-term. The commercial usability of sustainability solutions could be enhanced by including businesses in the development process
Opportunities	Threats
• **Realization of efficiency potentials under consideration of rebound effects on the micro-scale**: A user-centric development process can increase user acceptance and be crucial for product success • **Potential to connect different strands of research**: Research at the interfaces between sustainability, innovation and user-integration already exist but are largely unconnected • **Existing international networks** of Living Lab research should be integrated • **Capacity development** at universities: Universities offer infrastructures for mediators between user needs, sustainability aspects and technological perspectives	• **Short-term thinking in business strategies**: The research design of Living Lab projects could insufficiently consider implementation perspectives • **Data security issues**: The sensitivity of data on consumption and behavioral patterns cannot be integrated sufficiently in Living Labs if not considered e.g., by implementing accompanying ELSA-assessments • **Compatibility of assessed micro-data**: Data sets from Living Labs might be incompatible with macro-data on the societal system, if data interfaces and assessment conventions are not defined (The compatibility with macro data is helpful when comparing potential environmental or social improvement of a specific case with reference values at the macro-level. If data sets are incompatible, comparisons are less meaningful.)

Source Geibler et al. 2014, p. 588

11.3.5 Discussion of Strategies and Options for Action (Results of Step 5)

The Living Lab project results as well as the discussions within the project consortium, including the participants of the expert workshops, exposed a number of strategies and options for action to integrate the approach into the German research and innovation system. The project consortium unanimously derived the following specific goals for the further development of a German research and innovation system based on Living Labs for Sustainable Development (for more detailed description see Geibler et al. 2014):

- The further development of a research infrastructure for user integrated development of sustainable products and services. The goal is the creation of sustainable innovations, which have been tested for systemic effects, in German, European and international cooperative research networks.
- Improved access to such a research and innovation system, especially for research institutes and companies who cannot maintain such an infrastructure themselves (e.g., small and medium sized enterprises).
- To enhance and speed up research and development of resource efficient, competitive and socially acceptable products and services that can contribute significantly to a system-wide reduction of resource consumption in the household or other fields of application, such as the point of sale.
- Improved networking amongst European researchers and execution of joint research projects.

Based on the results, the authors suggest strategies and tangible options to integrate Living Labs for Sustainable Development into the German research and innovation system (see Table 11.3). There are two basic strategies: a *structure formation within the research and innovation system* including networking and profile formation and a *funding program "Living Labs for Sustainable Development"*. While the first strategy aims at networking and profile formation to support the innovative capacity, participation and design competencies, especially of SMEs, the second one develops sustainability potentials in specific fields of application, e.g. by funding of lighthouse projects.

One suggested measure has been already taken up: The German Ministry of Education and Research is funding a demonstration project in Germany, see Box 11.1.

Table 11.3 Strategies and options to integrate living labs for sustainable development into the German research and innovation system (*Source* Geibler et al. 2014, pp. 591–592)

Strategies	Options
1. Structure formation within the research and innovation system	
Networking and profile formation of the so far poorly contoured Living Lab landscape	• Financial support of regional clusters with thematically complementary profiles. The funding should include lighthouse projects and projects conducted within clusters, as well as poorly institutionalized facilities for experimental research that do not use the term "Living Lab" • Networking and synopsis of results and experience gained in Living Labs, e.g., through conferences or workshop series • Creating links between German Living Labs and international partners, e.g., in EU research and innovation programs such as "Horizon 2020"
Promoting the innovative capacity of SMEs	• Targeted inclusion of SMEs and trades in Living Lab research and value chain-wide joint projects • Improve access to Living Lab innovation systems for SMEs
Strengthening participative processes	• Promoting participation in innovation processes through Living Labs, e.g., in sustainable urban development
Establish design competency	• Develop systems design competency through adapted education concepts for actors from science, business, local authorities and households, e.g., under consideration of milieu structures, socio-cultural approaches, social motions research and communication research
2. Program "living labs for sustainable development"	
Lighthouse projects – Use potentials – Initiate innovations – Demonstrate examples	• Foresight process for the strategic alignment of national and regional innovation initiatives based on network analyses of existing Living Lab clusters; analyses of actor relationships and thematic foci; analyses of potentials for individual fields of action under consideration of regional competition-relevant strategies • Invitation for tenders for integrative, inter-departmental lighthouse projects by the German research ministry or other ministries in a competition for Living Labs to promote creativity, innovation, reflection and multidisciplinarity Addressing specific fields of action, e.g., "user integration in life sciences and bio-economy", "IT security in home automation", "sustainability in lead markets" (key technologies, new materials), etc. • Funding of demonstration projects to lift the profile of the Living Lab approach with its potential to integrate non-technical aspects into the development of technologies and business models

(continued)

Table 11.3 (continued)

Strategies	Options
Basics of transition and innovation research – Transdisciplinary innovation research – Interdisciplinary action research	• Consolidation of empirical transition research regarding new approaches to innovation, e.g., at the interface between experimental approaches of action research and technology development • Further development of conceptional and methodological foundations regarding the application of Living Labs in ELSA analyses and for environmental and sustainability assessments, e.g., in product tests or business model development • Development of principles for sustainability oriented design research under consideration of interdisciplinary, self-reflective Living Lab approaches in the design process • Development of theories and concepts for inter- and transdisciplinary education/didactics
Accompanying communication – Raise the profile – Create awareness – Enable exchange	• Communication campaign for businesses, politics, households and research, e.g., at subject-specific events, which are so far conducted without consideration of sustainability aspects • Preparation of information materials and communication formats for stronger involvement of SMEs, the trades, business associations, consumer-, environmental-, and sustainability initiatives • Specify the concept of "Living Labs for Sustainable Development" and introduction of the concept as a brand • Launching of a competition for lighthouse projects for "Living Labs for Sustainable Development"

Box 11.1 Demonstration project in Germany

Demonstration Project INNOLAB: The Potential of LivingLabs in the Green Economy

The project INNOLAB assesses and demonstrates the potential of LivingLabs in the Green Economy. Based on close interaction with three German Living Labs, the project pursues the following objectives:

- Developing the methods for user integration and sustainability innovation in real-world laboratories, e.g. concerning rebound effects and obsolescence
- Demonstrating the potential of Living Labs through the study of sustainability innovations in the field of assistance systems and their diffusion in the key areas of sustainable consumption "living", "retail", "mobility"

- Strengthening of the Living Lab concept in the research and innovation system of a Green Economy through integrated roadmapping
- Developing the national and international network and transfer

The project is funded by the German Ministry of Education and Research under the framework programme "Research for Sustainable Development" (FONA) and within the funding programme "Sustainable Economy". More information can be found at the project's website: http://www.innolab-living-labs.de.

11.4 Summary and Conclusions

The status quo analysis of the German R&D landscape identified the potential of a future Living Lab research infrastructure for Sustainable Development, in which the identified facilities could be integrated. Although a landscape of several Living Labs already exists, sustainability is not a key consideration in the innovation processes yet. Especially resource efficiency solutions are hardly addressed. The same applies to explorative research approaches like open observation or participative co-creation. Furthermore, an established network between the labs and the innovation and sustainability arena is needed to achieve viable outcomes for sustainable development.

One of the main goals of the SusLabNWE project was to create such a unique new infrastructure where insights, co-development and validation of sustainable products, services, legislation and combinations of these take place directly with users in the complexity of their living environment and daily practices. At four locations in Europe (Rotterdam (NL), Ruhr region (DE), London (UK) and Gothenburg (SE) regional Sustainability Living Lab infrastructures were set up (Baedeker et al. 2014). These locations form an important part of the infrastructure to be developed further.

Some key conditions for a successful Sustainability Living Lab infrastructure can be defined as follows: "long time horizons for research projects, enabling open-ended innovation processes, reflexive learning and strong transdisciplinary, systematic sustainability assessments that are integrated into the innovation process, and the consideration of socio-cultural factors in user behavior and acceptance" (Geibler et al. 2014, p. 593). It became apparent, that Living Labs offer the opportunity to integrate sustainability research, design, innovation and technology studies, social sciences and cultural studies.

Although the German Living Lab landscape is still a heterogeneous field with soft contours and the research body is still limited, some tentative recommendations can be provided. For the first step it is suggested to improve structure formation in the German research and innovation system, aim at networking and profile formation, support SMEs to get access to the Living Labs, and foster transdisciplinary and participative processes. Then, in a second step, based on the exploration of markets and specific services of Living Labs for Sustainable Development in different fields of application, Living Labs for sustainability can be promoted in specific funding programs.

In Germany some steps towards fostering such a research infrastructure could be made in the course of SusLabNRW in the Ruhr area and will be followed in the future. This infrastructure is now being used in related projects of the project partners and extended to other fields of sustainability, making use of some of the conclusions drawn from the analysis presented here: using the developed infrastructures (e.g. for indoor-climate data logging and analysis at Hochschule Ruhr West in Bottrop), using the fostered methodological three-steps toolbox for development of sustainable product-service-innovations, seeking opportunities for further funding and development (cf. textbox demonstration project INNOLAB above) and aiming to establish a permanent consulting offering as a product-service-system based on project results. Those Living Labs, which are established in different fields of application, could serve as anchor points for the further development for the German research and innovation system.

Other promising steps are developed in the ongoing INNOLAB project: Beside an update of the living lab mapping by Geibler et al. (2014) an integrated roadmap will be set up with the aim to further strengthen the Living Lab concept in the research and innovation system of a Green Economy and to develop the national and international network.

Acknowledgments We are grateful to Moritz Linder, Joanna Behrend, Till Heller, Marco Hasselkuß and Aline Fink for their assistance in formatting and proof reading. We acknowledge the research funding received from the German Ministry of Research and Education for underlying research in two projects: "Nachhaltigkeitsinnovationen im Living Lab" (grant no. 16I1624 and 16I1625) and "INNOLAB – Living Labs in der Green Economy: Realweltliche Innovationsräume für Nutzerintegration und Nachhaltigkeit" (grant no. 01UT1418A-D).

Annex

(See Tables 11.4 and 11.5.)

Table 11.4 Evaluation criteria and scoring scales for the selection of product/technologies/services with high resource efficiency potential (*Source* Geibler et al. 2014, supplementary material)

No.	Criteria to evaluate product group	Scoring scale	
1	Observation environment	1 = Living space;	4 = Shop
		2 = Working space	5 = Region
		3 = Hotel/ Restaurant	6 = other
2	Key parameter of observation	1 = Energy	4 = Money
		2 = Water	5 = Time
		3 = Waste	6 = other
3	Possible to investigate in Living Labs?	1 = yes	0 = no
Environmental sustainability aspects			
4	Resource inputs/quantity, e.g., – Product/infrastructure requiring high resource inputs – Mass application – Rebound effects	0 = none	3 = high
		1 = low	4 = very high
		2 = medium	
5	Resource efficiency potential, regarding e.g., – Abiotic resources, biotic resources – Water, Energy – Other resources	0 = none	3 = high
		1 = low	4 = very high
		2 = medium	
Economic sustainability aspects			
6	Feasibility, e.g., – Technical feasibility – Technical know-how available in Germany – Economic viability – Market and societal acceptance	0 = none	3 = high
		1 = low	4 = very high
		2 = medium	
7	Economic importance, e.g., – Market potential – Level of innovation – Relevance for export – International importance – Dependence on finite natural resources	0 = none	3 = high
		1 = low	4 = very high
		2 = medium	
Social sustainability aspects			
8	Improving or maintaining health and quality of life, e.g., – Maintaining societal status – Active participation in society	0 = none	3 = high
		1 = low	4 = very high
		2 = medium	
9	Social relevance, e.g., – Job security and social security – Improving and maintaining equal opportunities – Participation opportunities – Adherence to social standards along the value chain (e.g., ILO standards)	0 = none	3 = high
		1 = low	4 = very high
		2 = medium	

Table 11.5 List of identified Living Labs in Germany and neighbouring countries (*Status* April 2012) (*Source* Geibler et al. 2014, supplementary material)

No.	Living lab	Research field	Homepage
Ambient assisted living			
1	Advanced Construction and Building Technology, Munich	Assistance systems for the demographic change	http://partner.vde.com/bmbf-aal/aal-steck-briefe/documents/advanced%20construc-tion%20and%20building%20technology%20%28acbt%29.pdf
2	Ambient Assistance for Parameters of LIFE, Lüneburg	Assistance systems enabling an independent life for elderly people	http://www.vde.com/de/technik/aal/steckbriefe/seiten/projektambientassistanceforparametersof-life.aspx
3	Ambient Assisted Living, Sozialwerk St Georg, Gelsenkichen	Technical aids for personal freedom and autonomy for people with need of assistance due to disabilities, illness, social problems and old age	http://www.ambient-assisted-living.org
4	Fraunhofer IESE, Ambient Assisted Living Environment, Kaiserslautern	Technology concepts for better designed and con-nected life at old age	http://www.iese.fraunhofer.de/de/projekte/med_projects/aal-lab.html
5	Ambient Innovation Robotics, Munich	Socio-technical solutions for demographic change	http://www.airtum.de
6	Assisted Living, Kaiserslautern	Suitable, subtle, invisible and easily usable house automation technology to support aging people	http://www.ami-kl.de/frame.html?de_inhalte
7	Bremen Ambient Assisted Living Lab, Bremen	Better living for old or physically disabled people	http://baall.informatik.uni-bremen.de/en/index.php/Main_Page
8	Forschungsinstitut Technologie und Behinderung (FTB)—AAL, Wetter/Ruhr	Assistance systems for aging people	http://ftb-esv.de/wohnung3.html
9	FZI Living Lab Ambient Assisted Living, Karlsruhe	Technologies and service concepts for an inde-pendent life at old age	http://aal.fzi.de
10	Technologie-Zentr. Informatik und Informationstechnik, Bremen	Research area: Ambient-Assisted-Living	http://www.tzi.de
11	Assisted Living in Kaiserslautern	Assistance systems for aging people	http://www.assistedliving.de

(continued)

Table 11.5 (continued)

No.	Living lab	Research field	Homepage
12	LivingLab Schwechat (Austria)	Assistance systems for aging people	http://www.ceit.at
13	Living Tomorrow, Vilvoorde (Belgium)	Experimental space for new living concepts (i.e., mixing of functions), adapted to elderly's needs	http://livingtomorrow.com/en/home
69	SRG Senior Research Group, TU Berlin	Better user-friendliness of technology for the elderly	http://www.srg-berlin.de
Smart home and other building-related activities			
14	CIBEK, Limburgerhof	Intelligent products for residential and other purpose buildings	http://www.cibek.de
15	Connected Living, Berlin	Showroom for "connected" living	http://www.connected-living.org
16	Distributed Artificial Intelligence Living Lab, Berlin	5 topics—health, living, cloud computing, security and network infrastructure	http://www.dai-labor.de
17	Fraunhofer-inHaus-Zentrum, Duisburg	Development of new solution for rooms and buildings	http://www.inhaus.fraunhofer.de
18	Future Care Lab, RWTH Aachen	Health care	http://www.openlivinglabs.eu/livinglab/future-care-lab
19	iHomeLab Luzern (Swizzerland)	Smarter living	http://ihomelab.ch/index.php?id=20
20	Philips Home Lab Eindhoven (Netherlands)	Intelligent technologies	http://www.noldus.com/default/philips-homelab
21	Smart Environments, Berlin	Care applications in various settings	http://www.fokus.fraunhofer.de/de/fokus/index.html
22	SmartHome der Universität Bundeswehr München	Energy efficiency in households	http://smarthome.unibw-muenchen.de
23	SmartHome Paderborn e.V	Connected, intelligent living	http://www.smarthomepaderborn.de
24	Haus der Gegenwart, München	Smart environments, e.g., in the office, car and home sectors, as well as personal living and working spaces	http://www.allmannsattlerwappner.de/#/de/projekte/detail/76/

(continued)

Table 11.5 (continued)

No.	Living lab	Research field	Homepage
25	Futurelife Haus, Hünenberg (Swizzerland)	Development and research of connected smart sensor- and measurement systems to increase energy efficiency, living quality/comfort and security in residential and commercial buildings	http://www.haus.de/PH2D/ph2d.htm?snr=6850
26	e-Wohnen der Zukunft—Projekt 4, Berlin	Products and services for the intelligent and connected home	http://www.e-wohnen-der-zukunft.de
27	Intelligentes Haus IQ150 der FU Berlin	Experimental residential building	http://intellihaus.mi.fu-berlin.de/pmwiki/pmwiki.php
28	Demonstrations-Wohnung in Berlin	Experimental residential building	http://www.telematicspro.de
29	Das Mediale Haus, Nürnberg	show room, energy-efficient living incl. mobility	http://www.das-mediale-haus.de
30	Smarter Wohnen in Hattingen	Prototype of an "intelligent house" of the future	http://www.ims.fraunhofer.de/news/detailansicht/article/smarterwohnen-nrw.html
31	T-Com Haus, Rheinau-Linx	Energy-efficient living	http://www.weberhaus.de/fileadmin/weberhaus_de/redaktion/Downloads/haeuserheute_0205.pdf
32	Smart Home in Leipzig	Connected, intelligent living	http://www.swl.de/web/energie21smart/Smart-Home/Smart-Home.htm
Smart home and other building-related activities			
33	Adaptive Home Control, Winterthur (Swizzerland)	Connected, intelligent living	http://www.adhoco.com
34	Social Media Experience and Design Lab, Siegen	Connected living	http://www.openlivinglabs.eu/livinglab/smedl-social-media-experience-and-design-lab/http://www.wiwi.uni-siegen.de/wirtschaftsinformatik/praxlabs/
35	In-HAM vzw, Gits (Belgium)	Connected, intelligent living	http://www.in-ham.be/index.cfm?n01=default&lang=en

(continued)

Table 11.5 (continued)

No.	Living lab	Research field	Homepage
36	Pilotprojekt "Wohnen mit LON" (Local Operating Network), Gütersloh	Integrated house automation for comfort, energy saving and security optimisation	http://www.mh-software.de/newsletter/200507/tab.pdf; http://www.lonmark.de
56	Energieeffizienz-haus-Plus, Berlin	Efficiency-House-Plus with electro-mobility	http://www.bmvbs.de/DE/EffizienzhausPlus/effizienzhaus-plus_node.html
57	LivingLab ETH Zürich (Swizzerland)	Construction and sustainability in all dimensions	http://www.livinglabproject.org/partners/eth-zurich
66	Universal Home, Essen	Home-related developments in the areas of social environment, living systems, resource loops, flexibility and mobility, and economy	http://www.universalhome.de
67	Tobit intelligentes Haus, Ahaus	Showroom for connected, intelligent living	http://www.tobit.com/login/showTopicPerson.asp?ID=2068&sum=f722a2f92969f7ec68f031a0c94a2c73&LoginID=199-60001
Infrastructure			
37	DFKI et al. Innovative Retail Laboratory (IRL), Saarbrücken	Intelligent shopping environment	http://www.innovative-retail.de
38	AIT Austrian Institute of Technology GmbH, Wien (Austria)	Innovative infrastructure solutions	http://www.ait.ac.at
39	Car2go, Ulm	Option to rent cars (Smart) in minute-long intervals. Run by Daimler	http://www.car2go.com/ulm/de
40	DriveNow, Munich	Option to rent cars (Mini) in minute-long intervals. Run by BMW	http://www.drive-now.com
41	e-mobility Stuttgart	Services to integrate electronic vehicles into vehicle fleets of all categories	http://www.elektromobilisiert.de
71	SAP Future Energy Center (Karlsruhe)	A collaborative platform that demonstrates how the future of sustainable energy generation, distribution, storage, and utilization can be supported by innovative IT	http://global1.sap.com/corporate-en/innovation/living-labs/future-energy-center.epx

(continued)

Table 11.5 (continued)

No.	Living lab	Research field	Homepage
74	SAP Future Public Security Center (Darmstadt)	Create user-oriented solutions for civil protection and emergency response	http://global1.sap.com/canada/press.epx?pressid=10492
Infrastructure			
75	Metro Future Store, Tönisvorst	Innovations for retail	http://www.future-store.org/internet/site/ts_fsi/node/mgroup_fsi/Lde/index.html
Medicine			
42	Hamburg Living Lab, Hamburg	Biomechanics, implant technology and dental technology	http://www.demos-deutschland.de/node/41/project/demos-idea/living-lab-hamburg
43	Peter L. Reichertz Institut für Medizinische Informatik, Hannover	Assisting medical technology	http://www.plri.de
ICT (Information and Communication Technology)			
44	Fraunhofer Fokus, Berlin	Future applications of ICT	http://www.fokus.fraunhofer.de
45	Fraunhofer IGD, Darmstadt	Interaction and visualisation concepts	http://www.igd.fraunhofer.de
46	Fraunhofer ISST, Dortmund	Development of standards, architectures and concepts	http://www.isst.fraunhofer.de
47	Living Lab Vorarlberg (Austria)	Construction of long-term stable, complex systems in ICT	http://www.livinglab-vorarlberg.at
48	Schweizerisches Produktivitäts-Institut AG, Bern (Swizzerland)	Supporting the direct and indirect sustainability of products by applying the Living Lab method in the development and use phase	http://www.ipch.ch
49	The Virtual Dimension Center, Fellbach	Introduction and implementation of modern ICT solutions	http://www.vdc-fellbach.de
50	ViRaL Cooperation Lab, Sankt Augustin	Network Virtual Reality, Virtual Engineering, 3D-Simulation and 3D-Visualisation + Technology transfer	http://www.fit.fraunhofer.de; http://www.cooperation-lab.de

(continued)

Table 11.5 (continued)

No.	Living lab	Research field	Homepage
51	DFKI Smart Factory, Kaiserslautern	network of platforms and activities to facilitate research, development, innovation and market validation of new cooperation technologies in real-life environments	http://www.dfki.de/web/living-labs-de
76	European EPC Competence Center (EECC), Neuss	RFID technology and Electronic Product Code (EPC) research	http://www.eecc.info/index.php?article_id=106&clang=1
Food/Agriculture			
53	Danone LivingLab, Haar	Danone Institute for Nutrition and Food research	http://institut-danone.de/content/start.php
54	Frascati Living Lab, Frascati (Italy)	Intelligent systems/processes (incl. ICT + SatCom applications) in agriculture (viticulture)	http://www.c-rural.eu/index2.php?option=com_docman&task=doc_view&gid=98&Itemid=83
55	C@R Living Labs	Project cluster from Frascati Living Lab (I), Cudillero Living Lab (ES), Soria Living Lab (ES)	http://www.c-rural.eu
68	Restaurant of the Future Wageningen (Netherlands)	New meals, meal preparation methods and service systems, observation of eating habits	http://www.restaurantvandetoekomst.wur.nl/UK/
Work environment			
59	Knowledge Workers Living Lab, München	Research and development to support knowledge work	http://www.openlivinglabs.eu/livinglab/knowledge-workers-living-lab/www.cetim.org
61	Fraunhofer IAO Office Innovation Center, Stuttgart	Future of office/work equipment, organisation etc.	http://oic.iao.fraunhofer.de
62	SmartFactory, Kaiserslautern	Intelligent factory environment	http://www.smartfactory.de
63	SustLabs Universität Duisburg-Essen	How is sustainability perceived and implemented in office work routines	http://www.uni-due.de/bena/sustlabs.shtml

(continued)

Table 11.5 (continued)

No.	Living lab	Research field	Homepage
64	YOUSE GmbH, Berlin	YOUSE helps research consortia, producers and developers to develop user-friendly products and services	http://www.youse.de
70	OrgLab, Duisburg/Essen	Organisational development and learning for innovative and responsible behaviour in businesses and society	http://www.orglab.org
72	Sap Future Factory Initiative, Dresden	Foster co-innovation, research, and development for the manufacturing industry	http://global1.sap.com/corporate-en/innovation/living-labs/future-factory-initiative.epx
73	SAP Future Retail Center, Regensdorf	Foster research and development in retail, trade, and logistics	http://global1.sap.com/corporate-en/innovation/living-labs/future-retail-center.epx
Region			
52	iRegion Karlsruhe	Internet-based and mobile technologies for regions in the net economy	http://www.iregion.de
58	Neighborhood Lab [Design Research Lab UdK Berlin]	Fostering of networking to exchange knowledge etc. in urban environments	http://www.neighborhood-labs.org; http://www.design-research-lab.org
60	Mobile City Bremen	Communication solutions in Bremen, esp. for SMEs	http://www.technologiezentren-bremen.de/de/mobilecity_mobilecity_uebersicht
65	BLISS- Better Lighting in Sustainable Streets, Kaiserslautern, Eindhoven (Netherlands), Leuven (Belgium), St. Helens (UK)	Sustainable street lighting	http://www.bliss-streetlab.eu/about_bliss/

References

Baedeker, C., Bahn-Walkowiak, B., Bleischwitz, R. Kolberg, S., Mont, O., & Stengel, O., et al. (2008). Survey on Consumption Behavior and Its Driving Forces. European Environment Agency, European Topic Center on Resource and Waste Management: Copenhagen, Denmark.

Baedeker, C., Greiff, K., Grinewitschus, V., Hasselkuß, M., Keyson, D., & Knutsson, J., et al. (2014). *Transition through sustainable Product and Service Innovations in Sustainable Living Labs: Application of user-centred research methodology within four Living Labs in Northern Europe*. Paper for presentation at the 5th International Sustainable Transitions (IST) Conference, August 27–29, 2014 Utrecht, The Netherlands.

Bringezu, S., van de Sand, I., Schütz, H., Bleischwitz, R., & Moll, S. (2009). Analyzing global resource use of national and regional economies across various levels. In S. Bringezu & R. Bleischwitz (Eds.), *Sustainable resource management—Trends, visions and policies* (pp. 10–51). Greenleaf Publishing: Sheffield, UK.

Bringezu, S., Schütz, H., Pengue, W., O'Brien, M., Garcia, F., & Sims, R., et al. (2014). Assessing Global Land Use: Balancing Consumption with Sustainable Supply. A Report of the Working Group on Land and Soils of the International Resource Panel. United Nations Environmental Program: Nairobi, Kenya.

Buhl, J., Echternacht, L., & Geibler, J. v. (2015). Rebound-Effekte – Ursachen, Gegenmaßnahmen und Implikationen für die Living Lab-Forschung im Arbeitspaket 1 (AP 1.2a) des INNOLAB Projekts. Wuppertal Institute, Wuppertal.

Druckman, A., Chitnis, M., Sorrell, S., & Jackson, T. (2011). Missing carbon reductions?: Exploring rebound and backfire effects in UK households. *Energy Policy, 39*, 3572–3581.

Fischer-Kowalski, M., Swilling, M., von Weizsäcker, E.U., Ren, Y., Moriguchi, Y., & Crane, W., et al. (2011). Decoupling Natural Resource Use and Environmental Impacts from Economic Growth, A Report of the Working Group on Decoupling to the International Resource Panel; United Nations Environment Program: Nairobi, Kenya, p. 2.

Geibler, J. v., Rohn, H., Schnabel, F., Meier, J., Wiesen, K., & Ziema, F., et al. (2011). Ressourceneffizienzatlas. Eine internationale Perspektive auf Technologien und Produkte mit Ressourceneffizienzpotenzial. Retrieved November 4, 2015 from http://epub.wupperinst.org/frontdoor/index/index/docId/3902.

Geibler, J. v., Erdmann, L., Liedtke, C., Rohn, H., Stabe, M., & Berner, S., et al. (2014). Exploring the potential of a german living lab re-search infrastructure for the development of low resource products and services. *Journal Resources, 3*, 575–598.

German Federal Ministry of Education and Research (BMBF). (2012). *Bundesbericht Forschung und Innovation 2012*. Bundesministerium für Bildung und Forschung BMBF: Bonn, Germany.

Jackson, T. (2005). *Motivating sustainable consumption: A review of evidence on consumer behavior and behavioral change*. London, UK: Policy Studies Institute.

Lettenmeier, M., Rohn, H., Liedtke, C., & Schmidt-Bleek, F. (2009). Resource Productivity in 7 Steps. How to Develop Eco-Innovative Products and Services and Improve their Material Footprint; Wuppertal Special No. 41; Wuppertal Institute for Climate, Environment and Energy: Wuppertal, Germany.

Lettenmeier, M., Liedtke, C., & Rohn, H. (2014). A production- and consumption-oriented reference framework for low resource household consumption—Perspective for sustainable transformation processes of lifestyles. *Resources, 3*, 488–515.

Liedtke, C., Welfens, M. J., Rohn, H., & Nordmann, J. (2012). LIVING LAB: User-driven innovation for sustainability. *International Journal of Sustainability in Higher Education, 13*, 106–118.

Liedtke, C., Baedeker, C., Hasselkuß, M., Rohn, H., & Grinewitschus, V. (2015). User-integrated innovation in Sustainable LivingLabs: An experimental infrastructure for researching and developing sustainable product service systems. *Journal of Cleaner Production, 97*, 106–116.

Lin, I.-H., Chang, O. H., & Chang, C. (2014). Importance of sustainability performance indicators as perceived by the users and preparers. *Journal Management and Sustainability, 4*, 34–45.

Münz, R., & Reiterer, A. F. (2009). *Overcrowded world? Population explosion and international migration* (1st ed., p. 324). London, UK: Haus Publishing.

Pierson, J., & Lievens, B. (2005). Configuring living labs for a "Thick" understanding of innovation. *Ethnographic Praxis in Industry Conference Proceedings, 1*, 114–127.

Rockström, J., Steffen, W., Noone, K., Persson, Å., Chapin, F. S., Lambin, E. F., & Foley, J. A. (2009). A safe operating space for humanity. *Nature, 461*(7263), 472–475.

Rohn, H., Lang-Koetz, C., Pastewski, N., & Lettenmeier, M. (2009). Identifikation von Technologien, Produkten und Strategien mit hohem Ressourceneffizienzpotenzial—Ergebnisse eines kooperativen Auswahlprozesses. Arbeitspapier zu Arbeitspaket 1 des Projekts "Materialeffizienz und Ressourcenschonung" (MaRess). Ressourceneffizienzpaper 1.2.; Wuppertal Institute: Wuppertal, Germany,

Rohn, H., & Leismann K. (2012). Ergebnisse des Arbeitspakets AP 2.1 "Identifizierung von relevanten Anwendungsfeldern für NachhaltigkeitsLivingLabs", Interner Ergebnisbericht im Projekt "Nachhaltigkeitsinnovationen im Living Lab". Friedberg, Faktor 10—Institut für nachhaltiges: Wirtschaften.

Saurat, M., & Ritthoff, M. (2013). Calculating MIPS 2.0. *Resources, 2*, 581–607.

Sorrell, S. (2007). The rebound effect: An assessment of the evidence for economy-wideenergy savings from improved energy efficiency. A report produced by the Sussex Energy Group for, UK Energy Research Centre.

Spangenberg, J. H., Femia, H., Hinterberger, F., & Schütz, H. (1999). *Material flow-based indicators in environmental accounting*. Luxembourg: Office for Official Publications of the European Communities.

Stabe, M., & Schnalzer, K. (2012). *Ergebnisse im Projekt „Nachhaltigkeitsinnovationen im Living Lab"; Ergebnisse des Arbeitspaketes AP 1: „Status Quo Analyse zur deutschen Infrastruktur der Innovations- und Laborforschung [Bestandsaufnahme]", Interner Bericht*. Stuttgart: Fraunhofer Institut für Arbeitswirtschaft und Organisation IAO.

SUSLAB (2015). The SUSLAB Project. Retrieved November 1, 2015 from http://www.suslab-nwe.eu.

Talwar, S., Wiek, A., & Robinson, J. (2011). User engagement in sustainability research. *Sci. Public Policy, 38*, 379–390.

Welfens, M. J., Liedtke, C., & Nordmann, J. (2010). Sustainable consumption: Between unsustainable reality and people's willingsness to act. *Proceedings of the Knowledge Collaboration and Learning for Sustainable Innovation ERSCP-EMSU Conference*, Delft, The Netherlands, 25–29 October 2010.

Wiesen, K., Teubler, J., & Rohn, H. (2013). Resource Use of Wind Farms in the German North Sea. The Example of Alpha Ventus and Bard Offshore I. *Resources, 2*, 504–516.

Zweck, A., Bachmann, G., Luther, W., & Ploetz, C. (2008). Nanotechnology in Germany: From forecasting to technological assessment to sustainability studies. *Journal of Cleaner Production, 16*, 977–987.

Part IV
Living Labs and User Engagement

Chapter 12
In-Situ and Mixed-Design Interventions

Natalia Romero Herrera

Abstract Sustainability in living and working contexts aims to design innovations that are appropriated by users in their daily life activities. Appropriation is a dynamic process that acknowledges the complexity of practices in the adoption of technologies. It involves adaptation of the technology and its intended use as well as the practices that are affected by it. In Living Lab settings the innovation process is user-centric, meaning that is driven by users, their practices and the process of appropriation. This requires an active involvement of users at all stages of the design process: for gathering insights, ideation, co-designing, experimentation and evaluation. When the focus is on active involvement of users, qualitative methods are central in the design process. Qualitative methods support a wide spectrum of user involvement, from been observed to actively self-report experiences and practices to inform the design process. The more active the involvement of users and the more complex their context, the more effort and skills are needed from them. If no clear incentives for participation, active involvement becomes a burden and does not sustain overtime. In-situ and mixed-designs interventions support user centric, situated and integrated design research practices. In-situ and mixed tools take the form of mixed data probes and in-situ interventions that facilitate user involvement in the activities of data collection and interpretation. In this chapter, the SusLab Toolkit is presented as an implementation of in-situ and mixed-designs interventions in the context of heating practices at home.

Keywords Technology appropriation · Daily life practices · In-situ methods · Mixed methods · Sustainability living lab

N. Romero Herrera (✉)
Delft University of Technology, Delft, The Netherlands
e-mail: n.a.romero@tudelft.nl

D.V. Keyson et al. (eds.), *Living Labs*, DOI 10.1007/978-3-319-33527-8_12

12.1 Introduction

Sustainable domestic innovations are designed to co-exist within the dynamic complexity of daily life practices. If home living practices are ignored during the design process, solutions are doomed to failure (Spaargaren 2011). An example is homes that monitor and provide feedback for saving energy, where inhabitants often question and neglect the advice given on their consumption even if it has been tailored to their specific energy practices (Gram-Hanssen et al. 2007). One reason to explain this is the assumption that energy consumption actions are driven by rational energy saving decisions. This assumption ignores aspects that are central in people's everyday lives such as convenience, fashion, culture, and wellbeing. In addition, several research studies have shown clear evidence that technological improvements alone do not reach the expected impact on energy reduction (Majcen et al. 2013). Home residents go through a continuous process of appropriation in which technology use, technology itself and related practices are adjusted to fit users' lifestyles (Dourish 2003). Understanding this process could help address phenomena such as the rebound effect, a human reaction to technical improvements, where a decrease in the perceived costs of energy services may result in an increase in demand. Therefore there is a need to develop methods that go beyond monitoring objective factors around energy related practices by including subjective factors such as users' comfort (Guerra-Santin and Tweed 2014).

Sustainability Living Lab (SLL) offers an infrastructure to support a user-centric and contextual innovation process for the design of sustainable practices (Liedtke et al. 2012). Future users in their living context are involved as key actors in communicating and discussing their values and lifestyles: by means of reporting and reflecting on experiences around their daily life practices (Romero et al. 2013). The reports and reflections are included as rich input to support ideation and experimentation of new strategies and to evaluate the appropriation and impact of such strategies in people's daily life. For this, SLL adopts a multidisciplinary approach to incorporate user research into design practices and vice versa connecting the process of collection, analysis and interpretation of data with the design activities of understanding, ideation, experimentation and evaluation of concepts.

How to support users' involvement is then key to facilitate this process. Reporting and reflecting on daily life practices are not trivial activities. Practices are routines highly dependent on context, on other practices and on time, so they are not always easy to track and to have a clear overview, therefore making difficult to quantify and qualify their impact.

Mixed-designs are research constructions based on Mixed Methods Research (Creswell and Piano 2011) that propose the integration of qualitative and quantitative methods for the study of complex settings. In Chap. 2 The emergence of Living Lab Methods a mixed approach for SLL is described that implements mixed tools to support data collection (mixed self-reporting), data analysis (mixed data probes), and data interpretation (mixed design activities). Applying mixed tools in design research activities aims to facilitate the involvement of users at different stages of the design process: gathering insights, co-design (ideation), experimentation and evaluation of sustainable innovations.

This chapter describes the implementation of four mixed tools developed as part of the SusLab Toolkit, in collaboration with the SusLabNWE partners (www.suslab.org). The SusLab Toolkit (Harinxma et al. 2014) was designed to support a SLL infrastructure focused on home heating practices. This chapter presents the SusLab toolkit designed to support in-situ and mixed-designs interventions and to enable user-centric and contextualized design research activities.

The toolkit provides an infrastructure for in-situ self-reporting and reflection of practices based on Experience Sampling Method (Hektner et al. 2007), Daily Reconstruction Method (Kahneman et al. 2004) and Sensor Networks. Mixed tools extend the toolkit by the implementation of mixed-designs to facilitate users in the activities of describing and contextualizing practices in a quantified and qualified rich and reliable manner.

The chapter starts by briefly introducing the methods involved in the development of the toolkit, followed by a description of an implementation of the toolkit in the context of home heating.

12.2 Mixed Designs

Mixed-designs are formulated within the paradigm of Mixed Methods Research (Creswell and Piano 2011). They are based on three basic ways to mix data from qualitative and quantitative methods: by *connecting*, where one data source builds on or follows up on the other; *by merging*, where the results from quantitative and qualitative methods are compared or related to; or *by embedding*, where the results of one method is explained by the other (see Fig. 12.1).

Mixed-designs are defined and characterized by three factors: the order in which qualitative and quantitative methods collect data, whether sequential or concurrent; the priority given to quantitative or qualitative research, whether the same for both or prioritizing one over the other; and the stage of integration of data, whether in data collection, analysis, interpretation or a combination of them. Figure 12.1 shows the most used mixed designs. Connecting data supports *sequential explanatory* and *sequential exploratory* designs. The first uses qualitative data to explain and to interpret the main findings coming from quantitative results, whereas the second one uses quantitative methods to explore a phenomenon that has been identified in a qualitative study. Merging data supports *concurrent triangulation* where two concurrent data collections are used to confirm, corroborate or cross-validate a study. Embedding data supports *concurrent nested* designs where a predominant data collection is extended by adding a second data collection to gain a broader perspective. Two other designs, *sequential and concurrent transformative,* help employ the methods that best serve a theoretical perspective. In particular, *sequential transformative design* data is integrated to better understand a phenomenon or process that is changing as a result of being studied.

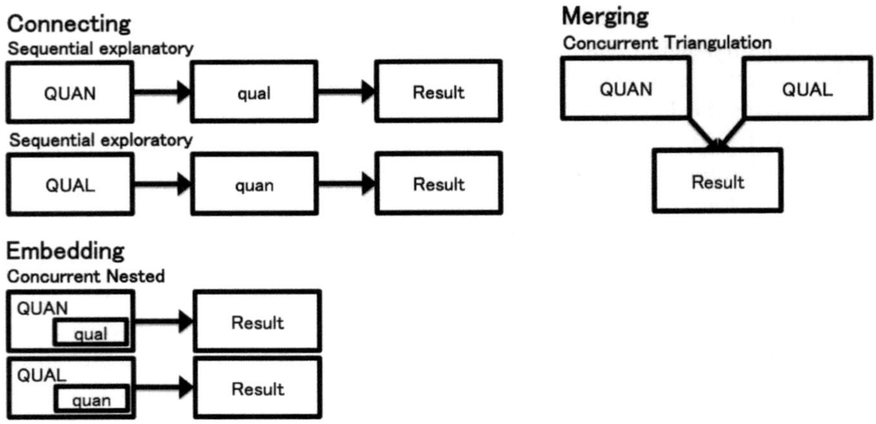

Fig. 12.1 Three ways of mixing quantitative and qualitative data. Notation: a predominant method is symbolized in capitals; in the absence of a predominant method both approaches are equally represented in the results (Creswell and Piano 2011)

12.3 In-Situ Methods

For most daily life practices, people are unaware of their impact on sustainability. The more routinary the practice is the harder to describe them and explain their impact. Unexpected high-energy bills are one example that illustrates this in real life situations. Nevertheless, participants involved in user-centred design research studies are expected to provide deep reports and reflections on their daily life practices. Traditional methods such as questionnaires and interviews are memory biased by the ability of users to recall relevant practices and experiences. Besides, users also lack an accurate overview of their practices and its impact. In-situ methods support users reporting and reflecting on daily life practices in their social, physical and temporal context. The reports can be automatically enriched by time of the day and other contextual data (see sensor networks below). Experience Sampling Method (Hektner et al. 2007) implements different protocols to trigger users report their experience at several moments in time. Connecting experiences and practices, while keeping sampling a lighter activity for users, is implemented by combining ESM with Daily Reconstruction Method (Kahneman et al. 2004) using sampled experiences as inputs to reconstruct more detailed information on practices (Khan et al. 2008). Sensor networks can inform ESM protocols based on contextual variables and provide richer input information to DRM.

12.4 Sensor Networks

Sensor networks are also discussed as relevant techniques to contextualize the user practices. In understanding the sustainable impact of practices, sensor networks can provide information otherwise invisible for users. For example, monitoring energy consumption as well as contextual variables such as indoor climate (CO_2, temperature, humidity) and other social (presence, movement, etc.) and environmental parameters (light, sound, etc.) can be integrated as feedback or used to inform the protocol of in-situ methods. Sensor network technologies have been adopted rapidly in Living Lab setups, as they can provide unobtrusively and to relatively low costs objective information around people's behaviour and patterns. On the one hand, they support large-scale and longitudinal collection of data; on the other hand, the knowledge gathered from sensor data alone is not sufficient to gain rich insights into real life situations (Mulder et al. 2005).

12.5 The SusLab Toolkit

The SusLab Toolkit supports design research practices in Living Lab settings. Designers, researchers and users interact with different layers and sources of data to generate user centric and contextualized insights, ideas, concepts, prototypes and evaluations. The toolkit provides a set of tools that implement in-situ and mixed-designs interventions to support design activities with high involvement of future users in Living Lab setup. The tools are developed to collect subjective and objective data in relation to heating practices, thermal comfort sensation and indoor climate variables. The in-situ and mixed-designs interventions are designed to integrate quantitative and qualitative data collection with users. The toolkit supports several designs: sequential explanatory mixed-design to facilitate the data collection of experiences and practices around heating; concurrent triangulation to integrate the analysis of objective and subjective quantitative data related to impact of heating practices; sequential exploratory mixed-design to guide the generation of quantified and qualified insights of practices and the ideation of solutions through deeper and focused analysis of practices; and sequential transformative implementation to enable experimentation and assessment of technology appropriation and emergence of new practices. The following sub-sections include descriptions of each mixed-design, which are summarized in Table 12.1.

Table 12.1 Implementation of in-situ interventions and mixed-designs using the SusLab Toolkit

Mixed designs	In-situ interventions	Purpose
Sequential explanatory	Comfort dial: sampling experience based on quantitative self-reports on thermal comfort Heating diary practices: reconstruction of practices based on qualitative self-reports on heating practices, clothing, and physical activity) Comfort dials connects to diaries of practices	Facilitates users to reconstruct qualitative reports on (heating) practices on the basis of quantitative self-reports on (thermal comfort) experiences
Concurrent triangulation	Sensor box merges with comfort dial = mixed probes	Visualizes objective (indoor climate) and subjective (thermal comfort) impact of (heating) practices
Sequential exploratory	Mixed probes: integrated visualizations of quantitative objective (sensor boxes and energy meter) and subjective (comfort dial) to integrate impact on sustainability and comfort Mixed probes connects to interviews Mixed probes connects to co-design	Interview sessions provide deep and focus explanations of the impact visualized by mixed probes Co-design sessions translate insights from mixed probes into contextualized requirements
Sequential transformative	In-situ interventions to provoke reflection or to evoke certain experience by guided user interactions with artefacts	Evaluates the process of appropriation of technology and the development of new practices

12.5.1 Sampling and Constructing Practices—Sequential Explanatory

Living Lab participants, as future users, can report their heating practices at home using the heating practices diary (see Fig. 12.2), a digital interface that invites users to reconstruct and qualify their heating practices on a daily to weekly basis. The heating practice diary is based on DRM. Acknowledging the challenges of recalling relevant practices even if reporting on a daily basis, the diary is connected to the inputs of the comfort dial. The comfort dial (see Fig. 12.3—left) is a personal input device, based on ESM, that users can carry with them to input their experiences on thermal comfort. Every time users experience a change in comfort they can report it on a seven scale from very cold to very warm by turning the comfort dial left or right. The inputs are colour coded in a range from blue to red and displayed in the timeline of the heating practices diary with a vertical peak. Inputs of the comfort dial can then be connected to the descriptions provided in the diary, as they can be used as memory anchors to describe the practices around those moments.

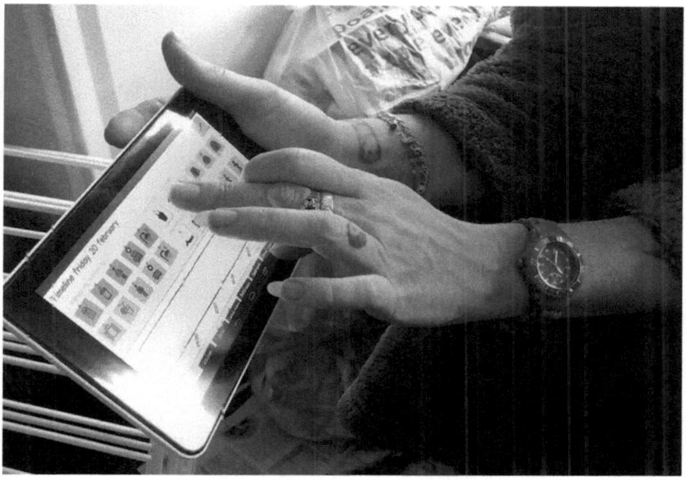

Fig. 12.2 A householder using the heating practice diary, in Dartford, England

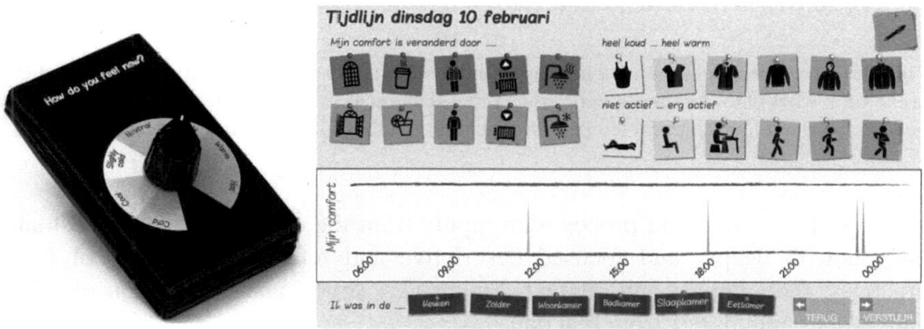

Fig. 12.3 *Left*—comfort dial; *Right*—heating practices diary: color coded peaks in the timeline represent inputs from comfort dial, the icons are draggable to indicate heating and cooling practices (*top-left* of the timeline), level of activity and clothing (*top-right* of the timeline) and location (*bottom* of the timeline)

12.5.2 Connecting Needs and Impact of Practices—Concurrent Triangulation

Users' quantitative and subjective data from the comfort dial can be visually merged with quantitative and objective sensor data on indoor climate and contextual data including temperature, humidity, and CO_2, resulting in mixed data probes (Fig. 12.4—bottom). Mixed data probes provide a first layer of integration of users' heating needs (e.g. too cold, too hot) and the quantified description of the heating context (e.g. low temperature, high temperature). Mixed and in-situ interview sessions are implemented as instances for future users and researchers

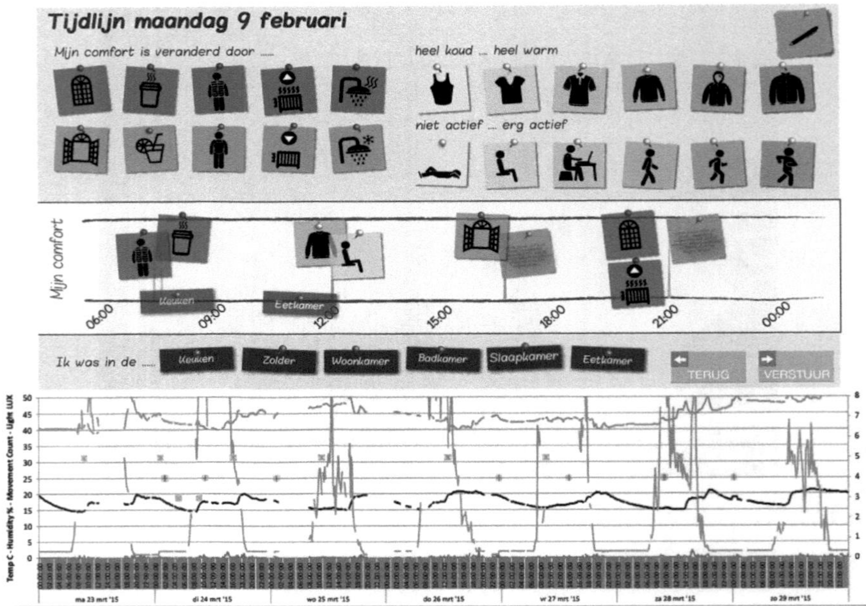

Fig. 12.4 *Top*—instantiation of a heating practice diary; *Bottom*—instantiation of a mixed data probe of 1 week: the *dots* represent comfort dial data (scale 1–7 at the right) and *black*, *green*, *pink* and *purple lines* represent temperature, humidity, light and movement respectively

to merge the mixed data probes with inputs from the diaries and to connect such insights with deeper and focused descriptive and reflective exercises of users' practices, their impact and context. Figure 12.4 shows an instantiation of mixed data probes and diaries from real data.

12.5.3 Ideation of Solutions—Sequential Exploratory

Mixed Data probes and the insights gather from the reflective exercises can be connected to co-design sessions supporting future users and designers to generate solutions that integrate social and sustainability values of practices. Mixed probes can help users identify relevant practices, study their context and reflect on their personal and sustainable impact. Co-design practices become well supported by rich and relevant data, bringing context further in the ideation and definition of requirements.

12.5.4 Experimentation of Technology Appropriation—Sequential Transformative

In-situ interventions can be developed to activate users reflect on and assess their practices and experiences when interacting with prototypes. For instance, by implementing online merging of comfort dial and sensor box data, interventions in the form of visualizations, questions or tasks can be triggered and prompted to users. This enables users to track and assess their appropriation process of the new technologies in their daily life.

12.6 The SusLab Toolkit Generic Setup

A generic setup is proposed to optimize user involvement and data richness across the three levels of integration represented by the three rings in the Mixed approach introduced in Chap. 2: The emergence of Living Lab Methods. The setup encompasses data cycles. A data cycle defines the frequency and schedule of data collection and of the mixed-design practices described above. The length of a cycle and number of cycles deployed in one study depends on the project resources and main objectives. As a rule of thumb it is suggested to set the length of a cycle to a minimal of 3 weeks and a maximal of 3 months, including at least 2 cycles. For example, to study heating and cooling practices in one year, it is recommended to implement 4 cycles (one per season) with a length of 3 months each.

Figure 12.5 illustrates a study setup with 4 cycles of 6 weeks each, with a total of 24 weeks of sensor data collection, 8 weeks of self-reporting data implemented in groups of 2 weeks at the end of each cycle, and co-design sessions of 1 day at the beginning of each cycle (by combining activities of interpretation of mixed probes and ideation of solutions (see Sects. 12.5.2 and 12.5.3 respectively).

A cycle starts with an in-situ user session. For the first cycle this session is more intensive than the ones in the following cycles, with the main goal to get an initial view of attitudes, practices, context (location, social) and use of technology. In the following cycles, this session invites users to work with mixed probes (see Sect. 12.5.2). This activity is located in the inner level of the mixed approach (see 'intake' in Fig. 12.5).

In the verifying level (outer ring) the primary data collection comes from sensors. Depending on the level of intrusiveness, the stability and cost of the sensor network, it is suggested to implement a continuous protocol of data collection for every cycle. In Fig. 12.5 this are represented by the white dots in the outer ring.

For the exploring level (middle ring) user involvement is in the form of self-reporting and self-reflection. To minimize users' burden and fatigue in long-term studies, it is recommended that every cycle applied self-reporting activities in the last one-third of the total length of the cycle. Therefore if the cycle is set to 6 weeks the self-reporting activity takes place in the last 2 weeks of the cycle.

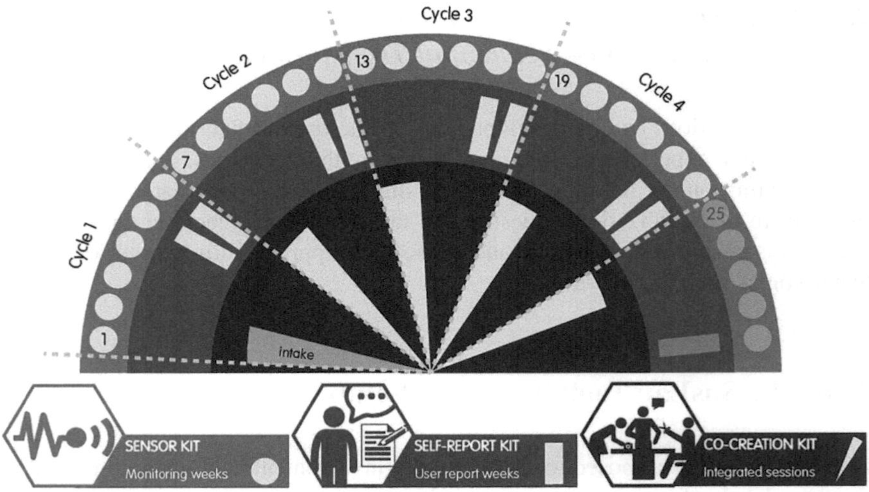

Fig. 12.5 Example of a study setup using the Suslab Toolkit

This activity hosts sampling and constructing of practices or experimentation of technologies depending on the progress of the study (see Sects. 12.5.1 and 12.5.4 respectively).

In the experimenting level (inner ring) co-design sessions are implemented usually in the form of group in-situ sessions at the end of a cycle using the inputs from the other levels (see Sect. 12.5.3).

12.7 Conclusion

The presented tool aims to empower users by actively collecting, analyzing and integrating rich data to qualify and quantify practices and their impact on daily life. Users then become active collaborators in design research activities to explain current practices, explore new ones and transform practices by appropriating new technologies.

Acknowledgments The development and deployment of the SusLab Toolkit have been funded by the SusLabNWE Interreg European program.

References

Creswell, J., & Piano, V. (2011). *Designing and Conducting Mixed Methods Research*. Sage Publications.

Dourish, P. (2003). The appropriation of interactive technologies: Some lessons from placeless documents. *Computer Supported Cooperative Work (CSCW), 12*(4), 465–490.

Gram-Hanssen, K., Bartiaux, F., Jensen, O. M., & Cantaert, M. (2007). Do homeowners use energy labels? A comparison between Denmark and Belgium. *Energy Policy, 35*.

Guerra-Santin, O., & Tweed, A. C. (2014). In-use monitoring of buildings: An overview and classification of evaluation methods. *Energy and Buildings, 89*, 176–189.

Harinxma, W., Romero, N., & Keyson, D. (Eds.) (2014). SusLabNWE Brochure. http://suslab.eu/fileadmin/suslab/Images/SusLab_brochure_2014_1.0.pdf.

Hektner, J. M., Schmidt, J. A., & Czikszentmihalyi, M. (2007). *Experience sampling method: Measuring the quality of everyday life*. Sage Publications.

Kahneman, D., Krueger, A. B., Schkade, D. A., Schwarz, N., & Stone, A. A. (2004). A survey method for characterizing daily life experience: The day reconstruction method. *Science, 306*(5702), 1776–1780.

Khan, V. J., Markopoulos, P., Eggen, B., IJsselsteijn, W., & de Ruyter, B. (2008). Reconexp: A way to reduce the data loss of the experiencing sampling method. *Proceedings of the 10th International Conference MobileHCI* (pp. 471–476). ACM.

Liedtke, C., Welfens, M. J., Rohn, H., & Nordmann, J. (2012). LIVING LAB: user-driven innovation for sustainability. *International Journal of Sustainability in Higher Education, 13–2*, 106–118.

Majcen, D., Itard, L., & Visscher, H. (2013). Actual and theoretical gas consumption in dutch dwellings: What causes the differences? *Energy Policy, 61*, 460–471.

Mulder, I., Hofte Ter, G. H., & Kort, J. (2005). SocioXensor: Measuring user behaviour and user experience in context with mobile devices. *Proceedings of Measuring Behavior* (pp. 355–358).

Romero, N., Almahmud, A., Beella, S., & Keyson, D. (2013). Towards an integrated methodology to design sustainable living practices. *Proceedings of International Conference of Ambient Intelligence Ami'13*, Springer.

Spaargaren, G. (2011). Theories of practice: Agency, technology, and culture: Exploring the relevance of practice theories for the governance of sustainable consumption practices in the new world-order. *Global Environment Change, 21*, 813–822.

Chapter 13
Co-creation in Living Labs

Shea Hagy, Gregory M. Morrison and Peter Elfstrand

Abstract Living Labs are places for open innovation where co-creation is a method for addressing real-life issues through the attribution of knowledge from science and society, the latter being a form of transdisciplinary social learning. In a Living Lab the representatives from business, society and academia, as well as citizens, have different value perceptions and propositions, providing heterogeneity across the stakeholder value spectrum. This provides a rich set of ideas and values for co-creation which can be used for both the operational phase and the integral shaping and creating the design for the physical infrastructure of the Living Lab itself. The use of co-creation workshops are demonstrated for ideation amongst the stakeholders for the HSB Living Lab. This is exemplified in the development of the social washing room which will be prototyped and tested in a fit-for-purpose multifunctional design space.

Keywords Living labs · Co-creation

13.1 Introduction

Living Labs are co-creative by design and definition as the Living Lab concept offers an environment that aims to facilitate co-creation; an interactive platform for collaborative research where users play an active role (Rosado et al. 2014).

S. Hagy (✉)
Building Technology, Chalmers University of Technology,
Gothenburg, Sweden
e-mail: shea.hagy@chalmers.se

G.M. Morrison
Curtin University Sustainability Policy (CUSP) Institute, Curtin University,
Perth, Australia

P. Elfstrand
Tengbom Architecture, Gothenburg, Sweden

D.V. Keyson et al. (eds.), *Living Labs*, DOI 10.1007/978-3-319-33527-8_13

The Living Lab is appropriate for co-creation as it is issue-driven and exists within a rich, complex and contested real-world context (Carew and Wickson 2010); that of sustainable living. The theoretical basis for the co-creation methodology is transdisciplinary where the knowledge is generated in patterns across relevant disciplines and discourses. Consequently, the hermeneutic framework for the Living Lab can be defined as the attribution of knowledge from science in an issue-driven process (Max-Neef 2005; Hadorn et al. 2008) and which builds on the early ideas of Jantsch (1970).

While the interpenetration of epistemologies within relevant issues for society is relatively well agreed as characterizing transdisciplinarity (Carew and Wickson 2010), the role of practitioners for the co-creation of knowledge is new (Pohl 2008). The integral thinking process may involve practitioners being active in knowledge production or involve practitioners reacting to research conducted (Mobjörk 2010). The former is a process of social learning where researcher's explicit disciplinary knowledge and practitioners tacit knowledge may be linked to provide new hypotheses for further research or societal action (Baars 2010; Leys and Vanclay 2011).

Transdisciplinary social learning becomes a more stringent methodology in the Living Lab through co-creation. Co-creation is integral thinking between stakeholders to provide value leading to innovation (Franz 2015); this innovation process should provide more relevant products and services which are quicker to market. There are two key challenges in the co-creation process. Firstly, the fundamental difference between the nature of the knowledge resources provided and offered by the stakeholders and secondly, the value perceived in the co-creation outcome by each party (Hughes 2014). For the second challenge the question remains whether the very basis of the triple helix idea (the nexus of academia, business and society), which is increasing innovation through knowledge, is seen as a value for all parties. Value perception and outcome has a heterogeneity across the stakeholder spectrum and may include new research, technology adoption, behavior or practice change, or effect on public policy (Hughes 2014). Hughes (2014) argues that different value perceptions might be addressed by identifying consensus and innovation spaces within the co-creation process.

Current thinking is that involvement of citizens in the co-creation process provides a greater user-centred value and thereby shifts focus from a rational organizational basis to a more novel and creative process (Ind and Coates 2013; Leminen and Westerlund 2012), which is both iterative and reflective (Vicini et al. 2013). However, this also implies a greater emphasis on the social representativeness of the outcomes (Franz 2015). If the user is to gain value and meaning in the co-creation process, a greater emphasis on user adoption of innovation based on preference and needs is required (Mangyoku et al. 2014). By bringing together the transdisciplinary knowledge resources of stakeholders and users in an open co-creation process, it should be possible to provide the Living Lab with societal meaning in a structured innovation space. This high degree of relevance should also reduce the risk of market failure of the innovations developed (Leminen and Westerlund 2012).

13.2 Co-creation

Co-creation is a process that provides an opportunity for on-going interaction between partners, clients, and users, allowing collaboration and fostering innovation (Ind and Coates 2013). A Living Lab facilitates and promotes open innovation systems and initiatives through a co-creative platform for experimentation. Living labs are thus spaces of co-creation, which promote open innovation processes making co-creation an essential element within a Living Lab environment. This article considers the contemporary use of the co-creation process in Living Labs and its relevance for both bringing together academic, business and society around common problems, as well as shaping new concepts for potential prototyping.

13.3 The Theory Behind Co-creation

13.3.1 Co-creation. The Method of Choice for HSB Living Lab

The HSB living lab is built on the Chalmers campus as a meeting place for business, society and academia. The lab includes ten business and society partners as well as Chalmers as the academic partner. During the formation of the partnership it was realized that there was a need for a methodology to bring partners together onto common ground, and also to provide a creative space where new ideas for innovations and services could be generated. As a consequence, co-creation workshops (CCW) became the method of choice underpinning the HSB Living Lab. The CCW methodology will not only be used during the operational phase of the Living Lab but has been integral in shaping and creating the design and conceptual basis for the physical infrastructure itself (see Fig. 13.1).

13.3.2 Co-creation Workshop as a Tool for Innovation

Business-society-academia workshops were designed to generate ideas in an intensive co-creative environment. The aim for these co-creation workshops was to provide early innovation ideas for HSB Living Lab which may or may not lead to prototyping in the Lab. These ideas generated through the workshops were then used by the design and planning teams to create the program for the building and inform the design documents. During the operational phase of the HSB Living Lab, CCWs will continue to be drivers of idea generation and the evolution of the living environment, where the residents will themselves be given the opportunity to co-create aspects of their environment and engage with the industry and academic partners.

Fig. 13.1 Image from a CCW during the concept and design phase of HSB Living Lab

13.3.2.1 Ideation

The co-creation workshop is used as a concept and idea generator in the first stage of the Conceive- Design-Implement-Operate (CDIO) framework, which takes an idea all the way through from concept to use. Product, process, and system lifecycle development and deployment are key elements of a CDIO program as defined by the CDIO Initiative (CDIO 2015). CDIO is considered an appropriate context for engineering education. Integrating co-creation into this framework through CCWs in a Living Lab environment can be used as well as a bridge between academia and industry. CDIO is a model of an entire product, process, and system lifecycle. The Conceive stage includes defining customer needs; considering technology, enterprise strategy, and regulations; and, developing conceptual, technical, and business plans (CDIO 2015). In this stage, the co-creation workshop methodology can be used as an effective tool in the process, creating an environment, in which knowledge and skills are taught, practiced and learned. A physical Living Lab infrastructure, then, provides the opportunity to implement and operate these concepts in an iterative design process. In the case of the HSB Living Lab and CCWs facilitated by Chalmers University, this is expanded from being applicable to engineering and education as set out by the CDIO initiative, to integrating multiple disciplines and linking education with industry, to a transdisciplinary methodology (Hadorn et al. 2008; Mobjörk 2010).

13.3.2.2 Transdisciplinarity

A transdisciplinary methodology primarily connects a diverse range of disciplines but can also be seen as a way to bring society, business and academia into a common space. As described by Hadorn et al. (2003), transdisciplinary research for sustainability strives to investigate problems on descriptive, normative and operational levels and it produces both systems and target knowledge. The complex nature of sustainability creates a need for a third type of knowledge. A transformative knowledge can create the necessary conditions and strategies for changing undesired processes. Given that sustainable development involves societal problem-solving based on research, the knowledge held by non-academic actors has an important role to play throughout the process of knowledge production (Hadorn et al. 2003). Here lies the functional importance of co-creation and CCWs, creating thematic based arenas to enable transdisciplinary collaboration. This is of more specific relevance when relating to sustainability science and sustainable development, which is the focus of the HSB Living Lab.

The complex nature of sustainable development requires knowledge from a wide range of disciplines which can then be used to develop and test this knowledge in real-life context i.e. through a living lab methodology. Here, the users' knowledge is taken into account. The CCW is the method used in the HSB Living Lab environment, "… to transcend the boundaries between scientific disciplines," and "… open the academic research process to actors in public bodies, business and civil society and go beyond purely academic definition, analysis and interpretation of research problems" (these quotes are from Hadorn et al. 2003).

13.4 Co-creation Methodology in Practice

13.4.1 Overview

This section describes the overall process from collaboration of academia-business-society to working with stereotypes and finally developing concepts.

13.4.2 Preparing the Nature of the Co-creation Challenge

The planning of CCWs in itself becomes a co-creative activity as inputs from others outside one's discipline and/or competence are necessary to successfully plan a workshop. The theme or topic of the workshop can be specified and chosen, however when planning a CCW it is crucial to leave flexibility built into the schedule and plan.

There are many possibilities regarding the duration of a CCW. The authors have planned and run numerous variations of CCWs, from full-day workshops, to half day workshops, and 2-day workshops, evaluating their own pros and cons. The researchers have planned now to hold a workshop on 2 separate days with a 1 week interval. From past experience, we know that it can be difficult to bring together the right people for a workshop as they would need to take 2 full consecutive days from their other daily activities. Further, CCWs are very intensive and providing space between intensive sessions can help to foster creativity and keep participants excited and fresh.

Preparation of a CCW requires much planning, often a few months in advance. Careful planning of the theme and gathering relevant participants from a diverse range of industries and backgrounds is crucial in order to deploy and run a successful CCW, one example is the Next Generation Clothing and Laundry Workshop in February, 2014 (see further sections).

13.4.3 Co-creation Workshop Process

The most basic tools of a CCW are sketching tools such as paper, pens, clay, wood, cardboard, glue and computer programs. Having a wide diversity allows participants to choose the medium/media they are most comfortable with or excited about using, and enables participants to explore the topic at hand from different points of view through mixed media.

Another type of tool that we use in the CCW is called brain writing. This has proven to be extremely useful and important both in idea generation and in setting and fostering group dynamics and communication. In brain writing, facilitators craft one or more statements or questions related to the topic/theme of the workshop. This is meant to be an intensive, rapid generation of views about the topic where deep reflections are not the focus. Brain writing, while helping to start the ideation process, also allows for participants who may be less likely to speak in a group discussion to get their views out and read by everyone in their group, helping to mitigate issues with certain personalities dominating the conversation in later discussion sessions.

Brainstorm sessions often begin with the statement that there are no rules, this is important to provide the basis for innovative ideas and creativity. In a CCW there is a need to specify and create limitations and frames to narrow focus yet still allow for flexibility. Scheduling as well as the formation of the teams is an important aspect of preparing the proper environment for creativity. One such way to do this is through what has been termed brain swarming. This is done by forming teams where some of the members are familiar with one another and some are newcomers as well as having the teams or groups within the CCW remain intact throughout the entire workshop. Ideal group size for the breakout sessions is recommended to be between three to seven persons, and odd number groups are desirable. As the members become familiar with one another they tend to become more open, apt to share ideas and productive.

When planning the agenda of a CCW the tempo and pace must be considered, i.e. having long presentations, and/or long group sessions can stagnate the process. This tempo can be achieved through careful pre-planning of the agenda and integrating flexibility into the schedule where the facilitators are able to read the atmosphere of the participants and change the program accordingly.

13.5 The Washing Room Example in HSB Living Lab (Next Generation Clothing and Laundry Workshop)

The decision was made to hold a 2-day co-creation workshop dealing with the everyday human activity of washing. This was carried out with the help of researchers from the RCA in London (who ran the co-creation process) and was co-located at Chalmers University of Technology in Gothenburg, Sweden and the Johnson Space Centre in Houston, Texas. The participants communicated via video link.

The background for this workshop stems from the HSB Living Lab project which is a physical asset in the Climate-KIC BTA flagship. The lab has 12 business and societal partners to date. Electrolux, a multinational appliance manufacturer, and the housing association, HSB, have an interest in cutting edge innovation combined with issues of social inclusion (loneliness being a particular problem in Sweden). Meanwhile Chalmers University researchers have an interest in the sustainable design of technology connected to human behavior in everyday life. NASA and Rice University became engaged, being interested on the sustainability of washing on the planned missions to Mars. The co-creation workshops are seen as a central aspect of the BTA Living Lab network as they bring together researchers and stakeholders into the prototyping space. The aim is to provide early ideation that can then be accelerated by entrepreneurs or the partner companies.

The 2-day workshop was carefully planned beforehand with interviews of housing association residents, astronauts and others providing a common platform for the workshop groups. The workshop was met with much enthusiasm from the delegates in Sweden (Electrolux, HSB and other company employees, Chalmers students and researchers, architects—including the person who later designed the HSB Living Lab) and at NASA (NASA engineers, Rice students and researchers).

It became clear that washing does not necessarily involve only the traditional washing machine. Some interesting ideas included the wasketball (a basketball loop in which clothes can be thrown through with a sensor to detect whether the clothing actually needs washing—effectively turning the act of washing into a game- see Fig. 13.2) and a designer T-shirt dispenser with cubicle for changing (useful for those who need to change T-shirt after cycling).

Two ideas stood out. The first was the refreshment cabinet whereby clothes can be refreshed either through connection to the home ventilation system and/or through low energy UV LED. We do not know whether the residents will use this

Fig. 13.2 Workshop team (Peter Elfstrand, Charlotte Farrouch, Michail Mavromatis) presenting Washketball concept

and therefore Electrolux will install cabinets on each floor of the Lab for testing through research projects.

The second idea was recently the focus of a national (Swedish) press release by Electrolux. This involves the social washing room. The larger Electrolux washing machines have now become much quieter and allow the possibility of social spaces around or adjacent to the machines. This differs from the cellar washing rooms of Swedish housing associations that used to be standard. The Lab will be designed to allow a large multifunctional space where different layouts can be tested and will connect to a student design space where alternative furniture can be built. A prototype (See Fig. 13.3) was presented as an exhibition at the major political meeting (Almedalen) in July 2015 and will move into the living lab with the students as the Lab opens in February 2016.

13.6 Conclusion

CCWs provide a means to:

Engage a transdisciplinary team, bringing academia, society and business into a Living Lab. This becomes a neutral space with common respect for the tacit

Fig. 13.3 Render of Multifunctional laundry (*Photo* Tengbom Architecture)

knowledge held by business and society and the stringent scientifically based knowledge held by academia. Further, the presence of students brings intergenerational aspects into the process.

Ideate, revealing early common concepts and ideas that can be developed for prototyping in a Living Lab. This is important for a Living Lab as it keeps people in the knowledge-innovation-business pipeline active.

References

Baars, T. (2010). Experiential science; towards an integration of implicit and reflected practitioner-expert knowledge in the scientific development of organic farming. *Journal of Agricultural and Environmental Ethics*, published on-line.

Carew, A. L., & Wickson, F. (2010). The TD wheel: A heuristic to shape, support and evaluate transdisciplinary research. *Futures, 42*, 1146–1155.

CDIO: Conceive Design Implement Operate (2015). *Home.* http://www.cdio.org/. (21 September 2015).

Franz, Y. (2015). Designing social living labs in urban research. *info*, *17*(4), 53–66.

Hadorn, G. H., Pohl, Ch., & Scheringer, M. (2003). Methodology of transdisciplinary Research. In: Encyclopedia of Life Support Systems (EOLSS) (Eds.), UNESCO and EOLSS Publishers. Paris: UNESCO.

Hadorn, G. H., Bradley, D., Pohl, C., Rist, S., & Wiesmann, U. (2008). Implications of transdisciplinarity for sustainability research, *Ecological Economics*, 119–128.

Hughes, T. (2014). Co-creation: Moving towards a framework for creating innovation in the Triple Helix, *Prometheus: Critical Studies in Innovation*, pp. 1–13.

Ind, N., & Coates, N. (2013). The meanings of co-creation. *European Business Review, 25*(1), 86–95.

Jantsch, E. (1970). Inter- and transdisciplinary university: A systems approach to Education and Innovation. *Policy Sciences, 1*, 403–428.

Leminen, S., & Westerlund, M. (2012). Towards innovation in Living Labs networks. *Int. J. Product Development, 17*(1/2), 43–59.

Leys, A. J., & Vanclay, J. K. (2011). Social learning: A knowledge and capacity building approach for adaptive co-management of contested landscapes. *Land Use Policy, 28*, 574–584.

Mangyoku, M., Aragall, F., Serra, A., Sanchez, L. and Vuontisjarvi, H. (2014). 'IDeALL: Investigating design-for-all and living-lab methods for engaging users in value co-creation'. *2014 International ICE Conference on Engineering, Technology and Innovation* (pp. 1–8), IEEE, Bergamo.

Max-Neef, M. A. (2005). Foundations of transdiciplinarity. *Ecological Economics, 53*, 5–16.

Mobjörk, M. (2010). Consulting versus participatory transdisciplinarity: A refined classification of transdisciplinary research. *Futures, 42*, 866–873.

Pohl, C. (2008). From science of policy through transdisciplinary research. *Environmental Science and Policy, 11*, 46–53.

Rosado, L., Kalmykova, Y., Hagy, S., Morrison, G., & Ostermeyer, Y. (2014). A living lab co-creation environment exemplifying Factor 10 improvements in a city district. *Journal of Urban Regeneration and Renewal, 8*(2), 171–185.

Vicini, S., Bellini, S., & Sanna, A. (2013). User-driven service innovation in a smarter city Living Lab. *International Conference on Service Science* (pp. 254–259).

Chapter 14
Participatory Drawing in Ethnographic Research

Flora Bowden, Dan Lockton, Rama Gheerawo and Clare Brass

Abstract The chapter reports on a participatory drawing research study conducted by the Royal College of Art within the SusLabNWE project. It sought to explore people's notions of energy and to visualise their ideas and associations relating to it. The study is framed within the context of the broader ethnographic research tools that were employed by the SusLabNWE consortium. The study was conducted in three phases with visitors to the Helen Hamlyn Centre for Design's Life Examined exhibition at the Royal College of Art in September 2013; with students participating in the UK ArtScience Prize at The Silk Mill, Derby in April 2014; and with visitors to the Victoria and Albert Museum Digital Design Weekend in September 2014. Participants were offered drawing materials and asked to respond to the question: *What does energy look like?* In this chapter we discuss the outcomes of the research process, we analyse the images that were created and we explore what they tell us about the participants' ideas about energy and what this could mean for energy visualisations.

Keywords Drawing · Participation · Energy · Visualisation · Ethnographic research

14.1 Introduction

Drawing Energy was a research study that sought to explore people's perceptions of energy, an often-intangible concept that is ever-present in our daily lives. It was conducted jointly by the Helen Hamlyn Centre for Design and SustainRCA

F. Bowden (✉) · C. Brass
SustainRCA, Royal College of Art, London, UK
e-mail: flora.bowden@gmail.com

D. Lockton · R. Gheerawo
Helen Hamlyn Centre for Design, Royal College of Art, London, UK
e-mail: dan@danlockton.co.uk; dan.lockton@rca.ac.uk

D.V. Keyson et al. (eds.), *Living Labs*, DOI 10.1007/978-3-319-33527-8_14

and was an opportunity to bring together their distinct philosophies and research approaches to explore the emerging territory of inclusive environmental design. In the context of this research, this meant understanding the ways in which people use and think about energy in contemporary British society and how energy itself might be represented in new ways in order to help people to engage with it in more personally-relevant and environmentally conscious ways.

Inclusive design and sustainability often have different starting points and deal with different scales. Inclusive design usually focuses on people's needs and capabilities at the domestic scale, while sustainability embraces complexity and systems thinking, addressing systemic change. The research methodology employed in Drawing Energy negotiated a space between the two, bringing together people's aspirations and perspectives with the context of socio-political mandates and changing infrastructure and technologies.

Drawing Energy was a qualitative investigation of energy. Within the context of SusLabNWE, it built on the qualitative research methods shared by the project consortium, that are outlined in the SusLabNWE Research Methods Toolkit (Greene et al. 2013). It also worked alongside the quantitative research techniques employed by other project partner institutions to explore human understandings of energy, and offered a new lens through which to consider the ways in which people use energy.

14.2 Aim and Approach

The Drawing Energy study builds on research methods developed by both the Royal College of Art over the last two decades and on The SusLabNWE Research Methods Toolkit developed across the project partner institutions. The energy research began by conducting a series of ethnographic interviews in 2013 with people—mostly Londoners—in their own homes, to explore their ideas around energy, what it means to them and how it features in their daily life.[1] Within this engagement we used design probes and logbooks (a development of the diary study) to gain further insights, all of which contributed to the later development of our drawing research.

While the principal focus of the SusLabNWE project was on energy in the utility sense, the word 'energy', with numerous definitions in English (Merriam-Webster 2015), is a concept that resists easy visualisation and is not synonymous with any singular or distinct visual icon. We therefore decided to expand the frame of reference for our research to encompass fully all the meanings the term holds and explore its significance for our interviewees.

[1] The householders interviewed included social housing residents and owner-occupiers who were interested in monitoring their energy use.

The research process itself was people-centred, with the designers and researchers seeing householders as collaborators and contributors, not just 'test subjects' (Gheerawo and Bichard 2011). This was important: it is beneficial to work with real people as nothing can really replace the value of this process (Warburton 2003). This moves projects from being perhaps an ego-centric expression of design expertise, to having social relevance and value for the end user.

Drawing Energy was conducted with visitors to the Helen Hamlyn Centre for Design's Life Examined exhibition at the Royal College of Art in September 2013; with students participating in the UK ArtScience Prize at The Silk Mill, Derby in April 2014; and with visitors to the Victoria and Albert Museum Digital Design Weekend in September 2014.

14.2.1 Ethnographic Research Methods

Our research began with ethnographic interviews in which we discussed people's relationship to energy in the home and their broader thinking around it. We utilised a range of design probes and a logbook (a variation on a diary study), which for example, asked people to report which other words they associated with the word 'energy' and asked them consider and rate different methods of visualising energy, from an individual display on their phone or computer, to public displays, calculating the energy use in a residential area.[2] These methods were outlined for the context of the project within the SusLabNWE Research Methods Toolkit, which set out the agreed qualitative research approaches that each partner institution would employ in their investigation.

The toolkit comprised a set of twelve qualitative research methods to be used in the lab and in the field. These included questionnaires and diary studies to gain initial insights into a participant's ideas about energy and its relation to everyday activities, as well as design probes delve deeper into the issues. It also incudes co-creation and prototyping, to support the development of new designs, from a people-centred perspective, and lastly, the Toolkit also details in situ tasks, designed specifically for the lab setting. The Toolkit therefore offered a broad set of methodologies that facilitated our research at different stages of it.

In Drawing Energy we created logbooks, which were intended to gather insights into participants' 'thoughts and actions within the context of daily life practices' (Romero Herrera 2013), related to energy consumption. While the design probe, which is often called a 'cultural probe' was 'intended to reveal further insights into a participant's behaviours or motivations concerning their domestic energy use' (Bowden 2013).

---

[2]The design probes and logbook are ethnographic research methods detailed in The SusLabNWE Research Methods Toolkit.

These methods worked well for the exploratory nature of our research. By utilising these approaches we sought to uncover nuances in the ways in which participants think about energy, the associations they make with it in everyday life, and their responses to different forms of communicating energy use.

14.2.2 The Research Process

Early on in the research we found that people's mental models (Johnson-Laird 1983; Gentner and Stevens 1983) and notions of energy were very diverse (as confirmed by some other research, e.g. Rupp 2013) and that their associations were much broader than heat or electricity. One participant told us that to describe energy to a child he would say it is '[a] force. Something that creates change, or motion, or action.' We also consistently found, that the invisibility of energy was a significant attribute for the people we spoke to.

Another participant said to us:

> I think I worked out that through gas and electricity every year, the average house gets the equivalent of a bit over three tons of coal delivered completely silently and without any mess. And go back a hundred years ago and everyone would have a really good quantitative understanding of how much energy they used because they had to physically shovel the stuff. So, that made me stop and think.

The initial research interviews revealed that the energy's invisibility was a defining characteristic, and one that might be closely connected to understanding our own environmental impact through energy use. This was by no means a new finding (e.g. Burgess and Nye 2008)—and 'keeping energy use visible' is central to the thinking behind home energy monitors (Hargreaves et al. 2013)—but it is one that has often been addressed in design through leaping straight to new interface designs (Froehlich et al. 2010) without exploring the issue further in terms of the meanings, social and ecological factors of everyday lived experience (Mazé and Redström 2008; Strengers 2011; Hamilton and Hinshelwood 2014) and the stories around these (Mourik and Rotmann 2013; Lockton et al. 2014).

To investigate the questions that energy's invisibility might present, we decided to undertake a drawing study to explore energy in new ways, and to use the drawing process to uncover the associations people make with this immaterial entity. We developed a visual research method, which social scientists might term a 'participatory visual method' (see for example, Gubrium and Harper 2013; Mitchell 2011) in which we asked people to respond, through drawing or writing on paper, to the question: *What does energy look like?* As Gray et al. (2010) suggest, '[w]ords become more challenging to visualise as they become less literal', and energy, as a form of dynamism, power, force or activity, might be considered 'an idea that isn't anchored to an object in reality' (Brown 2014). We reasoned that this method could help us to explore people's mental models and perceptions of energy, and of the infrastructures or meanings connected to it. Participatory drawing research has been used before to explore people's understandings of abstract or invisible concepts, for example Bibace and Walsh (1979) and Nemeroff (1995) explored notions of germs and illness, while Qualter (1995) and Devine-Wright et al. (2009, 2010) have explicitly looked at conceptions of electricity generation and the National Grid.

14.2.3 About the Study

The drawing study was conducted in three different stages and contexts, and it is important that we consider some of the factors in the research that may have informed the different outcomes produced.

The studies in the Life Examined exhibition and the Digital Design Weekend were both held within broader exhibition contexts and the participants were therefore all engaged with cultural events and design. Beyond this, there was no classification of their age ranges, occupations, or backgrounds. The students in the ArtScience Prize were the only group of a particular age range: they were all teenagers (13–18 year olds) working on art and design ideas inspired by the theme: 'Energy of the Future'. All of the participants could be said to have an interest in art or design, and this could have influenced the collection of images.

As the study was conducted in three different phases and locations there were inherent differences in the way each was structured. In the Life Examined exhibition participants drew on an angled board, at the V&A participants drew on tables, while at the ArtScience Prize many students chose to draw with their paper on their laps. At the V&A completed drawings were also hung on the wall, so participants could see some examples of earlier work.

All three strands of the study were conducted in either workshop or exhibition contexts, environments in which participants were encouraged to explore and create, which valued new ideas and even future visions. It is possible therefore that these conditions encouraged participants to express new ideas for what energy could be, or what they would like or expect to see in future, rather than to illustrate their experienced realities.

We acknowledge that the sample groups who took part in the study are not necessarily representative of British society as a whole, but we see this project as a way of uncovering individual views and ideas about energy that are not normally publicly expressed.

14.3 What We Found

From the three phases of the drawing study emerged a collection of 180 images of 'energy', presenting a diverse, multi-faceted and highly personalised picture of this often intangible and amorphous subject. In each of the three drawing studies participants spent as long or as little time as they liked creating their drawing, however, we did not ask people to complete surveys or questionnaires as a part of the study, so our analysis is purely based on our interpretation of the images produced.

It is important to note that this was not a study of people's drawing skills or observational drawing proficiency (Kozbelt and Seeley 2007). By asking participants to draw a physical representation of the invisible we were asking them to take part in a conceptual drawing exercise and as such, the study intends focus on the ideas, thought processes, emotions or experiences that the drawings seek to communicate.

In studying the drawings we began a process of clustering, or categorising all the pieces in the collection. These were quite fluid groupings at first that offered different lenses through which we could view and think about the material. We looked for relationships and commonality between pieces, but we also considered the divergence or contrast in what the drawings seemed to communicate—those examples that counter one another, or which could exist in a category all of their own. The richness and subjectivity in this body of work means that we felt our groupings were by no means definitive, and in something of a kaleidoscopic way, the categories can shift, reorganise, create new constellations and suggest new ways of thinking about the drawings for each individual viewer. Here are the subjects we propose:

14.3.1 Nature/Culture

Many drawings depict the 'end points' of the power we use in everyday life: the light bulbs, plug sockets, batteries or wires that fill British homes today. They are products; part of the energy infrastructure and in a way emblematic of energy in contemporary culture. Time, convenience, connectivity and the domestic are important considerations in these images as they seem to reference the instantaneous, on-demand moment of connection we have with physical power on a daily basis. They also reference the human scale. Rather than consider the National Grid, or power stations, for example, they present the visible points of contact that people have with the energy system and show how the people behind these drawings experience energy in their immediate environment.

Quite in contrast to images of manufactured products and technologies, there are also many images of nature referencing plant or animal life, the elements and environmental conditions. There are images of the sun and lightning bolts, five drawing of waves, four trees and six flowers, which collectively seem to represent the full range of force and power within the natural world. The sun could be read as the original source of energy, and waves and lightning as powerful forces of nature. In fact, we see lightning bolts 13 times throughout the collection, and (at the time of writing) a lightning bolt is also the first image on Wikipedia's Energy page (Wikipedia 2015), so perhaps this particular aesthetic has been adopted into our collective conscious as a symbol for energy, for example, via its use in battery charging iconography. How might more widespread adoption of electric cars affect this?

This sets up an interesting duality in our collection of images: in those discussed so far we can start to see a contrast or tension between the wild, which are in many instances (although not all) large-scale and potentially overpowering forms of energy; and the harnessed or the tamed, which is often energy that has been captured to be of service to people. It is also interesting to consider the perspective of the image-maker—the person doing the drawing—how they relate to the forms of energy they have depicted. Some image-makers may have drawn energy as it relates to them and some may have drawn the sources of renewable energy that we capture. In that sense, they have illustrated a starting point in the energy system whilst others may have sought to depict energy in its purest form, entirely beyond the limits of the engineered, human-made energy infrastructure.

14.3.2 Abstraction

Whilst in the previous section we discussed drawings with a broad range of subject matter, what these images had in common is that they were all representational. However, many of the other drawings in our sample are abstract.

There are images that are very concerned with colour and form, but suggest no explicit connection to, or association with, the objects or entities around us. We see swirls, zigzags, amorphous shapes, lines and blocks of colour across all the sample groups. One image is all bright orange and red, whilst another uses layers of colour to create a wash of deep blues and purple. Perhaps these images seek to visualise the matter of energy itself, to materialise the invisible, rather than to depict those elements or artefacts that either embody energy, or allow us access to it. Or perhaps they directly address the formlessness of energy, the way it is not neatly articulated in a precise and defined shape and cannot be pointed to, but exists intangibly all around us. Through this comparison we can see that the previous set of images depicted how energy is contained, while these drawings might do exactly the opposite.

But even in this range of abstract images, we see diversity. One drawing from the Life Examined exhibition shows a simple horizontal blue line across the middle of the page, reaching almost to the edges—a single mark on which to focus our attention—while another from the same group of participants, depicts a knotted, frenzied and chaotic mix of lines of different colours and trajectories. Whilst one drawing seems to suggest calmness and clarity, and the other speaks of disorder and confusion, they could both refer to energy's ubiquity and constancy. So there are parallels to draw, even in this seeming divergence, and it is interesting to consider how similar principles can be expressed in very different visual forms.

14.3.3 Process Drawings

At the V&A, the third and final strand of our study, we were able to offer the participants a wider variety of materials to work with than we had been previously. In the first two studies participants used pens and pencils, but in the third we added to this chalk pastels and ink and brushes.

Over the 2 days of the study (in which time the tables became increasingly stained and the materials looked worn and less precious), participants became more experimental with their drawings and freer with the materials. By the end of the first day people began to produce what we could call 'action drawings' or works wholly concerned with the process of their making. Several people dripped ink from above the paper; one person used a brush to draw circles then blew the ink across the page—leaving the trace of their energy in the path of the ink. Another three people (who didn't know each other beforehand) collaborated on a drawing, each making marks in pastel for 5 s, simultaneously. The resulting piece is the evidence of their energy on the paper. These images are non-representational, but they are records of the energy that has been exerted in their production. They are concerned with the paper, materials and the action of making—the drawing is a three-dimensional object and the result of an energetic process, not a picture plane (Greenberg 1961).

There is great breadth in the themes addressed by the participants. The drawings do not simply address the issues of the energy infrastructure, or environmental concerns. Instead they show us a much fuller scope of the ways in which people think about energy. The diversity, the contrast, the unexpected and the anomalies all serve to broaden our thinking on this subject, rather than to narrow our definition of it.

14.3.4 What We Do Not See

As well as all the subject matter that the drawings do represent, there are many issues that they do not address, and we would like to briefly consider the significance of some of what has been left out.

Several drawings show energy sources or supplies, but these are nearly all renewable. Alongside the drawings of the sun and waves, we see images of wind turbines on eight occasions, and from the Life Examined exhibition we find a new proposal for harvesting rainwater. Electricity pylons, which are a common feature of the British landscape, and a much longer-standing and established infrastructure than wind turbines—the first pylon was erected in 1928 (National Grid 2014) and the first wind turbine in 1991 (Nixon 2008)—do not feature at all.

The prevalence of renewable energy in the drawings is also intriguing when we consider that it remains a minority energy source across Europe. In the UK, renewable electricity accounted for 18 % of the total electricity generated in the third quarter of 2014, which was an increase of 4.2 % on the previous year, but coal and gas accounted for 58.5 % of electricity generated (DECC 2014). The energy mix in Europe is changing: the UK's target is for 15 % of all energy consumption to be from renewable sources by 2020 (DECC 2011) and across Europe renewables are forecast to account for 16 % of total residential energy use by 2020 (E3M-Lab 2013). However, our energy supply is still heavily dominated by fossil fuels and this is scarcely represented in the drawings. The emphasis we see on renewables is likely not to be people's lived experience. Instead, the drawings may look to the future, to imagine what will or could be, rather than intending to show what is at present, or has been in the past.

Nor is the fierce political debate that we often hear—or media coverage about the costs of energy in the UK—depicted in the images. In recent years the cost of domestic heating has become increasingly expensive and it was the subject of much public debate (see for example, Boffey 2015; Massy-Beresford 2014) over the course of the drawing study. Yet these worries or unsettling realities are not represented in the images.

Another interesting omission from the drawings is numbers (and units such as kilowatt hours). When we talk about energy, in terms of what we use and how much it costs, we typically quantify it—even our domestic bills rely on this information. Energy suppliers usually communicate to their customers in measurements—and real-time quantitative feedback for householders is a major plank of UK energy policy (DECC 2009). But apart from equations, the drawings do not portray numbers, and we therefore see no reference to one of the principal ways in which energy is talked about, or to the idea of quantities at all. This raises questions about how effective or useful quantitative metrics are for people in thinking about energy. Might other modes of communication or explanation be more valuable in engaging people in a dialogue about energy consumption? And how could we support people in thinking about the political and ecological systems in which they play a part, rather than talk to them purely about the money they owe?

14.4 Reflections

The drawings present stimulating material for considering the ways in which people currently do think about and visualise energy, and a key learning from the project was the broad scope and conceptual connections that people make beyond heat or electricity. What people have drawn in terms of technology at least, leans towards the future rather than towards the historic, but in other ways the results are more divergent. We find that across the sample the definitions of energy are varied. The drawings link to personal history and emotions, but also to infrastructure and systems. They stretch across scales, from the human and engineered, to the natural and untouched; and they are concerned with the political and environmental, the aspirational and the unique, as well as the ubiquitous and the everyday.

Building on the ethnographic research methods set out in the SusLabNWE Research Methods Toolkit, we were able to utilise the shared project approaches to identify key areas to explore further. Through interviews, design probes and logbooks we were able to identify that there was a broad spectrum of different meanings about the idea of energy and that the issue of invisibility was a significant factor in people's understanding. This led us to develop the drawing study to delve deeper into the questions around it. We therefore see the drawing study as an additional ethnographic research method, which responded directly to the findings uncovered in the first phase of work.

Perceptions of energy are, of course, very subjective. But in presenting these varied and sometimes opposing views, we think this study has reflected on some of the enormous complexities in what we often experience as simple daily realities. What has been produced is a diversity of representation, and through the associations and interpretations, we think that the study presents an exploration of how the subject of energy is culturally constructed.

We also think that designers and the energy sector should consider the imagination and creativity with which the participants in this study approached drawing energy, and that these broader interpretations and modes of visualisation might be able to transform energy systems for the better.

Acknowledgments Drawing Energy was a research study conducted by the Royal College of Art within the SusLabNWE project. The discussion of this research work presented here was originally published in *Drawing Energy: Exploring Perceptions of the Invisible*, an RCA publication produced on completion of the study. In this chapter we present our drawing study in the context of the ethnographic research methods utilised in the wider European project consortium.

References

Bibace, R., & Walsh, M.E. (1979). Developmental stages in children's conceptions of illness. In G.C. Stone, F. Cohen & N.E. Adler Associates (Eds.), *Health psychology: A handbook*. San Francisco: Jossey-Bass.

Boffey, D. (2015). Complaints to UK energy watchdog Ofgem triple as household bills rise. *The Guardian*. Retrieved January 14, 2015 from http://www.theguardian.com/money/2015/jan/10/complaints-uk-energy-watchdog-triple-ofgem.

Bowden, F. (2013). Method 4: Probes. In C. Greene, F. Bowden & R. Gheerawo (Eds.), *The SusLabNWE Research Methods Toolkit* (pp. 23–25). London: Royal College of Art.

Brown, S. (2014). *The Doodle revolution: Unlock the power to think differently*. New York: Portfolio/Penguin.

Burgess, J., & Nye, M. (2008). Re-materialising energy use through transparent monitoring systems. *Energy Policy, 36*, 4454–4459.

DECC. (2009). *Smarter grids: The opportunity*. Retrieved January 25, 2015 from http://www.techuk-e.net/Portals/0/Cache/%28DECC%29smart%20grid_web.pdf.

DECC. (2011). *UK renewable energy roadmap*. Retrieved January 14, 2015 from https://www.gov.uk/government/uploads/system/uploads/attachment_data/file/48128/2167-uk-renewable-energy-roadmap.pdf.

DECC. (2014). *Energy trends: December 2014*. Retrieved January 14, 2015 from https://www.gov.uk/government/uploads/system/uploads/attachment_data/file/386812/ET_Dc_14.pdf.

Devine-Wright, H., & Devine-Wright, P. (2009). Social representations of electricity network technologies: Exploring processes of anchoring and objectification through the use of visual research methods. *British Journal of Social Psychology, 48*, 357–373.

Devine-Wright, P., Devine-Wright, H., & Sherry-Brennan, F. (2010). Visible technologies, invisible organisations: An empirical study of public beliefs about electricity supply. *Energy Policy, 38*, 4127–4134.

E3M-Lab: P. Capros, A. De Vita, N. Tasios, D. Papadopoulos, P. Siskos, E. Apostolaki, M. Zampara, L. Paroussos, K. Fragiadakis, N. Kouvaritakis, et al. (2013). *EU energy, transport and GHG emissions: trends to 2050*. Retrieved June 30, 2014 from http://ec.europa.eu/energy/observatory/trends_2030/doc/trends_to_2050_update_2013.pdf.

Froehlich, J.E., Findlater, L., & Landay, J.A. (2010). The design of eco-feedback technology. In *Proceedings of CHI 2010*, Atlanta.

Gentner, D., & Stevens, A. L. (1983). *Mental models*. Hillsdale, NJ: Lawrence Erlbaum Associates.

Gheerawo, R., & Bichard, J. (2011). Support strategy. In *New Design*, Issue 87, pp. 33–37.

Gray, D., Brown, S., & Macanfuno, J. (2010). *Gamestorming: A playbook for innovators, rulebreakers, and changemakers*. California: O'Reilly.

Greenberg, C. (1961). Modernist painting. In *Arts Yearbook 4*.

Greene, C., Bowden, F., & Gheerawo, R. (Eds.). (2013). *The SusLabNWE research methods toolkit*. London: Royal College of Art.

Gubrium, A., & Harper, K. (2013). Participatory visual and digital.

Hamilton, J., & Hinshelwood, E. (2014). Creative approaches to energy in a community context. In *Proceedings of RGS-IBG Annual International Conference 2014, 'Narrating Energy' session*, London.

Hargreaves, T., Nye, M., & Burgess, J. (2013). Keeping energy visible? Exploring how householders interact with feedback from smart energy monitors in the longer term. *Energy Policy, 52*, 126–134.

Johnson-Laird, P. N. (1983). *Mental models*. Cambridge, MA: Harvard University Press.

Kozbelt, A., & Seeley, W. P. (2007). Integrating art historical, psychological, and neuroscientific explanations of artists' advantages in drawing and perception. *Psychology of Aesthetics, Creativity, and the Arts., 1*(2), 80–90. doi:10.1037/1931-3896.1.2.80.

Lockton, D., Renström, S., Bowden, F., Rahe, U., Brass, C., & Gheerawo, R. (2014). Energy storytelling through annotating everyday life. In *BEHAVE 2014: 3rd European Conference on Behaviour and Energy Efficiency*, 3–4 September 2014, Oxford.

Massy-Beresford, H. (2014). Winter energy bills frozen as Big Six urged to cut prices. *The Express*. Retrieved January 14, 2015 from http://www.express.co.uk/news/uk/521648/Winter-energy-bills-frozen-as-Big-Six-urged-to-cut-prices.

Mazé, R., & Redström, J. (2008). Switch! Energy ecologies in everyday.

Merriam-Webster/Energy. (2015). *Merriam-Webster*. Retrieved November 3, 2015 from http://www.merriam-webster.com/dictionary/energy (*Methods*. California: Left Coast Press).

Mitchell, C. (2011). *Doing visual research*. London: Sage.

Mourik, R., & Rotmann, S. (2013). Task 24: Closing the loop—behaviour change in dsm: from theory to practice. International Energy Agency.

National Grid Timeline. (2014). Retrieved July 1, 2014 from http://www.nationalgrid75.com/timeline.

Nemeroff, C. J. (1995). Magical thinking about illness virulence: Conceptions of germs from "safe" versus "dangerous" others. *Health Psychology, 14*(2), 147–151.

Nixon, N. (2008). *The Guardian*. Retrieved July 1, 2014 from http://www.theguardian.com/environment/2008/oct/17/wind-power-renewable-energy.

Qualter, A. (1995). A source of power: Young children's understanding of where electricity comes from. *Research in Science and Technological Education, 13*(2), 177–186.

Romero Herrera, N. (2013). Method 2: Diary. In C. Greene, F. Bowden & R. Gheerawo (Eds.), *The SusLabNWE research methods toolkit* (pp. 16–19). London: Royal College of Art.
Rupp, S. (2013). Considering energy: E = mc2 = (magic·culture)2. In S. Strauss, S. Rupp & T. Love (Eds.), *Cultures of energy: Power, practices, technologies* (pp. 79–95). Walnut Creek: Left Coast Press.
Strengers, Y. (2011). Designing eco-feedback systems for everyday life. In *Proceedings of CHI 2011*, Vancouver.
Warburton, N. (2003). Everyday inclusive design. In J. Clarkson, R. Coleman, S. Keates & C. Lebbon (Eds.), *Inclusive design—design for the whole population* (Vol. 15, p. 254). London: Springer.
Wikipedia/Energy. (2015). *Wikipedia*. Retrieved November 3, 2015 from https://en.wikipedia.org/wiki/Energy.

Chapter 15
Actor and Network Analysis

Carolin Baedeker, Marco Hasselkuß and Johannes Buhl

Abstract To uncover social influence in personal networks on how to set up the heating system, heating behaviour and advice on saving energy, as well as the influence of a household's peer group in terms of norms and status, a mixed-methods social network analysis was conducted in the City of Bottrop, Germany. In order to analyse a household's embeddedness in social networks, interviews were conducted with around 23 households. Personal relations (friends, neighbours, relatives and peer groups) and relations to actors in the value chain of heating/space heating (i.e. craftspeople, manufacturers) were analysed. Results indicate that contacts with family and friends play a major role and that consulting agencies and consumer advice centres influence investment decisions on insulation, for example, to a great extent—due to the highly developed infrastructure of consulting in InnovationCity Ruhr—Bottrop. The consulting organisations established therein apparently function well and it shows that advice is also further diffused through ego-networks.

Keywords Actor analysis · Network analysis · Mixed methods

15.1 Introduction

Social network analysis (SNA) is both a theoretical position and a methodological set in social science. It is a widespread approach used in different disciplines to analyse a variety of relational phenomena such as friendship, advice or

C. Baedeker (✉) · M. Hasselkuß · J. Buhl
Wuppertal Institute for Climate, Environment, Energy, Doeppersberg 19,
42103 Wuppertal, Germany
e-mail: carolin.baedeker@wupperinst.org

M. Hasselkuß
e-mail: marco.hasselkuss@wupperinst.org

J. Buhl
e-mail: johannes.buhl@wupperinst.org

D.V. Keyson et al. (eds.), *Living Labs*, DOI 10.1007/978-3-319-33527-8_15

Table 15.1 Types of ties in SNA (Borgatti and Halgin 2011, 1170)

State-type ties	Event-type ties
Kinship ties (e.g. brother of)	Interactions (e.g. giving advice to, sending e-mail)
Other role-based ties (e.g. boss or friend of)	
Cognitive (e.g. knows)	Transactions (e.g. signing treaty with; making a sale)
Affective (e.g. likes or dislikes)	

organisational collaborations in innovation research (Borgatti et al. 2009). The research object is actors i.e. persons or organisations, and their relations with one another. The relations can be analysed as conduits through which e.g. information flows or social influence. Borgatti and Halgin (2011) differentiate the following types of ties (Table 15.1).

For the context of interest in this chapter, the social influence in personal networks was in focus to find out more about how far relations with different kinds of actors have an influence on heating and ventilation behaviour (analysed as social practices), saving energy, and how to set up the heating system, but also the influence of a household's peer-group in terms of norms and status. To analyse a household's embeddedness in social networks, a mixed-methods social network analysis (Hollstein 2014) was conducted. In the German location of SusLabNWE within InnovationCity Ruhr, Model town Bottrop (InnovationCity Management 2015) the network analysis was conducted in the course of Insight Research as part of the three-tier model of research in the SusLabNRW methodological toolkit. It aimed both to gain insights into the status quo of influence factors on heating practices and to inform design ideas of Product-Service System (PSS) development (Baines et al. 2007; Baedeker et al. 2014; Liedtke et al. 2015) in the phase of Prototyping.

Both personal relations (friends, neighbours, relatives, peer-groups), and relations to actors in the value chain of heating/space heating (i.e. craftspeople, often responsible for maintenance and set up of the heating system) are of interest in this context. Employing a network analysis in a mixed-methods-design provides empirical data on the actual relationships of households that may influence their heating practices. We combined formal and qualitative methods to explore these aspects. Using **quantitative** methods provides information on the structure of **ego-networks** (see Sect. 15.2). The **qualitative network analysis** provides insights into meanings and subjective interpretations of influence on people's own heating behaviour. Thus, the following questions were addressed: What do the social networks of households look like in the field of heating and space heating? Which relevant actors can be identified and what evidence for their influence can be found? How do households interpret these influences in their heating behaviour?

15.2 Methodology: Mixed-Methods in Social Network Analysis

While the roots of SNA lie in formal, quantitative methods, meanwhile also qualitative forms of network analysis and the combination of both become more prominent. According to Hollstein (2014), mixed-methods network analysis is defined by three elements: (1) both numerical and qualitative, textual data is gathered, while one type of data might be transformed into the other, (2) during analysis of networks and relations mathematical as well as interpretative strategies are used, and, (3) at least at one point during the research process of data gathering, analysis and interpretation the data or strategies of analysis are integrated, whereas integration means to systematically link or relate quantitative and qualitative data or strategies to each other. Using qualitative and interpretative strategies next to formal methods to analyse networks provides some advantages: dense descriptions of networks can be produced and network practices and interpretations of actors can be analysed as well as dynamics of networks can be shown (cf. also Baedeker 2012).

These features can vary in terms of the number of research strands in a research design and the time at which they are used. Drawing on Tashakkori und Teddlie (2003), Hollstein (2014) describes five types of designs: sequential, parallel, fully integrated, embedded and conversion designs. Prell (2012) describes possibilities of qualitative pre-studies to inform questionnaire design for later formal SNA, thus, representing a case of a sequential design.

15.2.1 Research Design in SusLabNRW

The purpose of actor and network analysis is explorative. The embeddedness of a household (*ego*) in social networks can have an influence on heating practices. This includes the following groups (*alteri*):

1. Friends, neighbours, relatives (peer-groups), etc.
2. Other people in the value chain of heating/ space heating (i.e. people responsible for maintenance and set up of the heating system).

The method is intended for gaining insights into households' networks in order to understand how these existing networks can be used to influence behaviours. Employing a network analysis in a mixed-methods design provides empirical data on the actual relationships of, and between, households that may influence their heating behaviour. Using quantitative methods provides information on the structure of the networks. This is coupled with qualitative network analysis from which insights into how meanings are constructed and subjective interpretations can be gained. Analysis requires the combination of qualitative and quantitative techniques to understand how contacts (called alteri in SNA) influence heating practices. Collecting ego-network data in this way is adequate for the topic of interest since it is not possible to specify network boundaries a priori and collect whole network data using e.g. the closed roster-technique as a list of contacts presented to participants.

We used network maps (Hollstein and Pfeffer 2010; Schönhuth et al. 2010) to gather qualitative and formal SNA data in parallel.

With this technique the interviewee in a qualitative problem-centred interview is presented with a graphical representation of his or her ego-network and asked to collaboratively fill in contacts with the researcher in a participatory design. Interviews and network map drawing were conducted with the help of VennMaker Software (Schönhuth et al. 2010) to interactively draw and visually analyse ego-networks. The type of contact (family & friends, neighbours and colleagues, handicrafts people and manufacturers of heating systems, organisations and consultants) were represented through different sectors of the network map; emotional proximity was shown by drawing *alteri* closer or farther from *ego* in the centre; and spatial proximity as well as strength of contact were shown through different symbols on a concentric network map with *ego* being in the centre.

Data collected from network maps was analysed using quantitative network analysis to understand how different households' networks compare e.g. in terms of size, predominance of strong/weak ties, or network structure. Important people in the value chain can be identified along with the starting points from which information spreads across the network.

In SusLabNWE and a subsequent project, EnerTransRuhr, 23 households (representing different groups, including lead users and non-lead users regarding their interest in dealing with heating technology or esp. smart home systems; inhabitants of one family dwellings and apartment buildings; different socio-economic attributes) were interviewed. Both personal relations (friends, neighbours, relatives, peer-groups), and relations to actors in the value chain of heating/space heating (i.e. craftspeople, manufacturers) were analysed. Network analysis provides empirical data on the actual relationships of households that may influence their heating behaviour. Social influence coming from network relations is often regarded as a strong factor in behaviour change (cf. Jensen et al. 2015 for a modelling approach). Which relevant actors can be identified and what evidence for their influence can be found? How do households interpret these influences in their heating behaviour? The results presented in the following indicate that contacts with family and friends play a major role and that consulting agencies and consumer advice centres influence decisions on investments in insulation, for example, to a great extent—due to the highly developed infrastructure of consulting in InnovationCity Ruhr. The consulting organisations that were established when the InnovationCity Ruhr was founded and engaging the local population to participate apparently functions very well, as reflected by the large numbers of energy consultations performed since then. It also shows that advice is also further diffused through ego-networks.

Before results from quantitative and qualitative SNA are presented and integrated, two examples of the collected network maps are discussed in greater detail (Figs. 15.1 and 15.2).

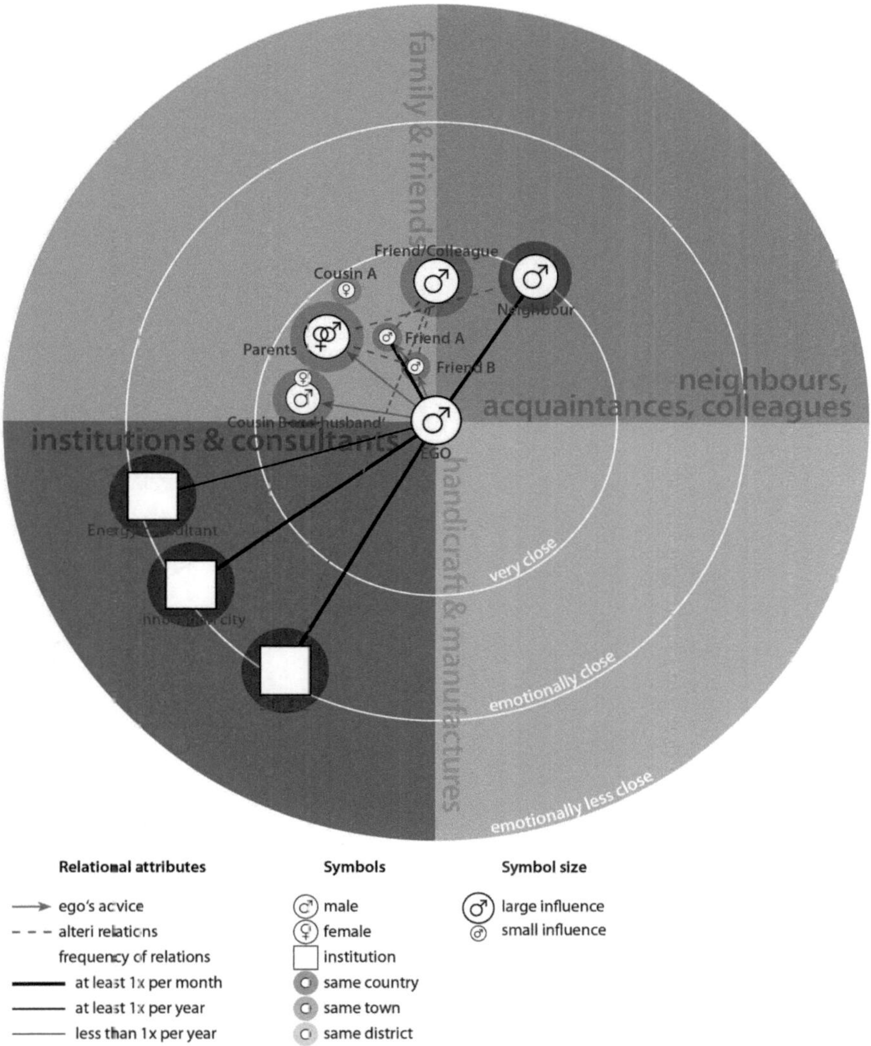

Fig. 15.1 Example of ego-network map household A, interactively drawn together with interview participants (using VennMaker Software, Schönhuth et al. 2010; modified own depiction based on these maps)

The network maps use the following symbols and sectors to represent ego-networks in the field of heating. The coloured sectors of the map symbolise the role of *alteri* named by *ego* during the interview. In the green sector family and friends were entered, the red sector covers neighbours, acquaintances and colleagues. The blue sector represents institutional actors like consultants and media, while the

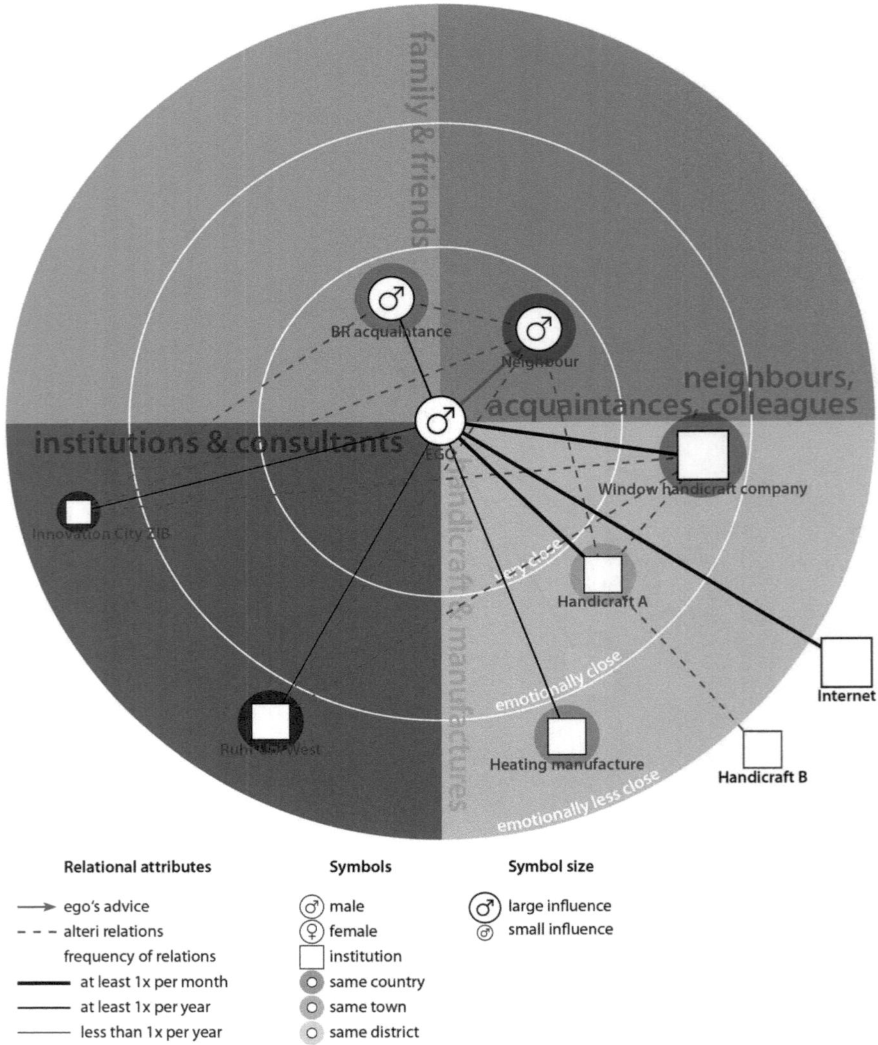

Fig. 15.2 Example of ego-network map household B, interactively drawn together with interview participants (using VennMaker Software, Schönhuth et al. 2010; modified own depiction based on these maps)

yellow sector stands for actors from manufacturers and maintenance. The symbols used for the *alteri* named by interview partners are circles and gender symbols for human actors and squares for institutional actors. The size of the symbol represents the strength of influence. Placing the symbols on the concentric circles represents the emotional closeness to ego and the kind of lines shows the frequency of contacts.

Looking at the two examples above, they show quite different structures of ego-networks in contacts about the topic of heating. In the example on the left side many emotionally close ties with family and friends were named as influential with regard to advice on or change of heating behaviour of ego, some of which have strong influence (bold lines). As we will show below this network map is in this way representative for many of the other cases. A contact to one neighbour has stronger influence on heating behaviour as well, and several institutional actors, namely energy consultants and representatives from InnovationCity and the SusLab-project were named to be influential was well. In contrast, the network map on the right shows that in this case most influence in form of advice on heating behaviour comes from manufacturers, maintenance companies but also some energy consultants. Possible kinds of changes induced by the influences were covered in the qualitative interviews. The participants mentioned e.g. ventilation behaviour, application of new technology (windows, smart home systems) or higher awareness.

15.3 Quantitative Description of Social Networks

During the projects 23 households with altogether 216 social contacts have been analysed. The social contacts have been described by characteristics circumscribing their influence on heating practices, their emotional, spatial and social proximity, the corresponding frequency of contact and sex. We asked participants with whom they talk about their "heating" and which "influence" they have. We considered heating practices to be the everyday heating routine of the respondents. We present the distribution of the *alteri* per dimension and dependencies between the dimensions with respect to their influence on heating practices.

As such the alteri are completely distributed to each dimension. That means that every dimension includes the complete number of alteri of the respondents. In each dimension the number of alteri sum up to a maximum of 216 social contacts. Some respondents brought up web and media to be a relevant source of information. We deliberately did not exclude them from our analysis, but ascribed those contacts a rather abstract and anonymous social role. In very few cases, some respondents did not name any contacts. Those are treated as missing contacts and do not show up in the final analysis.

The bar plot in Fig. 15.3 shows the number of alteri per specification of each dimension. Most of the social contacts live in the same city, are family or friends, but emotionally less close. The stated social contacts have a moderate to strong influence on heating behaviour.

We thus hypothesise that emotional, social and spatial proximity are positively correlated with the influence on heating practices. In order to indicate whether this holds true, we calculated corresponding correlations between the number of alteri in the relevant specifications of emotional, social and spatial proximity and their influence on heating behaviour. We calculated Pearson's r to get an idea of how

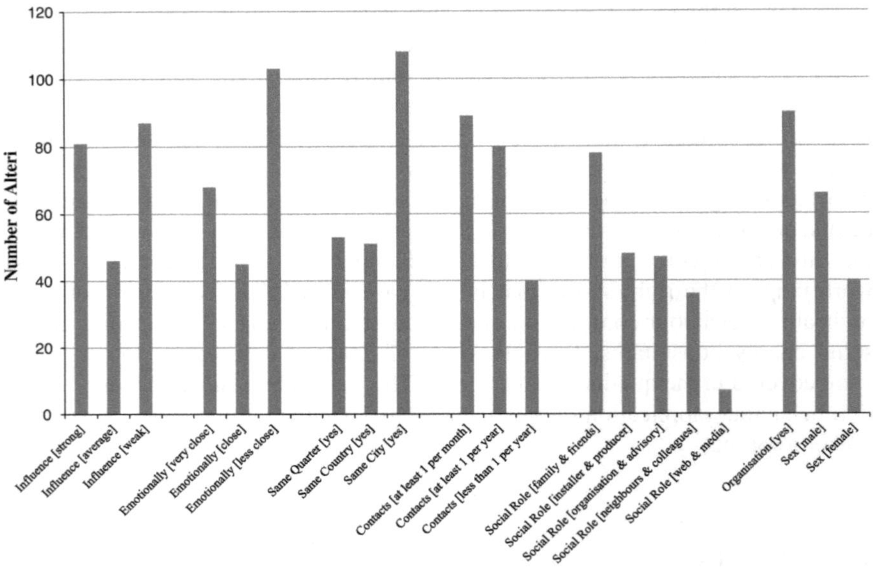

Fig. 15.3 Number of alteri per dimension and specification (n = 216)

the number of alteri with a strong influence depends on the number of alteri being friends and family, neighbours and colleagues, organisations and advisory agencies, being emotionally very close, staying in contact at least once per month or living in the same quarter. The correlation coefficient tells us the linear dependency between two variables. If one plots the number of alteri having a strong influence on heating practices against the number of alteri with specific characteristics as described above, Pearson's r resembles the slope of a simple linear regression between the two variables.

Table 15.2 shows the according correlations. We identified positive correlations except for the number of social contacts being neighbours and colleagues and the number of alteri having a strong influence. That means that the higher the number of social contacts being friends and family, organisations and advisory agencies, emotionally close with a frequency of contact of at least one time per month and living in the same quarter, the higher the number of social contacts with a strong influence on heating practices. The correlation between the number of friends and family and the number of social contacts having a strong influence is relatively high, whereas the correlation between the number of contacts in organisations and advisory agencies and the number of contacts having a strong influence is relatively weak. The negative correlation between the number of *alteri* being neighbours and colleagues does not imply that those have no or even a negative influence on heating practices. The coefficient rather indicates that the number of neighbours and colleagues have a moderate or weak influence on heating practices.

Table 15.2 Correlations of a selection of characteristics of social contacts with strong influence on heating practices

Characteristic of alteri	Pearson's r
Emotionally close	0.38
Contact at least 1 per month	0.36
Living in the same quarter	0.37
Friends and family	0.52
Neighbours and colleagues	−0.23
Organisations and advisory agencies	0.16

Note Pearsons r between -1 and 1

Eventually, the univariate and bivariate description of the social networks of respondents show that friends and family, social contacts that are emotionally and spatially close with having relative high frequency of contact show a strong influence of heating practices. However, the bivariate description does not reveal any interaction effects between the variables and thus offer a more causal interpretation of the findings. For instance, we don't know whether friends and family truly show the highest influence on heating practices or the influence of friends and family is moderated by its emotionally and spatially close relationship. That would be up to a more sophisticated multivariate analysis between characteristics of the alteri and the influence on heating practices or a closer look at the single networks in a more qualitative analysis.

15.4 Qualitative Analysis and Integration

In the following results from the qualitative interviews, conducted in parallel to using the network maps, are analysed more closely for the cases conducted in the course of SusLabNRW and integrated with findings from quantitative network analysis. As shown above influence from **family and friends** is usually high, both regarding the number of contacts (43 of 124 alteri named for the SusLabNRW cases) and strength of influence (21 of 43 with high influence) as these contact persons were often named as emotionally close. Often the interviewees stated that they talked to family and friends about topic of heating more generally. When family members or friends are planning to conduct renovations or construction works, or if some members have professional experience in related fields, then these persons were also contacted with regard to the topic of heating. In short, three motivations for talking about heating to family and friends were found:

- persons with professional experience in the family or among friends,
- persons with higher expertise in a certain related area in the family or among friends, or
- unfocused, casual talk.

Rather limited influence was found by the group of **neighbours and colleagues**, to which 22 of the 124 alteri altogether named in SusLabNRW-cases belong.

The strength of influence is also limited whereas 3 alteri of high, 9 of medium and 10 of little influence were named. The interviewees mostly engaged in conversations about the topic of heating with neighbours when experiences were exchanged or common interests were followed. With regard to direct or indirect neighbourhood effects (e.g. Johnston et al. 2005) it showed that technical measures and behaviour applied by neighbours are observed but any influence on personal decisions or heating behaviour is rather denied or only indirectly admitted. Heating is nevertheless considered a relevant topic between neighbours. Advice from this group of persons is mostly accepted when neighbours have some expertise in the field or work as handicraftsmen. Colleagues are more often contact persons on this topic if ego works in a field related to heating or if the contact is considered more close than just being colleagues.

Influence by **consultants or organisations** showed to be medium to high, where 29 of the 124 alteri named come from this group of actors with 11 stated to have high influence and 14 with medium influence on heating practices. Almost all of the participants are positive about institutional consultants and the related organisations are considered as trustworthy and good sources of more detailed information. In the InnovationCity Bottrop a consulting infrastructure was built up related to SusLabNRW and the project partner Hochschule Ruhr West (university of applied science). Under these conditions consultations were almost exclusively perceived as positive and InnovationCity and HRW were stated to have considerable influence on participants. The InnovationCity serves to give impulses, including consulting offerings on heating behaviour, to deal with the topic of energy efficiency and sustainability more consciously and closely and activities of InnovationCity cause that offerings of other organisations in the field of energy efficiency are more actively used.

There is a medium to high impact on the part of **maintenance companies and manufacturers**, 30 of the 124 contact persons belong to this sector of the network maps and their influence was named with 17 of large and 10 of medium impact. Manufacturers and maintenance are the executive body and especially give advice regarding upcoming investments, less in terms of heating behaviour. Both positive and negative experiences are portrayed. The choice of a craftsman is often stamped by the experiences of the environment (friends/family/neighbors) and is associated with geographical proximity (see Cluster proximity), which seems to be the decisive factor for households (time = money). In the Ruhr area however several cities are close to each other which also enables households to choose from other cities and, generally, the higher the reputation and specialisation of a company is it appeared that proximity becomes less important.

Media (print) and especially the internet play an important role next to personal contacts. Information is first searched for online and then the participants stated to seek advice by consulting agencies or maintenance companies and offerings by such agencies are found via print or online media. Nevertheless the influence was seen as rather low.

Taken together, three specific factors can be identified (Table 15.3).

Table 15.3 Three specific factors identified as important to heating behaviour, from the qualitative analysis

Trust/emotional closeness	As the quantitative SNA also showed, trust plays a very important role in implementation—craftspeople familiar with participants are chosen or recommendations by family and friends are crucial (also negative recommendations) "Trust in yourself": Own competences/ skills and (technical) knowledge in the subject area are—if available—emphasized and used
Motivation	Costs and comfort are the driving motivations for change (and also for participation in the research activities). On inquiry, also environmental aspects are named (rather suggestive) When participants' professional background is related to environmental protection, energy efficiency etc. environmental awareness as a motivation for behavior change is in the foreground
Change of heating practices	The majority of participants stated there had been changes in behaviour for themselves or in their network (7 of 11 cases in SusLabNRW) and concrete examples were named. These affect: ventilation behaviour, application of new technology (windows, smart home systems) and a different way of heating and higher awareness Reasons for change are taking personal action: professional advice was sought (InnovationCity or other agencies), getting professional advice at public events or as part of the research activities, technical novelties in the house which cause higher awareness and reflection of own behaviour (e.g. new windows or heating system) or problems that enforce change (e.g. mold formation)

15.5 Conclusions

The mixed-methods network analysis could show important insights into social influences on heating practices in the households, how and with whom participants mostly talk about or seek advice on heating-related questions. It showed that network analysis is an interesting and valuable method in Living Lab research. The analysis could be extended employing more sophisticated methods of formal SNA of ego-networks, such as structural holes analysis (Burt 1992) to show e.g. which ties could bridge the different clusters and thus improve flow of information. Furthermore, collected network data can also serve to inform agent based modelling approaches (see Jensen et al. 2015) to simulate the diffusion of energy-saving behaviour.

With regard to results that can inform Prototyping in SusLabs, the relevance of strong, emotionally close ties and the positive perception of institutionalised consultation infrastructures are especially relevant. As much as strong ties were named to have more influence on heating behaviour, trust appears to be decisive for the effect of consulting agencies. Product-Service System (PSS) design, e.g. the user-integrated development and testing of integrated sensoring and individually tailored energy consulting offerings, needs to take these aspects into account more thoroughly. So far the apparently high relevance of building up trust in good information and both emotionally and spatially close contacts are hardly considered for PSS design involving services close to the customer like e.g. consultation infrastructures. In the German test region this was successfully implemented in the InnovationCity, that is perceived as providing scientifically based, reputable information and succeeded to motivate people to participate in the conducted real-life experiments and often reflect personal routines more consciously.

Acknowledgments Part of this research (EnerTransRuhr: The German "Energiewende"—Development of an Integrative and Transformative Research Design in the Case of the Energy Transition of the Ruhr Area and North Rhine-Westphalia) was funded by the German Federal Ministry of Education and Research (BMBF).

References

Baedeker, C., Greiff, K., Grinewitschus, V., Hasselkuß, M., Keyson, D., Knutsson, J., et al. (2014). Transition through sustainable product and service innovations in sustainable living labs: Application of user-centred research methodology within four living labs in Northern Europe. In *Paper presented at the 5th International Sustainable Transitions (IST) Conference*, August 27–29, 2014 Utrecht, Netherlands.

Baedeker, C. (2012). *Regionale Netzwerke: gesellschaftliche Nachhaltigkeit gestalten - am Beispiel von Lernpartnerschaften zwischen Schulen und Unternehmen*. München: Oekom-Verl.

Baines, T.S. Lightfoot, H.W. Evans, S. Neely, A. Greenough, R. Peppard, et al. (2007). State-of-the-art in product-service systems. *Proceedings of The Institution of Mechanical Engineers. Part B-Journal of Engineering Manufacture, 221*(10), 1543–1552.

Borgatti, S. P., Mehra, A., Brass, D. J., & Labianca, G. (2009). Network analysis in the social sciences. *Science, 323*(5916), 892–895. doi:10.1126/science.1165821.

Borgatti, S.P., & Halgin, D.S. (2011). On network theory. *Organization Science*, *22*(5), 1168–1181. http://doi.org/10.1287/orsc.1100.0641.

Burt, R. S. (1992). *Structural holes. The social structure of competition*. Cambridge, Mass.: Harvard University Press.

Hollstein, B. (2014). Mixed methods social network research: an introduction. In S. Dominguez, S. Hollstein & B. (Eds.), *Mixed methods social networks research. Design and applications* (pp. 3–34). New York: Cambridge University press.

Hollstein, B., & Pfeffer, J. (2010). Netzwerkkarten als Instrument zur Erhebung egozentrierter Netzwerke. In *Unsichere Zeiten. Herausforderungen gesellschaftlicher Transformationen. Verhandlungen des, 34. DGS-Kongresses*.

InnovationCity Management. (2015). InnovationCity Ruhr. Blue sky, green city. Retrieved January 12, 2016 from http://www.icruhr.de/fileadmin/media/downloads/ICR_Imagebroschuere_neu_en.pdf.

Jensen, T., Holtz, G., & Chappin, E.J.L. (2015). Agent-based assessment framework for behavior-changing feedback devices: Spreading of devices and heating behavior. *Technological Forecasting and Social Change, 98*, 105–119 (2015). doi:10.1016/j.techfore.2015.06.006.

Johnston, R., Propper, C., Sarker, R., Jones, K., Bolster, A., & Burgess, S. (2005). Neighbourhood social capital and neighbourhood effects. *Environment and Planning A, 37*(8), 1443.

Liedtke, C., Baedeker, C., Hasselkuß, M., Rohn, H., & Grinewitschus, V. (2015). User-integrated innovation in sustainable LivingLabs: An experimental infrastructure for researching and developing sustainable product service systems. *Journal of Cleaner Production, 97*, 106–116 (2015). doi:10.1016/j.jclepro.2014.04.070.

Prell, C. (2012). *Social network analysis: History, theory and methodology*. Los Angeles/ London: SAGE.
Schönhuth, et al. (2010). VennMaker Software. www.vennmaker.com.
Tashakkori, A., & Teddlie, C. (2003). *Handbook of mixed methods in social and behavioral research*. Thousand Oaks, Calif: SAGE Publications.

Chapter 16
Design Participation in Sustainable Renovation and Living

Stella Boess

Abstract This chapter addresses resident participation in the renovation of sustainable housing. Such renovation efforts aim to reduce greenhouse gas emissions by reducing energy waste from heat loss. Resident behaviour after renovation is a key factor. The residents may, for example, continue to keep windows open in winter even though there is now a ventilation system. Aligning renovation processes with the residents' habits and preferences may therefore help reduce greenhouse gas emissions. No process framework currently exists that integrates resident participation with the renovation process. Design participation is a social design approach that seeks to support collaboration between the residents and the other stakeholders with design tools. This chapter shows how design participation reveals opportunities to innovate on the stakeholder process, as well as on technologies in the home. The examples arise from an education project in which design students collaborated with residents to address pre-, during and post-renovation needs as well as routine living. Each proposal reveals challenges and possibilities for the renovation process and for home technologies. The chapter maps the design participation examples onto the building management cycle and innovation issues in it. Overall, the examples reveal that there are still gaps to bridge between design participation thinking and the current participation and innovation processes in this field. While the latter tend to focus on agreements, being heard, and application of existing technologies, the examples presented here showcase the potential of exploration and joint discovery in promoting dialogue and innovation.

Keywords Design participation · Sustainable renovation · Social housing · Collaboration

---

S. Boess (✉)
Delft University of Technology, Delft, The Netherlands
e-mail: s.u.boess@tudelft.nl

D.V. Keyson et al. (eds.), *Living Labs*, DOI 10.1007/978-3-319-33527-8_16

16.1 Introduction

This chapter addresses the contribution of design participation in sustainable renovation of social housing. The chapter does this by identifying resident needs throughout the building management cycle and showcasing how early design concepts address them. The aim in generating such early proposals is to give residents' interests more presence in the process of sustainable renovation. In other words, the aim of presenting these design proposals is not *only* about new objects, spaces, or processes. Rather, the aim here is to highlight the benefits of the *proposing* itself. As stakeholders in sustainable renovation, we can benefit from such proposals in innovation processes. Because design participation is not yet a very common contribution to sustainable renovation, this chapter highlights the learnings that can be derived from design proposals, in the form of possibilities and challenges that arise from them.

The chapter draws on the author's experience of being involved in a research project that aims to develop a new technology configuration for sustainable housing renovation. The author has been peripherally involved for 18 months at the time of writing. The author has contributed research insights and conducted student design projects also intended to contribute. The author's background is primarily in methods development in participatory design of consumer products, systems and environments (e.g. Boess et al. 2011). The effort to realise design participation is part of social design as described by Margolin and Margolin (2002), where "the foremost intent is the satisfaction of human needs." Social design asks: "What role can a designer play in a collaborative process of social intervention?"

The author's involvement in this project led to learnings about the contribution of design participation to the context of building technology innovation. These learnings are presented at the end of the chapter.

16.2 Challenges in Participation

16.2.1 Participation in Sustainable Housing Renovation

Many recent publications have suggested that in sustainable renovation, an integral approach is needed that involves residents. It should start from the issues that concern them locally (Brouwer and Dijkstra 2010; Breukers et al. 2014). Pronk (2014) states that a combination of raising awareness, coaching, new business models, and intensive co-operation between stakeholders is required. Pronk also advises to adapt the communication with residents to the local circumstances. For Rotterdam, relevant local issues are for example flooded crawlspaces, security, energy-poverty, and senior citizens and their care. Besides taking into account local issues, the form of the process itself also influences successful participation. A good relationship with tenants should be fostered ahead of any renovation project, so that there is ease of communication at the moment that tenants start using the new energy systems such as ventilation or heating (Breukers et al. 2014). One of the key risks in sustainable renovation is that energy consumption remains high after a renovation because residents do not understand or trust the systems. The case in point is that residents often continue to open windows while the heating is on because they do not trust the ventilation system. No written instruction will change that. Only a good relationship beforehand and ample collaboration after renovation can influence this. Additionally the user interfaces of home systems such as ventilation and heating should be better designed.

16.2.2 Design Perspective

The currently available guideline documents for resident participation, such as the examples cited above, are careful and thorough, based on in-depth research and long-term experience. From a designer's point of view, they still lack a view on residents' experience of life beyond the functional aim of lowering the energy consumption of the resulting dwellings. There is little space for innovation in which users co-design any of those systems to fit their lifestyles, daily habits and preferences.

16.2.3 Design Participation

Design participation is, according to Lee (2008), defined as "bringing 'everyman' into the field of design" and has been so since the 1970s. Resident participation in architectural design was seen as an economic necessity and a chance for neighbourhood renewal, and it became part of government policies. Lee (2008) presents a categorisation of the possible relationships between stakeholders in change processes in residential architecture (Fig. 16.1). They range from organisation-led, where residents are only research participants, to resident-led, where organisations merely realise the ideas of residents. Because architecture is a complex stakeholder process, Lee's original term 'designers' has been replaced here by 'organisation'. By this is meant, the network of stakeholders professionally involved in a renovation. This includes, for example, architects, building companies, and housing corporations. The 'user' in Lee's original figure is here replaced by the 'resident', because this is the key thing that defines them as residents, they reside.

Drawing on the author' experience of current approaches to sustainable renovation from the perspective of Lee's (2008) overview, it can be observed that the current approaches are filled with urgency and are developing useful tools such as resident segmentation and service propositions (e.g. stroomversnelling.nl). They seek to minimise resident involvement by shortening the process and the period of actual renovation. Yet in their urgency they risk underestimating the necessary consensus and trust building that will promote the desired energy

Space of operation	Purpose of design participation	Relationship between organisation and residents	Role of organisational stakeholders	Role of residents
Organisation's space	The organisation's innovation goals	organisation residents	Masters/ authorities	Imagined residents/ representatives
Collaborative space	Collaboration	organisation collaboration residents	Co-designers/ facilitators	Co-workers/ partners
Collaborative space	Collaboration	organisation residents	Stimulators	Creative people/ advisors
Residents' space	The residents' own motivation	organisation residents	Craftsmen/ builders	Active clients

Fig. 16.1 Categorisation of relationships in renovation, based on the design participation typology according to Lee (2008). In Lee's original figure, 'organisation' is 'designers' and 'residents' is 'users'

reductions. This consensus and trust can only be obtained if the residents are accorded sufficient autonomy and participation, sitting at the table when the rules are made (Lee 2008). Usually, residents' wishes are represented in answers to questionnaires. Conversely, here we consider what is of value to residents and take their perspective: social housing is a context where people are at home, where they experience small and big life events and feel more or less rooted. The home is an important source of wellbeing and deep personal meanings (Chiu et al. 2014). Going beyond questionnaires, "user research could become more creative for all stakeholders including users if there is more professional designer involvement" (Lee 2008). What does and could sustainability mean to residents, and what has value for them, and how can we support this in design?

16.2.4 Contribution

This chapter explores the middle parts of Lee's (2008) categorisation: the possibilities for collaboration (Fig. 16.1). The chapter presents a set of early concept design proposals that resulted from co-design processes of designers with residents. The proposals span a range of design possibilities for this context. They serve to open up a space and inspire a broader and more design-oriented interpretation of the participation process than is currently the case. They are all small proposals each addressing a specific aspect of participation in sustainable renovation, in order to open up new spaces for consideration. Following each case, challenges and possibilities that it raises are presented. The intended contribution of this chapter is to present such preliminary and tentative proposals as an opportunity to learn how design participation can be integrated in innovation processes in sustainable renovation. The considerations presented are quite far removed from the way decisions are currently being made in this domain, as will be seen in the discussion. The integration remains a challenge yet to address.

16.3 Design Proposals

The proposals presented here were made by students in the 250-h project class Exploring Interactions in the Design for Interaction Master at the Faculty of Industrial Design Engineering at TU Delft. The projects were selected from two briefs:

- "My greener house: preparing for sustainable renovation. How can you as a designer support a change of people's living habits towards sustainability—while designing for *their* values and needs?" Out of 18 student projects responding to this brief, seven were selected for presentation here.
- "Promoting subjective wellbeing: appreciating the existing". Pohlmeyer developed this brief for the students based on her ongoing research into wellbeing (Pohlmeyer 2012, 2014).

Even though the first group of students was invited to design for residents' values and needs, not enough of those projects fully made a step towards a positive appreciation of residents' lives and how they changed before, through and after sustainable renovation. This is why an additional two student projects are included here that had responded to the brief posed and supervised by Anna Pohlmeyer.

16.3.1 Overview

The design proposals are presented in the following, divided into four themes: Before renovation, During preparation, Early use, and During use. Each of the items under the headings represents one student design project that will be featured further on below. They all revolve around the residents' perspective, here summarized as 'pleasant living at home'. In order to learn from the design proposals for the overall organisation of a renovation, the four themes are presented in alignment with the four steps of a renovation process as seen from the perspective of building management. Figure 16.2 shows such a commonly used

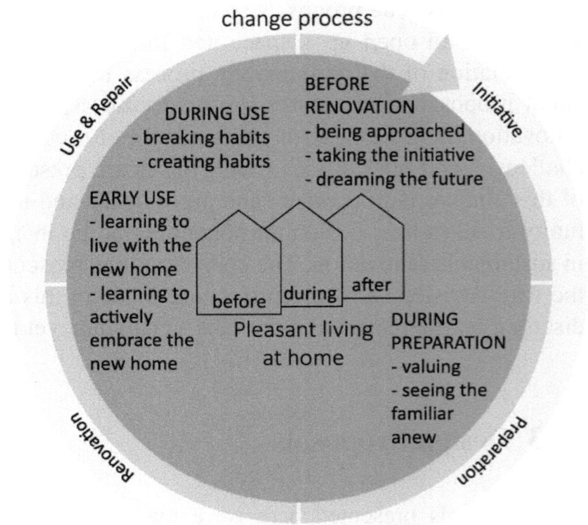

Fig. 16.2 A commonly used building management cycle of Initiative, Preparation, Renovation and Use and Repair, aligned with the four design participation themes Before renovation, During preparation, Early use and During use

building management cycle of Initiative, Preparation, Renovation and Use and Repair (Wamelink 2009) (Fig. 16.2). This set of design proposals is by no means complete—the figure immediately reveals that there are not many proposals for the actual renovation part. This is because it was not possible for the students to get access to this part of the process. In a new project, design participation in this particular part of the cycle could as yet be investigated.

16.3.2 Before Renovation: Taking the Initiative/Being Approached/Dreaming the Future

A building is in need of a renovation. Maybe as a resident you want to change something about your surroundings, but you need to engage the building owner. Or maybe a housing corporation wants residents to participate in choices about comfort, liveability or sustainability. How to support tenants in developing openness to possible changes?

16.3.2.1 Taking the Initiative with a Party-Set, A Proposal by Sofia van Oord (Fig. 16.3)

What if … you are a tenant, but you would like to change something about the house because you have the ideal of making it more sustainable, for example putting solar panels on the roof. Your landlord seems to show no interest in this direction.

Fig. 16.3 Party-set to have a party and get organised to realise a sustainable renovation

But what if … there was a way to make your neighbours enthusiastic, you could search for financing, and you could even convince your landlord?

Sofia found a group of students who did just that. They organised the financial support of former students, convinced the landlord and got the solar panels.

What if you had a party-set that enabled you to organise a party together with your neighbours and friends, and where you could brainstorm ideas together on how you could make your block more sustainable and how to get the money and support for it? You can use all the materials that you made at the party to send letters to your landlord and find funding. People love to do DIY (do it yourself), Sofia heard from residents.

A tenant Sofia worked with: "*Then we can build it ourselves, that's really nice!*"

Challenges

- for tenants it is not easy to realise something big—but it is possible.
- people often lack accessible and trustworthy information about sustainable building measures that are possible, affordable, make sense and provide increased wellbeing.
- many effective renovation measures only work on a larger scale, difficult for tenants to access.

Possibilities

- renovation is not just about the decision of buying technology, but also about doing something together and achieving impact.
- wherever there is motivation, it can be supported through community-building tools. An example is buurkracht.nl, which has been set up by an energy provider but presents itself as a grassroots organisation. Tenants' and environmental organisations can also help.

16.3.2.2 Being Approached on Corporation Initiative with a 'My House' Game, A Proposal by Rick Boellaard (Fig. 16.4)

What if … your housing corporation approaches you and proposes a renovation. You wonder what the benefit is, whether you will be saving money or whether it is affordable at all. Maybe you worry about other things that are more important to

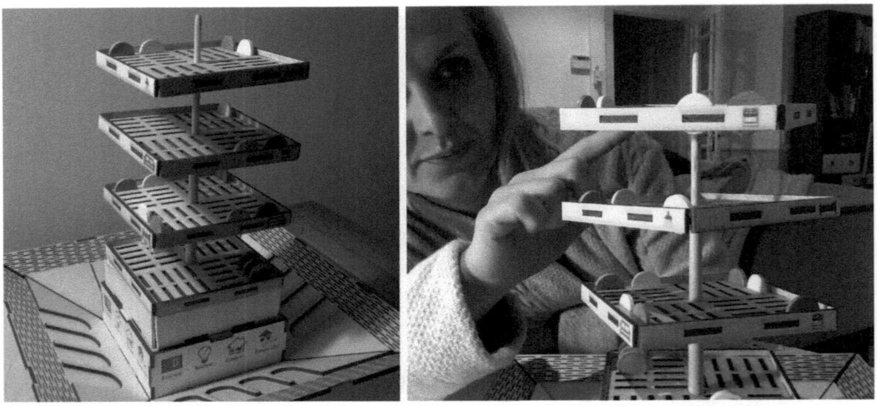

Fig. 16.4 'My house' game board to look ahead to the renovated home experience

you than your house, such as relationship or money problems, or you worry about furniture and other things in your house that have to be moved around in case of a renovation. A tenant Rick spoke with said: "*I feel like I lack the knowledge of making a good decision about the renovation*".

And a housing corporation wonders: "*How can we reach the tenants?*"

What if there was a playing board that the housing association made available. You could use it to see the new situation and be inspired by it. The game feels personal, enriching, and inspiring because you can apply it to your own living situation. In the game, you can lay pieces on a board that represent the technical possibilities of your house and your personal living situation. This way, you would be able to express your main priorities. You could also use this to talk about how the renovation would affect your personal situation.

Challenges

- the best process of approaching tenants is higly dependent on local circumstances and issues. A game can support it once initial collaboration has been established.
- the technical effects of nieuw house technology and systems are difficult to represent in abstract form.
- tenants tend to have little trust in housing corporations and their communication, especially if buildings have not been maintained well.
- many tenants would like to save energy and are perhaps environmentally conscious. But they find it more difficult to think in larger measures than e.g. switching lights off. Much more is possible, but that is difficult to imagine.

Possibilities

- For residents, their home is not just about energy. Make energy about the home and the neighbourhood, the quality of it and the quality of life in it.
- Devise communication means that bridge the technology and knowledge gap.

- Use participation moments to push the frame and think about the future as well as the present.
- When the aim is to achieve 'no regrets' (doing a renovation so well that it saves energy and money for decades to come), this requires a few thinking steps from tenants, and quite some time to get used to the idea. Next to actual model houses (and also *in* them), games, scenarios and related services are means by which residents can get more familiar with possible energy measures and their effects.

16.3.2.3 Dreaming the Future with an Extreme Scenario Game, A Proposal by Anton Garrigue (Fig. 16.5)

Sometimes it is useful to think in extreme ideas first in order to then come up with insights that are practical and close by. It is not just a house that makes up a home and a neighbourhood. Residents could learn to think of their neighbourhood and their city as something they can influence, even play with different ways it could be. A neighbourhood is not just a location but also a network of transport modes and these are intricately linked with people's experience of their home.

What if together we could develop a very futuristic scenario, in order to discover unexpected possibilities of how we could do things differently? For example, what if cars and bicycles were completely shared? Say there was only one type of car, and it carries bicycles. You can call it via an app, and while you are using it, others could take a bike from it or return a bike to it. You could switch your mode of transportation at any time. Apps provide coordination. The freedom to move around in each way you want at any moment is more valuable than ownership, and streets would gain all the space vacated by cars.

Residents that Anton spoke with, said when playing the game:

> I would tend to take my bicycle more with car sharing.
>
> Car sharing makes me feel more social, free and connected to the city.

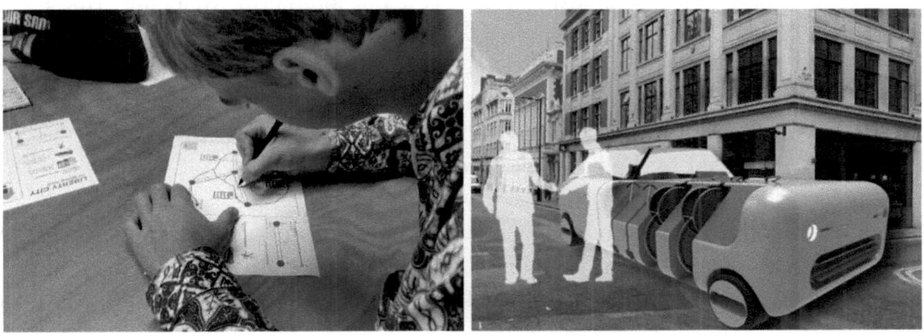

Fig. 16.5 Games to develop very futuristic scenarios together to learn from

Challenges

- Residents perceive the liveability of their surroundings not just in terms of their dwelling, but also of the neighbourhood. Often, cleanliness and usability of the neighbourhood even determines residential satisfaction. In any renovation process, therefore, it is useful to take those along.
- As citizens we are not used to thinking in urban scenarios. We tend to be busy with our own lives, and we perceive the city mainly in terms of our needs (Meesters 2009). But the street, the neighbourhood and the city change continually, also in people's experience. This means it is necessary to listen well to their concerns.

Possibilities

- Co-creation and city games are ways to start thinking in scenarios.
- New lifestyles and technologies bring potential for new ways to organise the city, for example from grassroots level.
- Every routine daily activity harbours the potential of change. The challenge for a neighbourhood or city is to find the right balance between supporting routine practices, necessary for survival and wellbeing, and variations or disruptions of these practices—that we also need.

16.3.3 During Preparation: Valuing/Seeing the Familiar Anew

Say you are living pleasantly at home—everything is as usual. Gradually you get so used to the things around you that you do not even notice them anymore. Without you even realising it, your experience of your daily life gets shallower. The two proposals presented next arise from the brief "appreciating the existing" given by Pohlmeyer (2012, 2014). We can become more aware of our home environment and neighbourhood, and discover the value for ourselves. This can, indirectly, lead to energy saving, as the examples will show. Wellbeing should always come in first place. After all, sustainability is about long-term wellbeing (Boess and Pohlmeyer 2016).

16.3.3.1 Valuing What You Have with Maintenance Kit 'RElove', A Proposal by Felix Marschner (Fig. 16.6)

Think of something that you have acquired and that you love, that makes you feel happy. For example, a favourite pair of sneakers. Gradually, though, they start to feel less new and give you less satisfaction. There are even other shoes you once

Fig. 16.6 Maintenance kit 'RΞlove your shoes' to foster appreciation

loved around still, that have perhaps gone out of fashion. Fixing them up again would cost you time and you already have so little. But getting rid of them does not feel right, either.

What if there was a way you could appreciate things you own more in your own home? And if this way also gave you calm and strength to lead your busy life? Say there was a toolkit that enables you to build an emotional connection to things you own. For example, your shoes. Would you wear them for longer, even if the fashion has changed? The shoe cleaning box contains a number of shoe caring products and tools. The tools are designed in such a way that the process is not quick but actually slow, and that it becomes a concentrated, detailed, almost meditative activity. Felix' research showed that people like doing that. He asked residents, after letting them try out the shoe cleaning box prototype for a week, how they liked it. They said:

> It felt purposeful and pleasant to do.
>
> It was slow but also more precise. It was a nice thing to do.

Challenges

- Sustainable technologies for the home often fail to result in the expected energy savings because people use them differently than expected or do not take care of them well.
- Few innovation ideas related to sustainability take into account the necessary care and devotion to keep domestic systems, but also any items you own, up and running and valuable.
- There is often a tendency to make things as fully automatic and maintenance-free as possible, and avoid any involvement of the user. After all, who would want to take care of and maintain an ugly ventilation system that has no inherent part in one's daily life?

Possibilities

- It is possible to create designs that do the opposite of the above ambition of being maintenance-free. Create designs that make it attractive and pleasant to take care of something. For example, by raising the value of a cleaning kit through aesthetic design. Furthermore, by devoting attention to the actual cleaning actions that a person will carry out.
- As in the example of the shoe cleaning kit, it may even be possible to make this a meticulous and painstaking activity, inviting you to prolong your engagement with the shoe and the activity. This way, it ceases to be a household chore to be squeezed in on the fly, and instead becomes an artisanal activity, in its painstakingness a source of satisfaction, pride and attachment.

16.3.3.2 Seeing the Familiar Anew by Rediscovering Your Everyday Route with 'Explordinary', A Proposal by Julia Mattaar (Fig. 16.7)

Say that you have lived in your house for a few years. You are more or less comfortable and happy there. You have your daily routes to the shops, the park or the local playground—in short, your neighbourhood. You have grown so used to how things are that you hardly notice anymore how things gradually change, or how the neighbourhood looks to you.

What if there was a way you could appreciate your environment more while out on the streets or on your way to work? Say there was a simple app that offers you a theme every day. It challenges you to find something in your environment that fits this category. The themes help you focus on things you would not normally see. You can also see pictures taken by nearby users for the same category. After completing 20 themes, you can order a set of postcards with a selection of the taken pictures on it. You can keep the cards as a physical memory, or send or give them away.

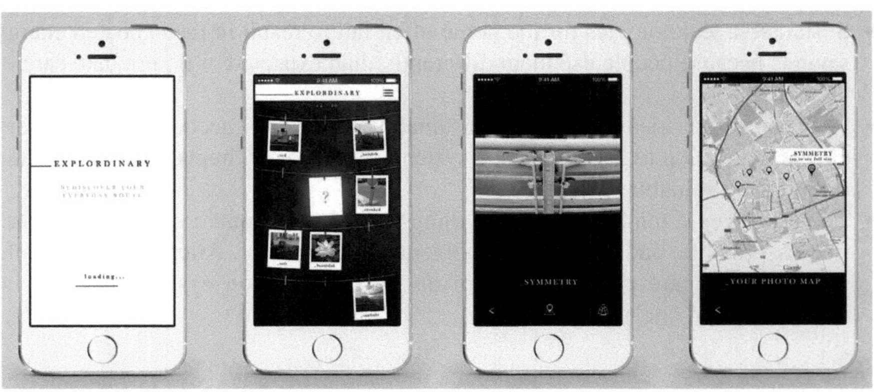

Fig. 16.7 App and postcard service 'Explordinary' to see the familiar anew

Challenges

- In daily life, we tend to accept things as they are. In fact, if you add something to a home or a neighbourhood, people sometimes experience it as unpleasant, because it requires them to invest thought into whether it is a good thing or not. To be on the safe side, complaining is then the safest initial reaction to make sure you do not get pulled into something you do not want. This is how complaining can come to be the most accepted, 'normal' style of communication on the dwelling or neighbourhood level.
- This form of communication is often, then, a disguise for the question for more information. However, it leaves little room for exploration because it fixates people in a yes/no pattern. Any dynamics of change will have a negative starting point, or a positive starting point that may feel enforced, for example if it is part of a marketing strategy.

Possibilities

- Reflecting on our environment helps us to open up our perspective on it, on our daily experience of our neighbourhood, our house and ourselves.
- Ideas that like the inspiration above, that enable people to see the well-known in new ways, should be introduced in an authentic way, and people should be able to choose them themselves and adapt them, and perhaps even create them themselves.
- Communication media have the potential of subtly introducing another paradigm of communication. If people have the opportunity to gain a perspective of their own and insights via a medium like this, they are better able to then contribute their own values and ideas to processes in which many parties have a stake. Such as renovations or other neighbourhood improvements.

16.3.4 Early in Use and Repair: Learning to Live with the New Home/Learning to Actively Embrace a New Home

Taking a moment to reflect on comfort and wellbeing in your home helps to make the new environment your own. It helps being more open towards the experiences in it, and it brings ideas on whether this is what you want, or what you would like to change. This is how you can create your very own comfort (Strengers 2014). A good moment to do this is when something recently changed about your house.

16.3.4.1 Learning to Live with the New Home with 'The Pebble', A Proposal by Justus Kuijer (Fig. 16.8)

What if … you get some new technology in your home, either you have bought it or it has been arranged for you. It feels new and pleasant, and it changes your life a little bit. For example, a new floor heating system. It is pleasant to the touch of

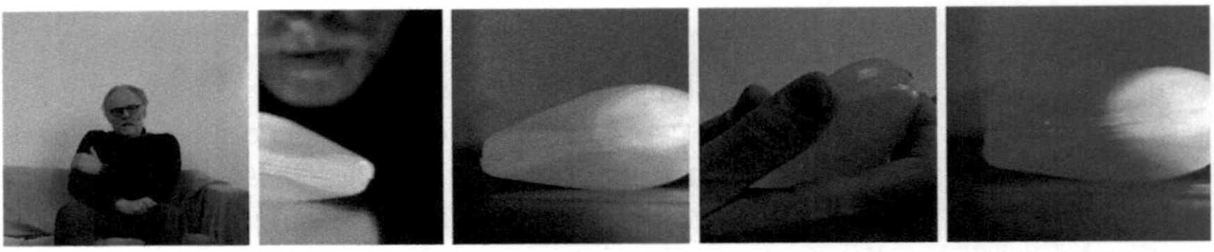

Fig. 16.8 'The pebble' enables you to see and feel what your heating is doing

the foot and you do not need to turn it up much to feel warm. But you are often not sure whether it is heating up or cooling down. The only place to see that is a box on the wall. So how can you develop your habits with it?

What if you could directly influence the warming up or cooling down of your new floor heating, and if you could see and do that from anywhere in your house? What if you had something that enabled you to see and feel what the heating is doing? The pebble lies in your house. When the heating is warming up, it glows red. When the heating is cooling down, it glows blue. When the heating holds the temperature, it glows green. This way, you not only get the information that it will be nice and warm very shortly, or nice and cool. You also have a tangible presence of the heating, just like people had a tangible presence long ago with a fire in a hearth. A resident about Justus' concept: "*Interacting with the pebble feels reassuringly familiar. I want to play with it, but because it is such a beautiful shape, I also like to just hold it and feel how it's warming up.*"

Challenge

- Domestic systems, especially comfort systems, do not always respond directly, because the underlying processes like warming up require time. So people end up turning up systems too high, because they have nothing tangible that they can integrate with their habits and needs.

Possibilities

- Make appliances and systems amenable to experience in terms of what they are doing and how you can influence your home environment. Make it visible and tangible.
- Design form and behaviour of systems in such a way that people can integrate them in their habits and preferred behaviours. For example, make it possible to carry something around in the house but also be easily found, make it intuitive to grasp, and give it a presence but also a passivity so that people can find their own ways of interacting with it.

16.3.5 Inspiration: The Welcome-Box … Is Your Personalisable Lamp

16.3.5.1 Learning to Actively Embrace a New Home with the Welcome-Box Lamp, A Proposal by Staffan Till (Fig. 16.9)

What if … you move into your new house, or a renovation has just been done, or it is just a special moment in your life. Perhaps you have done little with your home so far, or you are not attached to it because you don't know whether you want to stay there for a long time.

What if you are a resident, and a welcome-box was waiting for you or was handed to you by the building owner. In it, you find a lamp with a lampshade to help you get started in your new home. The lampshade is easy to adjust to your own preferences, and you can use it to create a pleasant light in your new house. A resident that Staffan spoke with:

> When I spend more time on personalising my home, I feel more attached to it.

Challenges

- Many home systems function in ways not directly accessible for residents. People who have little attachment to their house, or who have little affinity with manipulating items in the house (for example, not feeling competent), are not inclined to change anything about their house.
- A house needs some investment of care in order for people to feel ownership. People who have little attachment, have little motivation to take care of things well and create good living conditions for themselves. This can lead to people tolerating bad functioning of systems, or not even noticing if something is not pleasant.

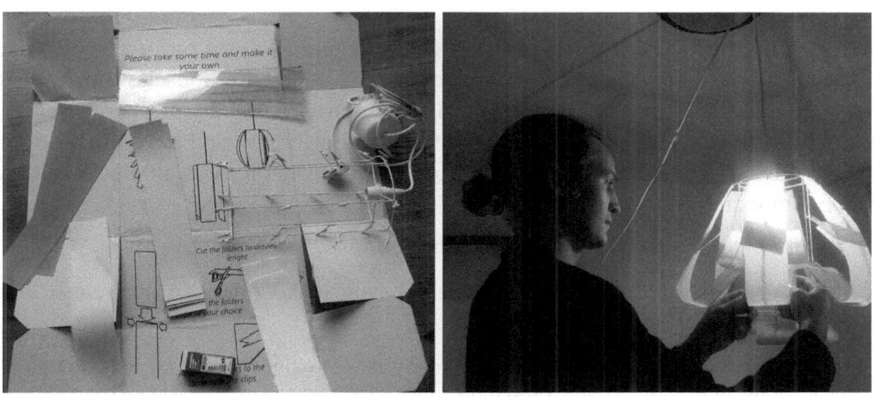

Fig. 16.9 The welcome-box that you turn into your personalised lamp

Possibilities

- Create opportunities for the adjustment of objects and systems in the home to one's desires, in order to foster attachment.
- Those adjustments should make a use value apparent, for example a more pleasant quality of the light or something that warms up. This in turn means that people will experience making adjustments as something positive, as a possibility they discover, that they could integrate into their habits. This can tip the balance towards noticing energy waste in their home (for example a heating that is on high for too long) and to adjust it.

16.3.6 During Use and Repair: Breaking Habits/Creating Habits

Much of living at home is habitual. Some habits come and go. Some slip away. It is a relationship that needs to be maintained, and sometimes tweaked. You like to live well, you also try to keep a check on costs, and it is not easy to always keep everything in view. Energy saving is also not the most important part of your life. It would be nice if that aspect of life took care of itself sometimes, or if it was pleasant to do in itself. After all, you can set an alarm to go off at a certain moment—why should other things not help you a little?

One might ask which role this plays in a renovation. One of the key issues in sustainable renovation is that actual energy savings turn out to be less than can be expected. A suspected reason is in people's habits. To address this, it is key to support people's habits already from the initiation stage of a renovation.

16.3.6.1 Breaking Habits: Designing Kids' Gaming into Daily Living with a Playful Lamp, A Proposal by Daniela Passa (Fig. 16.10)

Current sustainable renovation initiatives have begun to focus on households and household types, but not much yet on the needs of specific household members. Teenagers are a group that is often said to be unreachable about energy saving. Daniela went to speak to teenagers at school and at home, and developed design ideas with them. It turns out they like to use new technology to interact with the home. For example,

> I would love to control the light from my I phone, it would use less energy

Also, teenagers are still developing their habits and often do not realise that home living uses up resources.

> My mother asked us to consume less water, she was controlling us everyday to not turn it on too long, or to waste it. Now I still think of what she said and my behaviour changed.

Fig. 16.10 A playful lamp to break energy-wasting habits, co-developed with kids

What if the awareness-raising of energy use was built into objects of daily use in a playfully challenging way? For example, what if there was a lamp that uses the interaction to convey its meaning in a natural metaphor (tree shape). If the light is left on unnecessarily (the lamp can see that no one is in the room), it drops leaves. And it teases the user by staying off for a few minutes. It cannot be turned back on fully until you have given it back its leaves. It continues to excite and tease with its presence, giving teenagers the feeling they can finally touch energy with their hands. That creates a learning effect.

Challenges

- Some aspects of wellbeing in the home are connected to behaviour. Whether it is about living together in harmony as in the example (teenagers and their parents), or about energy waste, as with lights that stay on unnecessarily. A lot of research is currently being conducted into behaviour change and 'nudging' (Thaler and Sunstein 2008) or persuasion (Fogg 2002). It has turned out, however, that these approaches, embedded into products and services, often have only incomplete or shortlived effect.
- Nudges and persuasions are entertaining for a short time, but can soon be experienced as annoying or forcing. Even more so if they do not come from a source socially connected with the person being addressed. A parent is, but a persuasive product is not and is likely to be discarded. Good awareness of the social context is key in designing persuasive products.

Possibilities

- Activating techniques such as persuasive products can have a transferring effect on a person's beliefs and habits. They can contribute to character forming (Boon et al. 2015). For that it helps if the interaction is connected to people's social lives.
- Persuasive objects and services can punish or tease in a playful way. This enriches the interaction, just like a game of tag is fun (Fokkinga and Desmet 2012).

- The objects and coercive techniques themselves should be transient in people's lives. They should either be able to change with people or, like novelty toys, they should be easily acquired and easily passed on to someone else.

16.3.6.2 Creating Habits: Designing Support into Daily Choices with a Decision Tree for Fruit, A Proposal by Elske van den Ende (Fig. 16.11)

We tend to overestimate our abilities to make rational, informed choices about aspects of our lives. For example, we neglect to take care of our pensions, we eat things we do not intend to eat, and we get overwhelmed with food shopping decisions on a busy day. No amount of information can help here. Only hints that can be acted on at the very decision moment can ultimately sustain good decisions. Strong hints are for the short term, which is why the example given here is for a canteen campaign rather than for the home itself.

What if there was a beautiful, encouraging yet also teasing way to support you at the very moment of making a decision? For example, what if there was a display at the canteen presenting fruit. The most sustainable fruit (e.g. locally produced apples) is presented in the most attractive light, and you get cheers when you pick one of those. If you pick a non-sustainable banana (longest travel distance), however, you hear a booing sound.

Fig. 16.11 'Decision tree', a display for fruit that seeks to influence your choice

Challenges

- Only few people will themselves buy products to influence their habits such as home energy management systems.
- A conspicuous product like the 'decision tree' in the example teases and amuses, but can also feel annoying soon, making such products only suitable for temporary campaigns.
- If you accord this much initiative to a product, it needs to be designed carefully to prevent unexpected side effects. To give an example, that staff would start protesting against it as a nuisance.

Possibilities

- A good opportunity to introduce such stimulating products as the fruit decision tree is at locations such as supermarkets or canteens, as a part of advertising campaigns. New inspiration and also short-term provocation is more easily acceptable in such contexts. It might even become an inspiration for people to think up provocations of their own.
- Products such as the fruit decision tree can stimulate people to form new habits.

They have an initially engaging and fun effect, and create a lot of attention within a neighbourhood or work environment. If it is possible to change choices made once, this might also affect the choices people make at home, because of increased familiarity with this new possibility and intuitive and practical learning of the new choice.

- Products with 'a will of their own' also have the potential to become an accepted presence in a home or other environment if they are able to adjust to people's lives, including the changes and subtle variations in them. An option is to acquire such a product via a rent-construction, in which people do not become owners of products but use them as part of a service. That lowers the barrier that would come with a purchase, and enables people to pass such products on.

16.4 Discussion

This chapter has presented a number of design proposals. Their presentation serves to showcase the possible contribution of design participation in the building management process. The proposals are as yet quite far removed from the current forms of collaboration with residents in this domain, as we will see in the discussion. This integration remains a challenge yet to be addressed. It should be noted that a key aspect we hope to bring across here, is not the specific content of each proposal, but the value the proposals reveal for the stakeholder process in sustainable building renovation. We argue that learning from the challenges and possibilities in these proposals reveals this value. We articulate resident statements as well as challenges and possibilities that arise from the proposals as insights. We presented the proposals as exemplars of design participation (Lee 2008). We have shown that design participation is a vehicle for insights and decision-making in renovation planning. In the following, a short reflection on the relevance of the presented proposals to the processes in renovation practice is presented. In this, the author makes use of published accounts and evaluations of processes (e.g. Sijpheer et al. 2015) as well as first hand experience of a renovation technology innovation project with building industry stakeholders.

16.4.1 Gaps to Bridge

The gap between design participation thinking and stakeholder processes in sustainable building appears wide. The innovation process for sustainable housing technology allows little space to consider the various potential implications for the later use situation in depth, as was done in the proposals. A sense of urgency leads to the desire to make decisions quickly and not re-consider them. Decision-makers look towards abstract rather than concrete supports for decision-making. These are, for example, matrices, calculations, and technical drawings. Lee (2008) explicitly mentioned the gap between abstract and concrete as a key dimension separating designers (or planning professionals) from users. The abstract and concrete can be bridged, but this has not yet become part of the planning processes of the building industry stakeholders. Planning processes and participation processes are very different and hence difficult to connect.

16.4.1.1 Abstract Versus Concrete

There is little space in the building planning process for the use-oriented design thinking exemplified by the proposals presented here. Current co-design processes in practice tend to mean giving the user a limited number of choices to generate a *feeling* of involvement. During the innovation project the author witnessed, the discussions on technologies remain abstract and are conducted in a neither visual nor experiential way. An attempt to bring more of this into the stakeholder process via a set of workshops was graciously accepted but not taken any further afterwards. There is still a need for new tools and some rapprochement between disciplines for this process to get any closer to including design participation.

16.4.1.2 Consideration of the User Interfaces

During the entire process in the aforementioned building innovation project, the future user interfaces of home systems are never addressed. They are seen as an afterthought, to be provided by installation suppliers to finished homes. Individual project members address this issue occasionally, but it is not part of the structural process. There are careful and thorough resident participation processes in place (such as Brouwer and Dijkstra 2010), mentioned previously. However, these tend to follow the same paradigm of fixed technology to be used correctly, rather than technology being built up around residents' lives and ways of using technology as part of their lives. This is in spite of it being well known by now that key problems to achieve energy neutrality arise from use problems with interfaces (Breukers et al. 2014; Chiu et al. 2014).

16.4.1.3 Solution-Oriented Design

The stakeholder team in the innovation process experienced at first hand considers every solution as the definitive solution to be implemented. This might be due to project delays. However, the time pressure was applied from the start to 'just tell us what the users want'. It may be the way of thinking of the stakeholder team. This puts stress on the stakeholder process, since every decision carries much weight. The team is not taking a design thinking approach to interim solutions as learning opportunities, and opportunities for design participation.

16.4.1.4 Focus on Agreements and Being Heard Versus Exploration

Current participation processes (e.g. Brouwer and Dijkstra 2010; Breukers et al. 2014) tend to focus on *agreements* and *being heard*. The examples shown here have proposed a focus on *exploration* and *joint discovery*. This focus can help in developing dialogues on what we would want to change and how that change can be integrated into the concerns of our daily lives. What if we could have it easier and more pleasant and more secure in our lives—and what do we think is easier, more pleasant and more secure, in the first place?

16.5 Outlook

This chapter has suggested that *we can learn to evaluate design proposals in terms of the insights they provide on challenges and possibilities*. There are challenges ahead in integrating design participation in innovation processes in sustainable housing. Often, when discussing sustainability, we lack ideas and cannot get much further than solar panels or water stops for the tap. Rather than proposing and evaluating specific solutions, this chapter has sought to open up the discussion and promote a *broad joint exploration of possibilities*.

Acknowledgments Many thanks to the students Sofia van Oord, Rick Boellaard, Felix Marschner, Julia Mattaar, Anton Garrigue, Justus Kuijer, Staffan Till, Daniela Passa and Elske van den Ende for making their work available for this chapter. The pictures illustrating their projects are theirs. Thanks also Natalia Romero Herrera and Arnold Vermeeren for coaching the students and to Anna Pohlmeyer for the collaboration and allowing me to feature the work of two of her students, Felix and Julia, whose work responded to her brief on appreciation.

References

Boess, S., & Pohlmeyer, A. (2016). Designs with Benefits: hearth fire nights and bittersweet chores. *Proceedings of the 50th Design Research Society conference DRS, held in Brighton, UK* (in press).

Boess, S., Pasman, G., & Mulder, I. (2011). Making for participation. *Companion volume to the Participatory innovation conference PINC, held in Sonderborg* (p. 80).

Boon, B., Wever, R., & Quist, J. (2015). Beyond behaviour change: Technological artefacts and characterological development. *International Journal of Sustainable Engineering, 8*(3), 231–247.

Breukers, S., van Summeren, L., & Mourik, R. (2014). *Project BoB (Bewoners ontmoeten Bouwers) Eerst proces, dan prestatie.* Uitgebreide Samenvatting. ('First the process, then performance'). Study for Platform 31. Accessible via platform31.nl. Last accessed November 1, 2015.

Brouwer, J., & Dijkstra, A. (2010). *Bewonerscommunicatie bij duurzame woningverbetering.* Boxtel: Aneas. Retrieved November 1, 2015 from http://www.toolkit.nl/toolkit-bestaande-bouw.8122.lynkx.

Chiu, L. F., Lowe, R., Raslan, R., Altamirano-Medina, H., & Wingfield, J. (2014). A socio-technical approach to post-occupancy evaluation: Interactive adaptability in domestic retrofit. *Building Research and Information, 42*(5), 574–590.

Fogg, B. J. (2002). Persuasive technology: Using computers to change what we think and do. *Ubiquity* (December), 5.

Fokkinga, S., & Desmet, P. (2012). Darker shades of joy: The role of negative emotion in rich product experiences. *Design Issues, 28*(4), 42–56.

Lee, Y. (2008). Design participation tactics: The challenges and new roles for designers in the co-design process. *CoDesign, 4*(1), 31–50.

Margolin, V., & Margolin, S. (2002). A "social model" of design: Issues of practice and research. *Design Issues, 18*(4), 24–30.

Meesters, J. (2009). *The meaning of activities in the dwelling and residential environment: A structural approach in people-environment relations.* Series Sustainable Urban Areas, Vol. 27. IOS Press.

Pohlmeyer, A. E. (2012). Design for Happiness. *Interfaces, 92*, 8–11.

Pohlmeyer, A. E. (2014). Enjoying Joy. A process-based approach to design for prolonged pleasure. *Proceedings of 8th Nordic Conference on Human-Computer Interaction (NordiCHI'14).* ACM Press, pp. 871–876.

Pronk, M. (2014). Versnelling010—Acceleration in Rotterdam. Strive. Presentation held at *Symposium NeZer*, 23 Sept 2014. Retrieved November 1, 2015 from http://www.nezer-project.eu/download/18.1acdfdc8146d949da6d65ad/1419240585856/D7_4_StudyVisitReport_NL_Sep2014.pdf.

Sijpheer, N., et al. (2015). *Resultaten uit monitoring over: Tevreden bewoners.* http://energielinq.nl/resources/resultatenmonitoringbewoners.

Strengers, Y. (2014). Smart energy in everyday life: Are you designing for resource man? *Interactions, 21*(4), 24–31.

Thaler, R. H., & Sunstein, C. R. (2008). *Nudge: Improving decisions about health, wealth, and happiness.* Yale.

Wamelink, J. W. F. (2009). *Inleiding bouwmanagement.* VSSD.

Chapter 17
Supporting Iterative Research and Design Explorations in the Living Lab Context

Tomasz Jaskiewicz, David V. Keyson and Jantien M. Doolaard

Abstract This chapter presents an approach for organising research and innovation in the Living Lab context, where context research instruments, as well as conceptualised, developed and tested prototypes are integrated in one hardware and software platform (BOCS platform). The BOCS platform allows collecting of sensor and building management data, self-reporting of subjective information by users and providing feedback to users through a variety of channels. By this, the platform supports iterative cycles of context researching, co-creating, implementing and testing of solutions. The initial goal for the use of the platform is to enable creation of solutions aiding office occupants in improving their comfort while reducing building energy use. This goal is attained by enabling iterative identification and a gradual build-up of in-depth understanding of involved social practices, and incremental introduction and evaluation of ways to support the change of these social practices through monitoring, self reporting and feedback in office environments. The chapter outlines the organisation of the proposed process in detail. The approach is further positively evaluated based on the outcomes of a preliminary case study. It is finally suggested that in the future the approach may be applied to other Living Lab situations where complex challenges are faced and fast results are expected.

17.1 Introduction

A Living Lab can be broadly defined as "a real-life test and experimentation environment where users and producers co-create innovations" (ENoLL 2015). Living Labs enable and support "concurrent innovation" processes (Mulder and Stappers

T. Jaskiewicz (✉) · D.V. Keyson · J.M. Doolaard
Delft University of Technology, Delft, The Netherlands
e-mail: T.J.Jaskiewicz@tudelft.nl

D.V. Keyson
e-mail: d.v.keyson@tudelft.n

J.M. Doolaard
e-mail: J.M.Doolaard@tudelft.nl

D.V. Keyson et al. (eds.), *Living Labs*, DOI 10.1007/978-3-319-33527-8_17

2009). In such processes users play a central role and all other key actors are actively involved. In this way, concurrent generation of fundamental knowledge, improvement of technology and creation of innovative solutions are supported.

The common perspective towards living labs is to consider them as testbeds that support design researchers in developing deep understanding of the context and user interactions with existing and newly designed products and services. Although living labs have been approached as dedicated, space-bound facilities, considered to be labs where users can live for some period of time (e.g. Intille et al. 2005) there is an increasing trend to also treat real-world context situations as living labs, without strictly defining their physical boundaries and without requiring permanent facilities of a dedicated laboratory.

In this respect, Living Labs offer a unique opportunity to explore approaches for behaviour change in a broadly considered context, and to develop solutions that not only facilitate or discourage specific user activities in isolation from each other, but that holistically support change of entire, highly complex social practices, ensuring lasting behaviour change. However, combining living labs with "in the wild" research and design, brings a challenge of a lack of research and prototyping instruments in highly complex real-world contexts, and poses many organizational challenges in respect to time and resources that living lab studies require.

The approach, methods and tools presented in this chapter have been developed for the Building Occupancy Certification System (BOCS) project, which is part of the Building Technologies Accelerator (BTA) program of the European Climate Knowledge Innovation Community (C-KIC). The BOCS project aims to combine research with development of market-ready solutions that support reduction of CO_2 emissions from office buildings through occupant behaviour change. The BOCS platform has been developed to facilitate development of such solutions. The resulting approach presented in this chapter provides a new way for organising research, innovation and implementation of solutions in a living lab context, which allows parallel execution of research and innovation, providing new opportunities for addressing the challenges of social practice change. The approach is organised around the BOCS platform, which is a software and hardware platform supporting performing of objective and subjective measurements in context, providing feedback to users, while also allowing easy modifications and addition of features by employing a modular system architecture.

17.2 Social Practice Change for Energy-Conscious Office Occupancy

Presented approach assumes a broad perspective towards changing energy use practices. Social practices can be defined as "broad cultural entities that shape individuals' perceptions, interpretations and actions within the world" (Hargreaves 2011). In discussed research, social practices of occupants are considered not only in relation to their impact on energy use, but also in respect to occupants' comfort and well-being, as well as other, difficult to anticipate aspects of their daily work routines. Altogether, this provides a holistic perspective serving as a foundation for implementing lasting behaviour change, where designed solutions may become embedded into individual routines and habits of occupants and are enforced through social interactions.

Social practices can vary largely. Even within one office space, occupants can engage in very different activity patterns, originating from such factors as their work experiences, cultural backgrounds, family context or individual lifestyle preferences. At the same time, office occupants influence each other's practices, and oftentimes new practices unique to a given office emerge from social interactions among people. The ways in which office practices form and transform are complex and involve people's lives outside the offices. Therefore, in order to effectively support people in shaping their social practices towards higher sustainability and improved comfort, an in-depth understanding of the opportunities for change is needed. These opportunities may relate to the existing work organisation and social-cultural aspects, as well as involving the physical environment, including the building technology. An additional complication here is that social practices related to comfort and energy consumption may differ not only across short timescales of a workday or a weekly work rhythm, but due to varying weather and sunlight conditions radically change across seasons. Consequently, the required research, design and development of solutions aimed at supporting behaviour change in the context of social practices is time consuming. A typical sequence of steps would traditionally begin with surveying the context and identifying problems and opportunities, which in this situation requires studying of office practices during all seasons. The following step of designing one or more solutions, and testing them, traditionally can require another year of work, and may lead to additional research questions, requiring another extended context study, leading in turn to new design requirements.

17.3 Iterative Context Researching, Conceptualising, and Implementing Solutions Using the BOCS Platform

The presented approach aims towards researching office occupant social practices, as well as designing, developing and testing solutions for improving these practices, all of which are supported by a dedicated BOCS platform. The BOCS platform consists of sensor and self-reporting devices, optional feedback devices, online interfaces, and a back-end for collecting and processing gathered data. The BOCS platform combines aspects of a research instrument, namely a research toolkit for monitoring subjective and objective parameters of offices (including hardware devices such as sensor nodes, or self-reporting devices, as well as

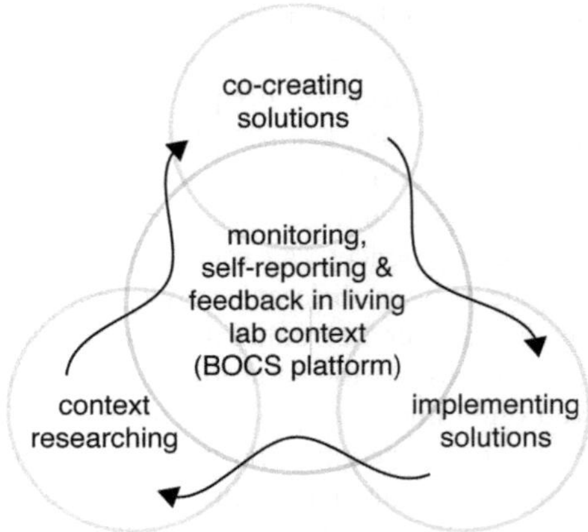

Fig. 17.1 The presented approach is organised around an iterative cycle involving context research, solution identification and solution implementation. The BOCS platform plays a central role as a research instrument and enabler of social practice transformation

software self-reporting interfaces), a kit of parts for exploring solutions for supporting occupants in changing their practices (including extensions to sensor nodes and self-reporting devices, and modular interface elements that can be added to the front-end interface), and the foundation of a system to be deployed as a long-term solution to support the target group in maintaining or improving energy and comfort-related office practices.

With the BOCS platform, user research and testing, generation of new solutions and their implementation are performed iteratively, as shown in Fig. 17.1. The BOCS platform can be seen as a continuously evolving prototype designed in context (Coughlan et al. 2007), driving change by permitting insights into the office context, supporting identification of new opportunities, their implementation and validation. Performing the process iteratively allows an ongoing validation of explored solutions and their continuous improvement based on otherwise impossible to predict effects that these solutions may induce on complex social practices of office occupants.

The BOCS platform has been based on the Living Lab infrastructure developed in the SusLab project (www.suslab.eu). However, as part of the BOCS platform, this infrastructure has been redesigned to support the system's flexibility and ability of fast adaptations regarding types of collected data, types of sensors used, types of self-reporting interfaces, and to support rapid development of context-specific screen-based, as well as physical feedback interfaces, as shown in Fig. 17.2.

The sensor and self-reporting device hardware has been developed using modular electronic components. These components in combination with a system of rapid-prototyped enclosures permit production of small batches of fully customized devices, attainable to be produced in the timeframe of several days. A similar approach has been taken for the online feedback interface in which "monitors" have been developed as modular interface elements, which can be assembled into a custom dashboard for each specific user, based on ad-hoc determined requirements (Fig. 17.3).

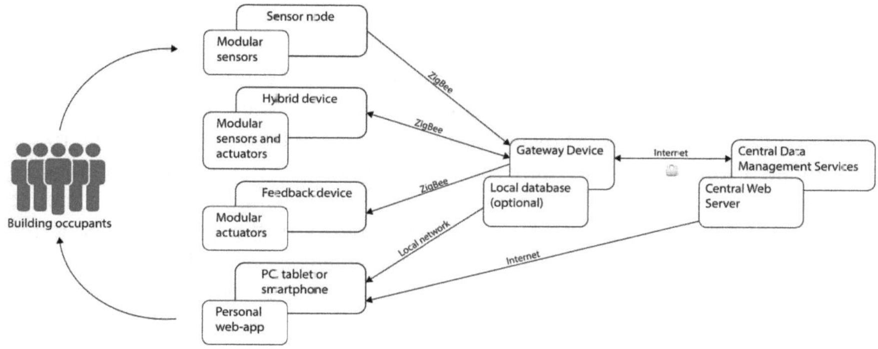

Fig. 17.2 BOCS platform system architecture supports the rapid addition, modification and removal of system components based on specific context requirements

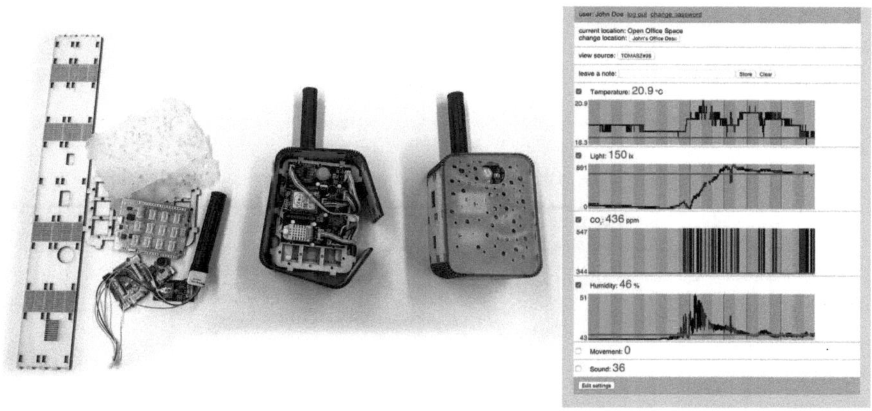

Fig. 17.3 The BOCS platform uses modular sensors and physical self-reporting elements (*left*) and modular screen interface elements (*right*) to allow fast prototyping of working and fully customized monitoring and feedback system solutions

17.4 Context Researching and Testing Solutions

Assuming an iterative approach requires the context researching and testing activities to be performed in a short timeframe, such that their repetition may gradually build up acquired knowledge through consequent process iterations. The aim of the research has been defined as identification and assessment of office practices that influence comfort and may cause wasting of building energy. The steps of the research activities in the process have been assumed to be the following: (1) identification of comfort-influencing and energy-use-impacting practices via observations, interviews and surveys, (2) analysis of the identified practices in relation to implemented solutions for climate sensor monitoring, self-reporting of subjective factors by office occupants, and feedback, if present, based on additional observations, interviews and surveys, (3) normalisation of the findings using a shared format for description and comparison of practices, (4) identification of practices that can be improved, and (5) evaluation of changes to practices and of the role of implemented solutions, if any were introduced in the previous cycle.

In step 1, a contextual questionnaire aimed at analysing key social practices and the subjective perception of comfort and energy in the workplace is administered. A timeline, which is part of the questionnaire is filled in by participants, as shown in Fig. 17.4. The step also includes encoding observations performed by a researcher in-situ and a post-observation interview format for verifying these observations.

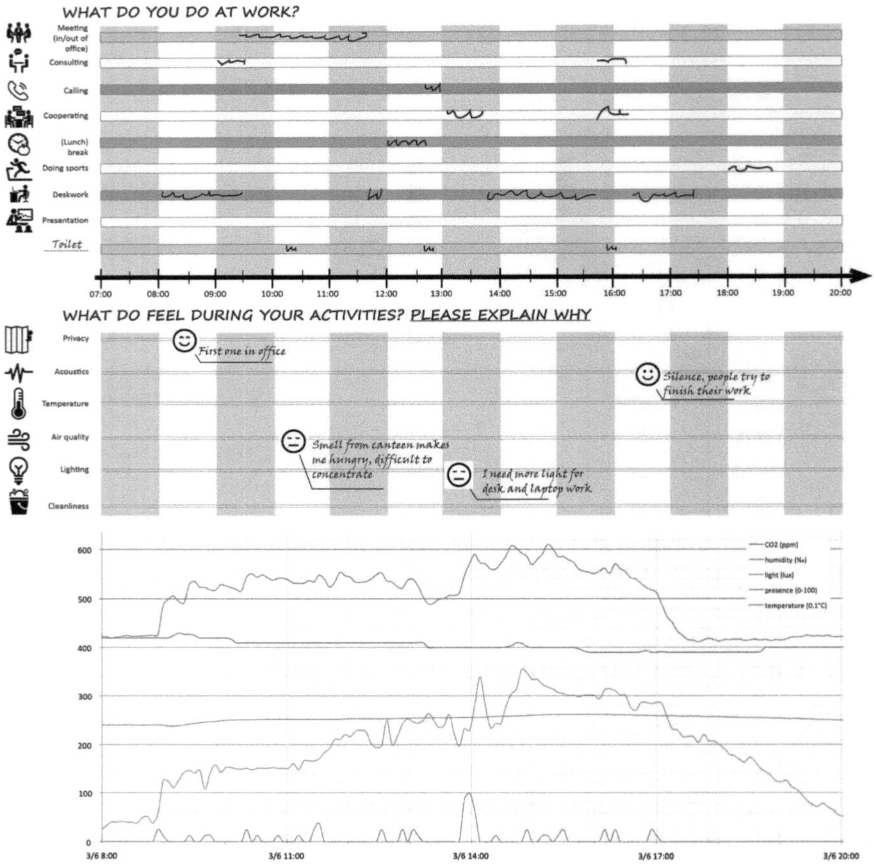

Fig. 17.4 An example of a timeline filled in by occupants as part of a contextual questionnaire (*top*), and indoor climate sensor readings from a typical workday, measured in one of the studied offices (*bottom*), through the correlation of which researchers and occupants can understand the reciprocal relationship between their comfort, actions and indoor climate parameters

The evaluation of comfort and energy use related practices (step 2) requires a more in-depth investigation than can be gathered by observations, interviews and questionnaires alone. Through traditional means such as qualitative interviews and videos, Bazley (2015) identified temporal and dynamic aspects of comfort. For example, comfort was found to vary depending upon the time of day and day of the week. To this extent, the in-situ self-reporting methods combined with quantitative sensor measurements in the office environment further expand the repertoire of current techniques for subjective comfort measurement in the workplace, including use of questionnaires, interviews and video analysis.

In the initial version of the BOCS platform, a combination of sensors measuring CO_2 concentration, humidity, sound levels, temperature, light intensity and movement were used. The combination of subjective and objective data contributes to capturing the influence that specific actions have on the office environment, how these actions build up practices of individual users and how these practices become social practices. This in turn allows to understand the relationship between participants' actions, social practices, participants' comfort and buildings' energy use (Lockton et al. 2013). To further analyze findings from the objective and the subjective measures (step 3), a template was developed which includes the materials, meanings and skills involved in the practice (Kuijer 2014). However, the template continues to undergo further revisions, towards developing a method of comparison of practices in respect to their contribution to comfort and energy saving. Initially such comparison has been performed through a time-consuming expert reviews, however in the future may be accelerated through the use of a standardised format (step 4). Evaluation of iteratively implemented solutions (step 5) begins from the second iteration on and involves the comparison of changes in practices across iterations, and comparison of practices between co-creating building occupants, solution testing building occupants and a control group of building occupants not involved in co-creating or testing activities. The repetition of the researching activities at least once per season, allows to address the variation of practices depending on seasonal weather conditions and associated social factors.

17.5 Exploring and Identifying Solutions

Identifying and analysing social practices in office environments and their relationship with occupants' comfort and buildings' energy consumption provides in-depth insights into the present situation, and allows to pinpoint specific aspects of these practices that offer opportunities for change. However, due to high intricacy of social office practices and their grounding in everyday habits, social relations and thinking patterns, an enforced modification of a social practice in a top-down fashion is difficult to achieve. Practices in office environments also involve more participants and are more heterogenous that practices in a household covered by other chapters. Additionally, attention of people to both comfort and energy consumption in office spaces is significantly lower than in case of households. As an alternative approach, co-creating a solution together with the occupants (Mulder and Stappers 2009) offers a possibility to take advantage of occupants as "experts in their own lives" in finding both practice improvement opportunities and concrete solutions to support them, as well as encouraging the involvement of participating users in the early adoption phase of the solutions.

The exploring and identifying solutions has been divided into (1) the preparation phase, (2) co-creation phase and (3) processing of co-creation results. Before the start of co-creation sessions (step 1), practices identified and analyzed during context research are translated into sketchy storyboards, consisting of four to six pictures illustrating typical moments in the day of typical observed participants in a typical office. The style of the storyboards is deliberately fast and informal. The following co-creation session (step 2) consists of four phases. The first "reflection" phase takes 15 min and begins by a short explanation of the storyboard to participants. Printouts of anonymised data from monitored offices are used to show how different indoor climate parameters correspond to activities identified as part of the practice, how they influence comfort in that practice and what impact they have on energy consumption. Participants with assistance of researchers are then encouraged to label the storyboard with post-its, in order to indicate (a) their own subjective feelings they associate with storyboard situations (yellow post-its), (b) objectively measurable aspects of the environment which they can explicitly identify (red post-its), and (c) opportunities they notice for improving comfort or reducing energy use. The second "opportunities" co-creation phase takes 45 min and is divided into two equal parts. In the first part participants discuss the opportunities identified and look for possible conflicts between energy use and comfort within these opportunities. In the second part participants are asked to identify possible positive and negative impacts that the opportunities for change may have on social relations in the office. During both parts occupants are requested to write down pronounced topics on post-its and organise them on corresponding flipchart pages. The third "concrete solutions" phase takes 30 min. In the first part of this phase participants identify what objective and subjective data from their office can support them in achieving the identified practice change. They use stickers to signify those parameters on the storyboards. Following this step, they receive a template print-out, stickers and post-its to work in sub-groups to make their own storyboard of a future scenario of their own office practices incorporating earlier identified opportunities for change, and including sensors and feedback devices supporting this change. The session ends with short presentations of these future scenarios to each other. This final storyboard serves as an initial blueprint and "contract" with participants towards turning envisioned scenario into solution to be further developed and tested, and participants committing to use these solutions. Following the co-creation session, participating design researchers revise the outcomes and provide additional descriptions based on performed observations (step 3). At this step earlier collected data from participating offices can be used to make a rough estimation of the impact of the ideas on the improvement of the work environment's physical characteristics and reduction of energy use.

17.6 Implementing Solutions

The setup of co-creation sessions leads to formulating a set of concrete future scenarios through which occupants commit to an attempt to change their office practices, and which indicate features that need to be implemented in a BOCS platform to support them in reaching this goal. In the current version of the BOCS platform, these features can encompass a range of objective and subjective sensing interfaces, as well as custom feedback web interfaces and devices, which occupants can involve in their social practices.

The process of translating the outcome of the co-creation session starts with (1) preparatory activities, followed by (2) the "prototyping sprint day" after which (3) engineering of specific features continues. The preparatory activities (step 1) involve integration of the context research findings with co-creation activities. Design researchers involved in both activities prepare a concise report and a presentation highlighting the most important findings. The "prototyping sprint" (step 2) takes an entire day and involves participation of design researchers, designers and engineers. Throughout the entire "prototyping sprint", quick mockups, demos and testing of existing features take place, to facilitate communication and integration of required features. However, despite the open format, the "prototyping sprint" follows a defined process consisting of three stages. In the first stage design researchers present the outcomes of preceding activities and highlight the requirements for adaptations and new features to be introduced in the BOCS platform. The following stage involves a structured discussion on the feasibility, implementation and design of requested features, using differently colored post-it to highlight raised questions and taken decisions, and grouping them in shared problem areas. In the last stage decisions towards the implementation plan are taken. Workload and planning for the following steps are also agreed upon. After the prototyping sprint starts the implementation (step 3), during which engineers and designers implement agreed upon solutions and test their functional operation. The total duration of this phase is limited to a 2-week timeframe, to ensure that co-creating occupants remember the details of co-created solutions and actively engage in testing them, upon the start of the next process iteration.

17.7 Closing the Loop

Finalising the solution implementing activities leads directly to the next process iteration, which starts with testing the implemented solutions in the context of use. The first iteration begins with the uninterrupted context, while the following iterations incrementally add solutions supporting occupant social practice change. Achieving practice change is a long process which in itself may require multiple months or even years to become fully embedded in people's daily routines. Office comfort related practices are also influenced by, and may vary depending on the time of year. In order to allow reliable evaluation of the effects of the implemented solutions on occupant practices, three occupant groups are involved in the study, working across similar, but not directly connected office environments. The control group is not affected by the study and is only observed and monitored using the BOCS platform. The testers interact with the iteratively updated devices, interfaces and services, but do not participate in the co-creation process reserved for the third group of participants. In this way, not only the effect of the developed solutions on the occupants can be evaluated, but also the effect of participating in the co-creation can be measured.

17.8 Pilot Study Summary

To evaluate the described approach and the BOCS platform, a study in three office buildings of a large international organisation has been carried out (Jaskiewicz and Keyson 2015). 33 offices participated in the study. 15 of these offices were located in one building (building A). Two groups of 9 offices were located in other two buildings (building B and C). All the offices participating in the study were selected to have a similar sunlight exposure and host between one and five occupants. The three buildings had different characteristics. Building A has been recently refurbished and its users have limited possibilities of controlling HVAC installations, but can open windows. Building B has older, manually controlled installations. Building C is an older building, where very limited climate controls are available and windows cannot be opened.

The research in the pilot study led to identification of two social practices with highest potential for improvement in respect to comfort and energy use reduction, namely "focused desk work" and "informal office meetings". General correlations could be found between qualitative objective data and moments when these practices were occurring, but the patterns were not conclusive to identify the occurrence of the practices based on data only. Nonetheless the outcomes of the preliminary research enabled the inventory and analysis of key social practices occurring in the studied buildings.

Three co-creation sessions were performed with participants from each of the three participating buildings. The sessions faced organisation challenges and low attendance of participants due to conflicting work obligations. Nevertheless, even in groups where only two or three participants attended, constructive solutions were developed. In all cases confronting participants with research findings triggered highly motivated discussions, and provided insights that were not captured in the research steps.

One example can be the role of perceived noise in the focused desk work practice in relationship to ventilating rooms. Despite low measured noise intensity, the fact that participants could understand and relate to casual conversations in the hallway caused them to close office doors and ventilate office rooms by opening windows. In that case co-creation session led to a scenario where participants leave the office doors open, but provide each other feedback about noise annoyances, with an aim to cluster social chats.

Despite setbacks encountered with the BOCS platform software, the prototyping was attainable in the designated time. The prototyping sprint has led to a blueprint of one self-reporting device that fitted all three future scenarios developed in the co-creation sessions. Two large buttons on the device allowed to indicate moments in which participants were noticing that their comfort is impacted by bad air quality or noise, as shown in Fig. 17.5. The device was additionally equipped with a slider, allowing the expression of occupant overall comfort level on a 5-step scale. The occupants were enabled to correlate their subjective observations with objectively measured data using a feedback display implemented as

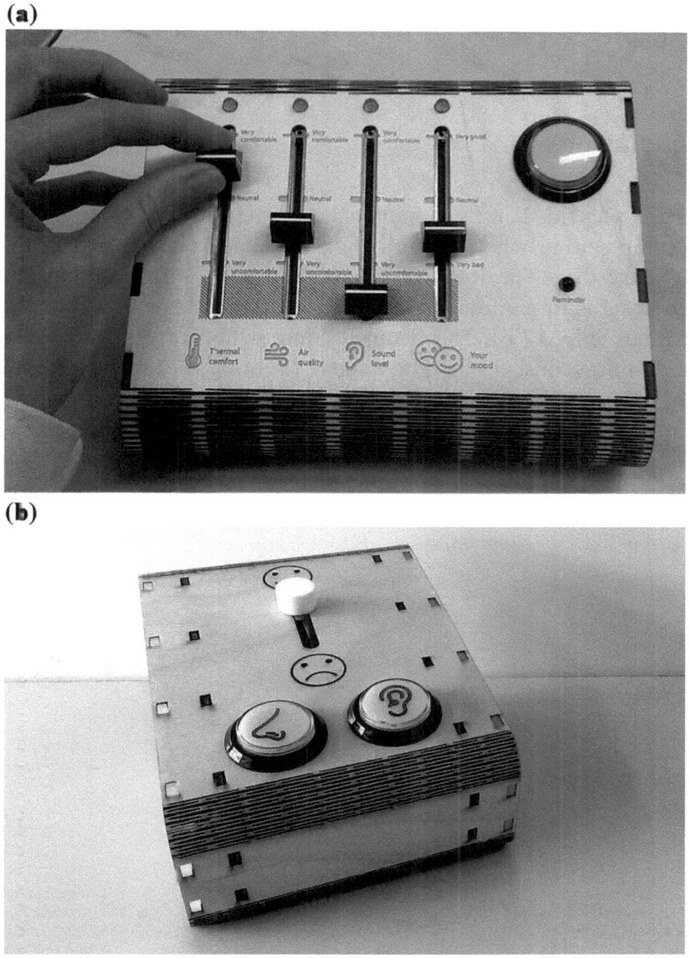

Fig. 17.5 Two different self-reporting tabletop devices developed based on the outputs of co-creation sessions

a web application. Subsequently, the prototyped devices, including subjective air and noise measurement were deployed among the participants in the experimental groups. This system is currently being tested by pilot study office occupants as part of a second design iteration.

However, the interviews with participants have also revealed that the time between co-creation session, ability to try out solutions and effectively bring them to a fully usable stage is too long, and they loose their engagement and motivation in bringing the envisioned change to life. This signifies that introducing a technical solution is only one aspect of the described process. The building occupants becoming early adopters of that solution, and using it to change their own lives are key to success of the approach. Further development of the approach may thus involve additional, smaller iterations to maintain participant engagement at high level.

17.9 Conclusions

The approach presented in this chapter demonstrates a novel way to innovate in the Living Lab setting when facing behaviour change challenges involving complex social practices. The iterative process allows co-creation and development of solution designs in parallel to ongoing longitudinal research on energy-related social practices (Fig. 17.6). The planned continuation of the study through more iterations promises not only improvement of initially developed solutions and creation of new ones, but also ongoing evaluation of users' responses to these solutions in order to better understand the individual behaviours and social mechanisms shaping the studied practices.

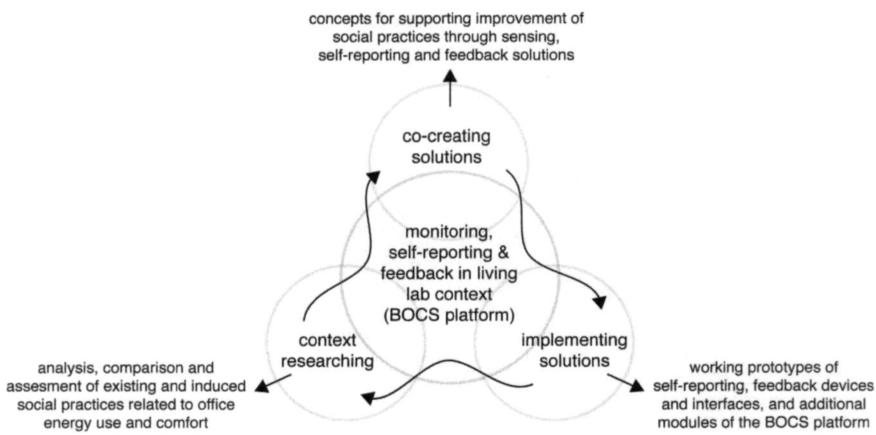

Fig. 17.6 Through iterations, the presented process leads to incremental generation of social practice insights, innovative concepts and technological solutions

Several drawbacks have been identified following the initial pilot study. The research methods have been optimised to allow their execution in three-month cycles. However, the context researching phase and implementation of co-created solutions both remain a time consuming activities, creating bottlenecks in the overall process, prohibiting further shortening of cycles from months down to weeks. A format for describing and comparing studied social practices is being further developed to support identifying relations between the skills, meanings and materials of practices (Shove 2007), and enable a more comprehensive analysis of qualitative data in the process. Although the co-creation process has been limited in order to cater to indicated time constraints of participants', the majority of participants have nonetheless not been able to attend the sessions due to other work priorities. It is hoped that the results of the first iteration brought back to the context will engage more participants to join the sessions during future iterations. Further revisions to the approach may also involve aspects of the co-creation process being performed by participants in small groups or individually at the time of their own convenience. Furthermore, the implementation of co-created solutions has also proven to pose additional difficulties. As outcomes of co-creation have ventured into unexpected areas, the need for developing features that have been not accounted for in the system architecture appears unavoidable, resulting in possible prototyping delays, or compromising of co-creation process outputs, for which better strategies need to be defined.

Despite the above shortcomings, the first iteration of the approach has shown to deliver results that triggered participants to reflect on their practices and initiate new practices in the work environment to improve comfort and reduce energy consumption. The answering of key research questions and verification of co-created solutions and their implementations requires several more iterations and from this perspective the approach cannot be evaluated at this stage. The intermediate results from the first iteration have delivered a wide range of outputs, as indicated by Fig. 16.6, promising successful continuation of the process. However, to maintain the engagement of occupants, the shortening of the duration of the cycle is recommended, along with potential introduction of additional, smaller in scope, iterations within one season.

The initial success prompts speculation about applying the presented approach, methods and tools to other living lab problem areas. One can imagine direct application of the approach to addressing energy-related occupant practices in other types of buildings without the need for substantial modifications of the approach and of the BOCS platform. Venturing further into speculation, using different sensors could allow addressing other problem areas ranging from healthcare to mobility. Yet, perhaps, the biggest future promise lies in studying social practices and finding approaches for their shaping across various domains, closely following and holistically supporting the full breadth of people's lives.

Acknowledgments This research was made possible by funding from the European Union Climate Knowledge Innovation program under the Built Technology Acceleration flagship. The pilot studies have been co-funded by the Amsterdam Institute of Advanced Metropolitan Solutions. The Waag Society has participated in the development of workshop formats, organisation of co-creation sessions and prototyping activities. Special thanks to Marc de Hoogh for database and web interface engineering, Martin Havranek and Richard Bekking for hardware engineering work, Stella Boess and Abhigyan Singh for support and feedback on the pilot study co-creation sessions.

References

Bazley, C. M. (2015). Beyond comfort in the built environment. PhD dissertation, Delft University of Technology.

European Network of Living Labs (ENoLL) Website. Retrieved November 2, 2015 from http://www.openlivinglabs.eu.

Coughlan, P., Fulton, Suri J., & Canales, K. (2007). Prototypes as (design) tools for behavioral and organizational change a design-based approach to help organizations change work behaviors. *The Journal of Applied Behavioral Science, 43*(1), 122–134.

Hargreaves, T. (2011). Practice-ing behaviour change: Applying social practice theory to pro-environmental behaviour change. *Journal of Consumer Culture, 11*(1), 79–99.

Intille, S. S., Larson, K., Beaudin, J. S., Nawyn, J., Tapia, E. M., & Kaushik, P. (2005). A living laboratory for the design and evaluation of ubiquitous computing technologies. In *CHI'05 extended abstracts on human factors in computing systems* (pp. 1941–1944). ACM.

Jaskiewicz, T., & Keyson D. V. (2015). Co-designing with office workers to reduce energy consumption and improve comfort. In *Behaviour, Energy and Climate Change Conference Proceedings*, eScholarship.org.

Kuijer, S. C. (2014). Implications of social practice theory for sustainable design. PhD dissertation, Delft University of Technology.

Lockton, D., Bowden, F., Greene, C., Brass, C., & Gheerawo, R. (2013). SusLabNWE: Integrating qualitative and quantitative data to understand people's everyday energy behavior. In *Behavior, Energy and Climate Change 2013 Conference*.

Mulder, I., & Stappers, P.J. (2009). Co-creating in practice: results and challenges. In Collaborative Innovation: Emerging Technologies, Environments and Communities, *Proceedings of the 15th International Conference on Concurrent Enterprising: ICE 2009* (pp. 22–24).

Shove, E. A., Hand, M., Ingram, J., & Watson, M. (2007). *The design of everyday life* (p. 192).

Chapter 18
Recruitment of Participants (Households in City District and Companies) for Insight Research and Prototyping

Kamil Folta, Dan Lockton and Flora Bowden

Abstract In the project SusLabNWE the integration of users in private households was a vital part of the concept and scientific approach because products and services need to be aligned to the user needs and fitted to their behaviour. In order to develop, design and implement innovative products that serve their purpose and are accepted by users, a high level of user engagement is expected. This chapter describes the recruitment of participants in two case studies within the SuslabNWE project.

18.1 Introduction: General Overview and Implementation in SusLab

Within SusLabNWE, the integration of experimenters into private households is a vital part of the project's concept and scientific approach. To get the most diversified insights and universal statements of results, it is on the one hand necessary to spread the project on different types of households. One point is a variation on demographic aspects (e.g. age, state of life, standard of living), but also on income differentials, which reflect the standard of living and especially the type of property, i.e. owner occupancy of rent.

K. Folta (✉)
Innovation City Management GmbH, Südring-Center-Promenade 3, 46242 Bottrop, Germany
e-mail: kamil.folta@icruhr.de

D. Lockton
Royal College of Art, London, UK
e-mail: dan@danlockton.co.uk

F. Bowden
SustainRCA, Royal College of Art, London, UK
e-mail: flora.bowden@gmail.com

D.V. Keyson et al. (eds.), *Living Labs*, DOI 10.1007/978-3-319-33527-8_18

On the other hand, the innovative products and services offered in the context of SusLABNWE (e.g. devices which help to safe energy and enhance comfort in terms of heating) need to be aligned with the users' needs and fitted to their behaviour in order to safeguard their application. To a greater degree some products even have negative effects because users show different behavioural patterns as engineers and designers had foreseen or predicted. In order to develop, design and implement innovative products which serve their purposes and are accepted by users in SusLab, a structure of LivingLabs and a high level of user engagement and preliminary investigation by the project managers is needed.

And at least it is a chance to explore some new technical opportunities and chances to implement technical assistances at highest technical level combined with visually attractive design. Following sections introduce the recruitment of participants in two SusLABNWE case studies.

18.2 Case Study Bottrop

The described differences and the initial position at all levels assume that recruiting appropriate private test persons adjust individual approaches. Therefore it is the responsibility of each partner to develop and implement a local volunteer program. The partners choose suitable instruments for activation and information, like information events, direct contact via personalized letters or contact forms via internet.

This is a prerequisite for creating a local test infrastructure for the topic "heating/space heating" in at least 50 homes. All participating households are located in the pilot area of InnovationCity Ruhr in Bottrop. The local involvement and recruiting for the project is coordinated by the Innovation City Management GmbH (ICM) for the German consortium consisting of Wuppertal Institute for Climate, Environment and Energy, Hochschule Ruhr West University of Applied Sciences (HRW) and ICM. The main benefit of that testing area is the existing sensitization of the inhabitants for unconventional and innovative projects regarding new technologies.

Since 2011, Innovation City Management has offered individualized energy consulting especially to private house owners through the Centre of Information and Advice (Zentrum für Information und Beratung—ZIB). In the past 5 years 2000 personal consultations have been carried out in total. This offer is absolutely cost free for the citizens of the pilot area and it is still much asked for. In 2015, 326 different consultations have been carried out by ICM's energy advisor/architect, every one of them for around one and a half up to 2 h. This service is available for everyone living in the city area offers energy efficiency advice to building owners, by analysing the energy consumption data of each building. For that a special online tool has been designed, which also can be used for a pre-review for every interest. Therefore the tool serves as an information basis for the interests and in addition it´s a tool for the energy consulter who gives qualified input to the software system to create an encompassing entry to each consulted building. With this data, personalized energy efficient retrofitting proposals are developed, implemented and customized to maximize efficiency for each individual unit. The advice is adapted to the individual financial possibilities combined with a special promotion for the total price of the modernization if someone lives inside the pilot area. The main advantage is due to the fact that subsidies spread more easily than in other existing programs for modernization by reducing administrative effort.

In parallel, informed citizens are motivated by advertising campaigns and targeted events and community workshops. These events are designed to engage the community and let citizens become part of a project and support its goals such as implemented in the SusLAB-project.

Additionally, Innovation City Management has organized networks of local craftsmen, architects and energy consultants and established a partner network for technology and process related innovation within the Ruhr metropolitan region. The industrial advisory board enables companies to develop and join projects. By means of this, major German energy related companies like Vaillant, Danfoss or RWE are involved, but also local and regional housing companies like VIVAWEST, GBB and Vonovia are members of the advisory board. The main benefit of this network is more than evident: By means of involving housing companies directly, thousands of test persons can be acquired.

For the SusLabNWE winter pilots ICM and HRW used this existing networks and initiatives to contact suitable households and asked them to participate and get involved in the project. The existing data represented he basis to analyse and then to find private landlords and tenants in flats and houses. They received a covering letter by mail from the HRW which described the project's background, the procedure and the benefits. Around 600 households received this letter and much more households replied than necessary.

The group of households involved increased after the first winter pilot by directly contacting further households. Another benefit was gained through media and newspapers: Positive reviews about the project spread quickly and numerous newly interested citizens approached the project. Apart from the media support the technical equipment improved.

Furthermore additional measurement toolkits were available to gather information and statistical data in the households. The main point is that the participants did not receive any kind of monetary incentive. But they provided their data containing detailed feedback about their individual behaviour and derived from that data individual potentials of savings.

18.3 Case Study UK

In the UK, the nature of the participants recruited, and the recruitment strategies used, varied throughout the SusLab project, depending on the goals of each part of the study. Overall, the primary aims of the first parts of the study were qualitative, while later phases aimed to integrate qualitative and quantitative data. In every case, the aim was better understanding of householders' use of energy in context, including a specific focus on heating practices and thermal comfort. This deeper understanding, requiring multiple home visits by researchers, necessarily meant smaller sample sizes.

In the first phase of our research in London, researchers from the Royal College of Art carried out home visits and interviews with householders, followed by a probe/logbook study (described in Lockton et al. 2013). Following methodologies developed in the context of inclusive design (Eikhaug and Gheerawo et al. 2010), in this work we focused on *lead users* in one form or another—people who have particular needs around, or interest in, energy use at home, and who are indeed often self-described 'edge cases'. Recruitment of the initial group of nine participating households was done through: (1) Tenants' groups with a focus on environment and sustainability, at two social housing providers in London (Poplar HARCA in East London, and Camden Council in north London); (2) Word-of-mouth suggestions of participants, by researchers who had previously worked on sustainability-related projects; (3) Postings and announcements in the London Internet of Things and Cleanweb meetup groups (online groups with in-person meetings every month). Participants were offered gift vouchers for Marks and Spencer, a slightly upmarket UK supermarket and department store.

Once potential participant availability had been explored, we arrived at nine households of a range of ages, backgrounds and family situations (with children, living alone, retired, etc.), living in London and south-east England: social housing tenants on limited incomes, some already part of existing programmes aimed at saving energy (via home energy displays and online monitoring), and some who have taken it upon themselves to cut their energy use without using any kind of display; people with medical needs which mean they use higher than average amounts of gas for heating; people with environmental motivations and people much more focused on cost; and people from the Internet of Things and Quantified Self communities, who have set up their own home energy monitoring systems for their own interest, and have incorporated using the systems into their everyday routines. Some of our 'early adopter' lead users could be in the vanguard of coming trends around technology use at home, but trends also represented in the group—such as ageing populations and more people living alone—would have other effects on energy use. The idea was that through learning from these interested users—understanding their routines, motivations and interactions with technology, we could both understand better the factors to concentrate on in subsequent research, and identify design opportunities for interventions that take account of the real contexts of everyday energy use. Some of these householders subsequently took part in our Home Energy Hackday and co-creation workshops.

A parallel phase of our research involved public engagement at events and exhibitions, which is not so relevant for this chapter; Chap. 14 on participatory drawing, and Chap. 23 on the Powerchord sonified energy monitor describe the participant contexts of this work.

For the final 'Winter Pilot' phase of the UK SusLab research, involving the installation of the SusLab Monitoring toolkit and use of the heating practice diary (see Chap. 12) in homes, the Institute for Sustainability, via its partners at Dartford Borough Council, recruited five households in the Dartford, Kent area (to the south-east of London).

In this study, the installation, configuration and monitoring of the SusLab equipment was carried out by researchers from Imperial College London. One household had to drop out as the broadband internet service present was incompatible with the SusLab equipment. As the recruitment was handled by other partners, the demographic and contextual information about householders was something the RCA researchers (Flora Bowden, Dan Lockton and Shruti Grover) elicited during interviews. The householders were all social housing tenants who were part of schemes to improve the energy performance of their homes, had broadband internet which would enable the sensor kit to be installed, and were happy to do the level of self-reporting required by the study. This home improvement work had either been completed in the preceding few years, or was about to be carried out, and it was considered that these householders were sufficiently interested in energy, and heating, to be motivated to take part in the study. The incentive in this case was to be able to keep the Android tablet used for the self-reporting, at the end of the study.

This is a consideration that is worth attending to when carrying out recruitment for studies which require substantial ongoing effort from participants (such as completing diaries or using apps)—participants who are motivated to participate to this degree, are likely already to be quite deeply interested in the subject, which does not necessarily make for a representative sample of the characteristics of the population as a whole.

References

Eikhaug, O. & Gheerawo, R. (Eds.). (2010). Innovating with people—the business of inclusive design. Norwegian Design Council.

Lockton, D., Bowden, F., Greene, C., Brass, C. & Gheerawo, R. (2013). SusLabNWE: Integrating qualitative and quantitative data to understand people's everyday energy behavior, in *BECC 2013: Behavior, Energy and Climate Change, 18–20 November 2013, Sacramento, CA*. Berkeley, CA: University of California eScholarship repository.

Part V
Sustainable Production and Consumption

Chapter 19
Analysing Social Milieus and Lifestyles—Their Contribution to a Better Understanding of Heating Practices

Johannes Buhl, Kathrin Greiff, Carolin Baedeker and Christa Liedtke

Abstract This chapter introduces empirical social research on social milieus, and why lifestyles are of interest to Living Lab research. We present the results of a milieu analysis for Living Lab research in Bottrop. We enrich the findings from the milieu analysis by describing the influence of lifestyles on heating practices, combining a representative statistical analysis for Germany with in-depth qualitative interviews of Living Lab participants. When it comes to heating practices, lifestyles expressing higher material wealth and socio-economic status show meanings associated with comfort or convenience rather than financial reasoning.

Keywords Social practices · Behavior · Sustainable consumption · Mixed methods

19.1 Theoretical Background

The analysis of lifestyles, social milieus and types of household[1] is gaining importance in an evermore experience- and event-driven society (Schulze 2013). Consumers increasingly focus on experience in everyday life. Demand in an experience economy manifests less in products people seek to buy and to have, but rather in services, activities and events people want to do, to make or attend to. In such economies, consumption is rather based on activities such as going to

[1]We consider social milieus to circumscribe a group of people who share a common lifestyle. In this sense, we refer to one's social milieu and lifestyle interchangeably in this chapter.

J. Buhl (✉) · K. Greiff · C. Baedeker · C. Liedtke
Wuppertal Institute for Climate, Environment, Energy,
Doeppersberg 19, 42103 Wuppertal, Germany
e-mail: johannes.buhl@wupperinst.org

D.V. Keyson et al. (eds.), *Living Labs*, DOI 10.1007/978-3-319-33527-8_19

cinemas, museums, amusement parks, attending concerts, theatre plays, going out, trips, travel, staying at hotels and going to restaurants. Being a member of clubs, such as sport clubs, are as much a thing people would do in an experience-based economy—voluntary and informal. In this respect, people do usually not seek experience for themselves, but while interacting with others. 'Experience' becomes a shared and social event. Consequently, as soon as people have found out successful strategies and ways to 'experience', those strategies and pathways serve as signals for social distinction, status and identity.

In this context, one may state that experience marketing is just another quality of service-oriented marketing. It is however not about products and services marketers strive to sell, but the experience itself is the primary goal of production and consumption. From an environmental perspective, experience-oriented lifestyles are not necessarily an immaterial way of living; on the contrary, seeking experience may come with resource intensive consumption patterns of products (e.g. ICT) as well as services (e.g. services in tourism). One might think that such "experience demand" (Schulze 2013, p. 98) is totally different from traditional demand sectors. However, experience markets comprise all products and services, the use of which is predominantly defined in aesthetic terms (nice, exciting, comfortable, stylish, interesting and so on). The concept of experience demand needs to focus on "inwardly oriented actions" (Schulze 2013, p. 101). Whereas outwardly oriented actions can be defined and measured clearly like money or product features, inwardly oriented actions refers to the inner life of the person like feelings, values, knowledge or skills. That is a challenge to new forms of design trying to adequately deal with newly emerging patterns of consumption in an experience-oriented society. Thus, when trying to understand who is taking part in such an experience society and why they do so, a conventional differentiation of societal groups into classes, such as income classes, might serve as a point of departure,[2] but insufficiently addresses social inequality in society. According to Pierre Bourdieu (1984), social stratum has replaced social class such that new forms of capital—like social and cultural capital—emerge next to economic capital.

More recently, sociological research seeks to understand how such patterns of capital translate into lifestyles and manifest in every day activities and social practices (see Chap. 3 about social practices). It is crucial to understand that lifestyles do not manifest in goods or products bought, but rather in how people use these goods in daily life and for what. It is important to grasp daily life as repeated social episodes constituting lifestyles. Then, lifestyles do not only depend on what kind of dwelling people live in (like the size of dwelling), but how and why people live in their houses as they do. In order to differentiate between lifestyles in households, one has to ask for and observe general preferences, attitudes and personal

[2]See e.g. Sinus (vgl. http://www.sinus-institut.de/loesungen/sinus-milieus.html); Sigma (http://www.sigma-online.com/de/SIGMA_Milieus) und sociodimensions (http://www.sociodimensions.com/files/milieus_2.pdf).

norms in their daily practices. The aim of our research on lifestyles in Living Labs was to develop and exemplify an approach to differentiating between households according to their social milieu. Based on such a milieu analysis, we tried to understand how lifestyles link to meanings of daily practices, understood here as heating practices. Thus, the identification of lifestyles might be used as a variable to explain practices and corresponding resource and energy use.

19.2 Methodology and Data

19.2.1 Milieu Analysis

The milieu analysis is an integral part of the insight research of the Living Lab approach as applied in SusLabNWE (Liedtke et al. 2015). In the Living Lab research in Bottrop[3] around 90 households participated in the study. In order to know what kind of households and what kind of lifestyles are responsible for what kind of heating at home, a model of social milieus for Germany has been used.

Households commonly share the consumption of energy services like heat, but members in the household do not necessarily share a common lifestyle. Lifestyles may differ according to roles in the household. The father might follow a different lifestyle than his adolescent son. Both share the heat in the living room and may have disputes about the temperature in it (for whatever reasons). Those issues are addressed in a network analysis (see Chap. 15 about Network Analysis). Thus, we assume that the household shares a common social milieu that is proxied by the individual characteristics of the head of the household or main income earner. In the end, we combine household and personal characteristics. A survey asking the participants not only for socio-economic characteristics of the participants consisting of data on household size, family status, income, education and age of head of household, but also for environmental and economic attitudes has been conducted (Table 19.1). The screening of households allowed us to compare responses to a representative lifestyle model for Germany (socio-milieus from Sociodimensions 2015), assigning social milieus according to social status, generation, age and family status. Table 19.1 shows the information gathered on the households in order to

[3]The Living Lab research in Bottrop comprises the results from the projects SusLabNWE and EnerTransRuhr. The project EnerTransRuhr follows up on the work done in SusLabNWE. We integrated data collection and data analysis of both projects in order to give a more comprehensive picture of the Living Lab research in the Ruhr area and Bottrop, respectively. Whereas the milieu analysis is foremost based on the data collection and analysis in SusLabNWE, the presented findings on the influence of lifestyles on heating behaviour is mainly done in EnerTransRuhr. We acknowledge the support in EnerTransRuhr by the German Federal Ministry of Education and Research through a grant from the funding priority 'Social–Ecological Research' (grant number 01UN1205A).

Table 19.1 Household and individual dimensions and according variables used in the survey

Household	Individual
Size of household (number of persons)	Socio-demographics (age and sex)
Number of children in household (below and above 18)	Family status (single, couples w/o children)
Housing type (single houses, multi family housing, home owner or rental)	
Household net income	Social status (education, occupation and occupational status for head and partner)
Energy consumption (costs)	Attitudes towards environmental protection and economic growth (choice of probable and desirable future scenarios)
Size of dwelling (in m^2)	

Source Own table based on Sociodimensions (2015)

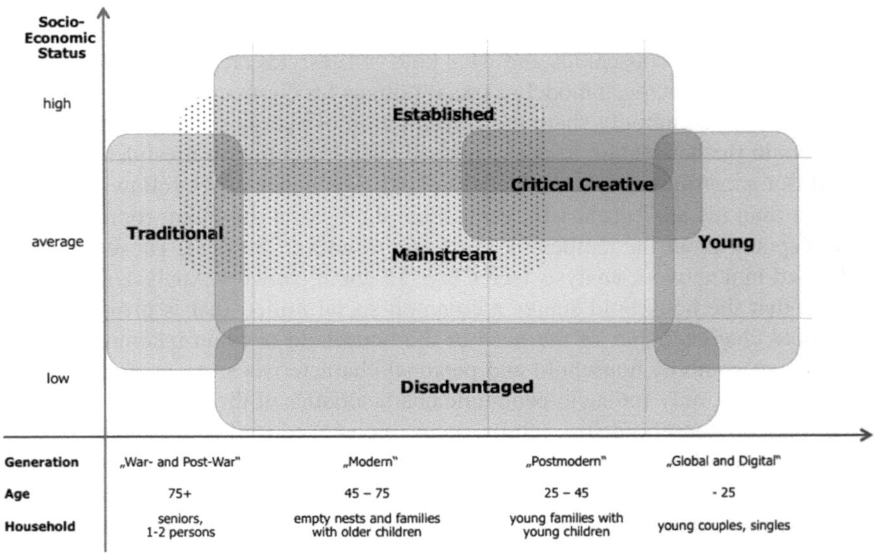

Fig. 19.1 Sample (dotted cloud) located in simplified Socio Milieus model. *Source* Own depiction based on Sociodimensions (2015), and according to FEA (2015)

derive their social milieu. Figure 19.1 depicts the respective social milieu in which the responding households have been assigned according to their scores.

Moreover, a cross-validation of the typology was possible based on a related socio-spatial geo-mapping of the social milieus of Bottrop by Microm (2015). The participants have been assigned according to their residential address in Bottrop, and their most probable social milieu has been derived. The two independently conducted milieu assignments eventually offered a triangulated assignment of the analysed households in Bottrop (Sociodimensions 2015). The geo-mapping allows us to integrate the findings from the research we conducted on a city level. It allows for analysing how social practices may spread and diffuse on a larger scale, within and between milieus—and within quarters and across cities (see Jensen et al. 2015).

19.2.2 Multivariate Regression Analysis

However, it remains unclear, to which extent lifestyles can explain (non) sustainable practices or not. What is the added value of knowing the lifestyle or the socio-milieu in addition to standard socio-economic variables? Therefore, we conducted a multivariate regression analysis of heating practices based on the action model from Liedtke et al. (2013, 2014).[4] We complemented the rather socio-psychological action model with information on socio-economics like age or income and lifestyles.

We used data from the study on "environmental awareness" in Germany for the year 2012 (FEA 2013). For the same study, the socio milieus have been employed for the year 2015. In 2012, the study relied on the milieu model from Otte (2005). Here, milieus are described by two dimensions as well, material status (or wealth) and modernity in terms of attitudes of lifestyles like traditional and religious or materialist and hedonist attitudes. Thus the model as such is very similar to the model applied for the respondents in the Living Lab research in Bottrop.

People have been asked if their household "practices" to keep the consumption of heating costs low. Thus we proxied heating practices by the household's routine to keep the heating costs low. We introduced financial meanings, as well as socio-psychological meanings like personal (e.g. feeling obliged to environmentally friendly behaviour) and social norms (e.g. feeling obliged to follow the behaviour of significant others, family and friends), as well as socio-demographics like age and sex, in addition to lifestyle dimensions (modernity and wealth). The statistical analysis is complemented by in-depth qualitative interviews with 18 participants of the Living Lab research in Bottrop. Here, the analysis is based on qualitative insight research from 6 households in SusLabNWE and 12 households in EnerTransRuhr. The qualitative research has been designed correspondingly in order to integrate the information collected. The extensive qualitative insights provide a deeper understanding of the meanings behind heating practices.

[4]The quantitative insight research has been conducted in EnerTransRuhr. The findings are given in Liedtke et al. (2014) as well.

19.3 Results

19.3.1 Identification of Lifestyles in a Milieu Analysis

The respondents of the Living Lab research in Bottrop are grouped according to their social milieu. The findings suggest that the sample is highly dominated by the social milieus of so called "established mainstream" lifestyles, whereas young and precarious milieus are highly underrepresented (Table 19.2).

The depiction of the findings within the graphic milieu model (Fig. 19.1, see also the Federal Environmental Agency (FEA) 2015 for another application of the milieu model) makes clear that the sample is composed of respondents with relatively high socio-economic status and of birth cohorts of the post-war generation (average age of respondents is 52 years) in "empty nests" (with an average of 0.7 children in the household).

The identification of lifestyles of the respondents enables us to derive more elaborated and more general understanding of the respondents' way of life, attitudes and values.

According to the milieu model applied from Sociodimensions (2015), "established empty-nesters" are typically satisfied with life, proud of their achievements, but open to eco-innovations. "Mainstream" households place high value on comfort and convenience, are relatively open to ecological arguments, thus show a relatively high personal norm towards environmental attitudes. They look out for reliable information (e.g. of product features) on which they can ground their sustainable behaviour. Accordingly, those milieus are grateful for information and consultancy respectively. At the same time, they show sceptical attitudes towards environmental efforts. Their perceived self-efficacy for environmental protection is relatively low. In this respect, "traditional" milieus are even more geared towards security and routines. They are sceptical about experiments in their lifes.

Table 19.2 Share of social milieus and deviance from population in and of sample

Socio milieus (sociodimensions)	Share in sample (in %)	Deviance from population (Bottrop = 100)	Deviance from population (Germany = 100)	Conclusion
Traditional Milieus	16.7	91	106	On average
Established Milieus	24.4	170	158	Overrepresented
Critical-creative Milieus	17.9	116	134	Overrepresented
Mainstream Milieus	34.6	143	140	Overrepresented
Disadvantaged Milieus	2.6	17	16	Underrepresented
Young Milieus	3.8	30	25	Underrepresented

Source Sociodimensions (2015)

In contrast, "critical creative" milieus are highly educated, open and tolerant, but expressing high aesthetic and qualitative expectations. They show a high environmental personal norm, they feel obliged to environmentally friendly behaviour. They are often pioneers when it comes to the adoption of eco-innovations such as e-mobility, or energetic retrofitting. The ascribed attitudes and values help us to better interpret the influence of lifestyles on heating practices.

19.3.2 The Influence of Lifestyles on Heating Practices

We briefly describe our findings and conclusions on the influence of lifestyles from a mixed methods research on heating practices (see Liedtke et al. 2014 for the results and Liedtke et al. under review). The statistical findings are derived from a cross-sectional, multivariate and stepwise logit regression of heating practices (i.e. their routine to keep their heating costs low) on financial variables like income, social and personal norms, socio-demographics like age or size of dwelling as well as lifestyle dimensions in terms of material wealth and modernity as an attitude. Otte (2005) refers to modern attitudes as hedonistic or traditional attitudes. He believes that those attitudes are closely linked to one's age. Older people are more likely to have a traditional or religious attitude than younger people, who, in turn, are more likely to have a hedonistic attitude.

We found that the material dimension (wealth) of the lifestyle is a significant predictor for heating practices. The higher the socio-economic status of the respondents, the lower the chance that respondents keep track of their heating practices in terms of costs. The influence of personal norms hints at the same direction. If somebody would rather have natural resources protected as scarce and valuable resources, he would have a greater chance of routinely trying to keep their heating costs low. The findings suggest that "established" milieus and milieus with higher socio-economic status need to be addressed not by financial motives (alone), but by other meanings as well, in order to intervene in potentially unsustainable heating practices.

This was the starting point for the in-depth content analysis of 18 interviews with participating households in the Living Lab research. A closer look at the statements of respondents with higher socio-economic status and energy consumption reveals meanings like comfort or social norms being more important for their heating practices than financial reasoning. We believe milieus with higher socio-economic status need to be addressed by attitudes as described in the milieu analysis. Those milieus are open to eco-innovations like e-mobility, energy efficient household appliances or energetic retrofitting, but they want to have their high aesthetic and qualitative expectations met. Eco-innovations need to meet aesthetic functions and address meanings like convenience and comfort, at the same time.

Moreover, they need to convey information in a reliable way. Eventually, the qualitative description of practices and their meanings in a structural content analysis served as a basis for prototyping an intervention design.[5]

19.4 Summary and Conclusion

The social differentiation of households according to socio-economic characteristics and lifestyles offers a more comprehensive picture of participants in the Living Lab research in Bottrop. By conducting a triangulated assignment of the households analysed within Bottrop it is shown that the sample is highly dominated by social milieus of "established mainstream" lifestyles, whereas young and "disadvantaged" milieus are highly underrepresented. Households of well-established "empty nesters" and of the "modern mainstream" represent the highest share, whereas the younger generation as well as "disadvantaged" households did not take part in this study. We studied the influence of lifestyles on heating practices in a mixed methods study combining a representative statistical analysis for Germany and in-depth interviews with participants of the Living Lab insight research in Bottrop.

The multivariate analysis of heating practices based on the action model from Liedtke et al. (2013, 2014, under review) suggests that for "established" milieus and milieus with higher socio-economic status, financial motives are not the primary motive behind their heating practices. The milieu analysis concludes that these milieus are open for eco-innovations, but want aesthetic functions and meanings like convenience and comfort to be met. The statements from respondents with higher socio-economic status point towards similar conclusions when it comes to heating. The statements suggest that meanings behind heating are associated with comfort and convenience rather than financial motives.

We conclude that a milieu identification and a social differentiation according to lifestyles is a promising tool for Living Lab research in order to deal with differing meanings behind practices according to a lifestyle's motives, norms, attitudes and values. The insight research introducing a milieu analysis offers a good basis to derive promising ideas for future intervention design and prototyping. The description of heating practices within a representative milieu model may help to potentially generalize effects of intervention designs, thereby depending less on individual contexts.

[5]Prototyping was not finished when this chapter was written. We decided to not show preliminary results and unfinished work.

References

Bourdieu, P. (1984). *Distinction: A social critique of the judgement of taste* (Harvard University Press).

Federal Environmental Agency FEA/UBA. (2013). Umweltbewusstsein in Deutschland 2012. Ergebnisse einer repräsentativen Bevölkerungsumfrage. Berlin. http://www.umweltbundesamt.de/sites/default/files/medien/publikation/long/4396.pdf).

Federal Environmental Agency FEA/UBA. (2015). Umweltbewusstsein in Deutschland 2014. http://www.bmub.bund.de/fileadmin/Daten_BMU/Pools/Broschueren/umweltbewusstsein_in_d_2014_bf.pdf).

Jensen, T., Holtz, G., & Chappin, E. J. L. (2015). Agent-based assessment framework for behavior-changing feedback devices: Spreading of devices and heating behavior. *Technological Forecasting and Social Change, 98*, 105–119. doi:10.1016/j.techfore.2015.06.006.

Microm Consumer Marketing. (2015). microm Geo-Milieus® auf Straßenabschnittsebene für die Stadt Bottrop (unpublished presentation and mapchart). Neuss.

Liedtke, C., Baedeker, C., Hasselkuß, M., Rohn, H., & Grinewitschus, V. (2015). User-integrated innovation in Sustainable LivingLabs: An experimental infrastructure for researching and developing sustainable product service systems. *Journal of Cleaner Production, 97*, 106–116.

Liedtke, C., Hasselkuß, M., Welfens, M. J., Nordmann, J., & Baedeker, C. (2013). Transformation towards sustainable consumption: Changing consumption patterns through meaning in social practices. In *4th international conference on sustainability transitions*, 19–21 June 2013, Zürich.

Liedtke, C., Hasselkuß, M., Buhl, J., Baedeker, C., & Welfens, J. (2014). Addressing consumption patterns through meaning in social practices—findings from a mixed methods analysis of heating practices. In *BEHAVE energy conference* 3–4 September 2014, Oxford.

Liedtke, C., Hasselkuß, M., Buhl, J., Baedeker, C., & Welfens, J. (under review). The role of social practices in transitions: heating by German households. *Environmental Innovation and Societal Transition.*

Otte, G. (2005). Construction and test of an integrative lifestyle-typology for Germany. *Zeitschrift für Soziologie, 34*(6), 442–467.

Schulze, G. (2013). The experience market. In J Sundbo & Sørensen F (Eds.), *Handbook on the experience economy* (pp. 98–122). Edward Elgar Publishing.

Sociodimensions. (2015). Wuppertal Institut – SusLab NRW, Haushaltsanalysen: Zielgruppendifferenzierung in der Innovation City Ruhr – Modellstadt Bottrop. Projektbericht (unpublished). Sociodimentions (Institute for socio cultural research), Heidelberg.

Chapter 20
Material and Carbon Footprint of Household Activities

Kathrin Greiff, Jens Teubler, Carolin Baedeker, Christa Liedtke and Holger Rohn

Abstract This chapter addresses the resource consumption and GHG emissions associated with household activities and household types. Over the course of 6 weeks 16 participating households were asked to provide data regarding their activities in the fields of housing, mobility, nutrition, waste, goods and appliances, tourism and recreation. This extensive survey enabled the authors to calculate the households Material and Carbon Footprint, representing the environmental pressure for certain household types and lifestyles. It was found that even households with similar soziodemographics differ highly in their overall impact as well as the shares attributed to the different fields especially for nutrition, housing and mobility. Two workshops were conducted where households were asked to identify possible short-, mid- and long time strategies for reducing their environmental impact (road mapping). Although not all households participated, it had become clear that many external factors prevent households from adapting their behaviour most notably in the field of mobility. However, the road mapping process also showed a high affinity of the volunteers towards lifestyle changes. Regarding the set of methods used in the study, the authors conclude that the approach is promising, but future research is necessary: amongst other potential improvements it would

K. Greiff · J. Teubler (✉) · C. Baedeker · C. Liedtke
Wuppertal Institute for Climate, Environment and Energy, Wuppertal, Germany
e-mail: jens.teubler@wupperinst.org

K. Greiff
e-mail: kathrin.greiff@wupperinst.org

C. Baedeker
e-mail: carolin.baedeker@wupperinst.org

C. Liedtke
e-mail: carolin.baedeker@wupperinst.org

H. Rohn
Trifolium – Beratungsgesellschaft mbH, Friedberg, Germany
e-mail: holger.rohn@trifolium.org

D.V. Keyson et al. (eds.), *Living Labs*, DOI 10.1007/978-3-319-33527-8_20

certainly be helpful to not only analyze the resulting environmental impact of households but also the circumstances that lead to the household's specific social practises and routines.

Keywords Households · Material Footprint · Carbon Footprint · Household activities · Roadmapping · Social practises · Life cycle assessment · MIPS

20.1 Background and Objectives

The global consumption of natural resources is still rising beyond the natural system's boundaries (Bringezu and Bleischwitz 2009; Giljum et al. 2014; Rockström et al. 2009; Schmidt-Bleek 2013; Ward and Neumann 2012). Thus, societal transformation towards sustainable consumption and production is a key challenge (Liedtke et al. 2012, 2015; Schmidt-Bleek 2009). Households play a major role in this transformation process (Lettenmeier et al. 2012, 2014). Within household consumption the fields of nutrition and mobility (next to housing) are identified as key fields of activity responsible for high natural resource use (Watson et al. 2012). Because all household activities, including those related to consumption, are interconnected by social practises[1] (Kuijer and De Jong 2011), an assessment of the whole household system is necessary to evaluate influencing factors, possible lifestyle changes and rebound effects (Brookes 1990; Buhl 2014; Buhl and Acosta 2015).

Next to the energy use and its greenhouse gas emissions (GHG) for heating and electricity, the natural resource consumption is used as an indicator for ecological sustainability (Huysman et al. 2015; Lettenmeier and Wuppertal Institut für Klima, Umwelt, Energie 2009; Liedtke et al. 2014; Schmidt-Bleek 2013). In this study, the indicator Material Footprint accounts for all material resources from nature (resource consumption), while the Carbon Footprint[2] is used for the global warming potential. This chapter presents an analysis of both indicators in the context of "Sustainable Labs North West Europe" (SuslabNWE). The analysis presented here is from the German part of the investigation (SuslabNRW). This part sought to identify the major influencing factors for the resource consumption of, and the greenhouse gas emissions by households. In that, it particularly took into account the corresponding social practises. In addition, the impacts in the field of heating have been compared to the whole resource consumption and the different household activities.

[1]Social practises can be defined as "a routinised type of behaviour which consists of several elements, interconnected to one another: forms of bodily activities, forms of mental activities, 'things', and their use, a background knowledge in the form of understanding, know-how, states of emotion and motivational knowledge." (Reckwitz 2002, p. 249).

[2]The Carbon Footprint in this chapter corresponds to the global warming potential in 100 years (GWP100a) and is expressed in CO_2 equivalents.

20.2 Methods

In order to identify influencing factors for ecological impacts by household activities, the corresponding Material Footprints and Carbon Footprints had to be calculated. Possible lifestyle changes were addressed in an additional road mapping process with the participating households.

The procedure consisted of three steps. Firstly, **(1) household monitoring** has been conducted. 16 households in the study area (the City of Bottrop) filled in a questionnaire about their housing-related activities. They were classified into the typology described above (see Chap. 19. Household Typology and social milieus) and monitored for 6 weeks. Parallel to this step the **(2) calculation and evaluation of the Material and Carbon Footprint** started. After interpretation of results a **(3) road mapping process** has been conducted: A researcher discussed the individual results with the households in order to create individual roadmaps for possible lifestyle changes. This process resulted in an analysis of drivers and barriers for lifestyle changes after two workshops.

20.2.1 Step 1: Household Monitoring

In order to calculate the household specific resource use and GHG emissions a detailed questionnaire was developed, which is based on a previous study in Finland (Kotakorpi et al. 2008). The questionnaire also contained supplementary questions regarding time use, expenditures and socioeconomic data in order to classify household types.

The households were contacted via press release through the Newsletter of the project partner Innovation City Ruhr. 16 households registered for participation. All households were differentiated into different household types, as described in Chap. 19. Household Typology and social milieus. The fields of activity monitored were housing, mobility, nutrition, waste, goods and appliances, tourism and recreation. The consumption was monitored for 6 weeks in October and November 2014. Each week the households received one questionnaire by mail to be answered in the following week. After the week they sent the questionnaires back to the researchers. Except for the field of goods and appliances (inventory of all possessions) that could be answered during 2 weeks, each field was enquired within 1 week.[3] No household abandoned the monitoring, but several households needed more time so that the last questionnaire was returned in January 2015. Figure 20.1 shows the fields of activities and describes the content regarding the gathered data.

[3]The questionnaires are not appended to this book because their total length runs to several dozen pages.

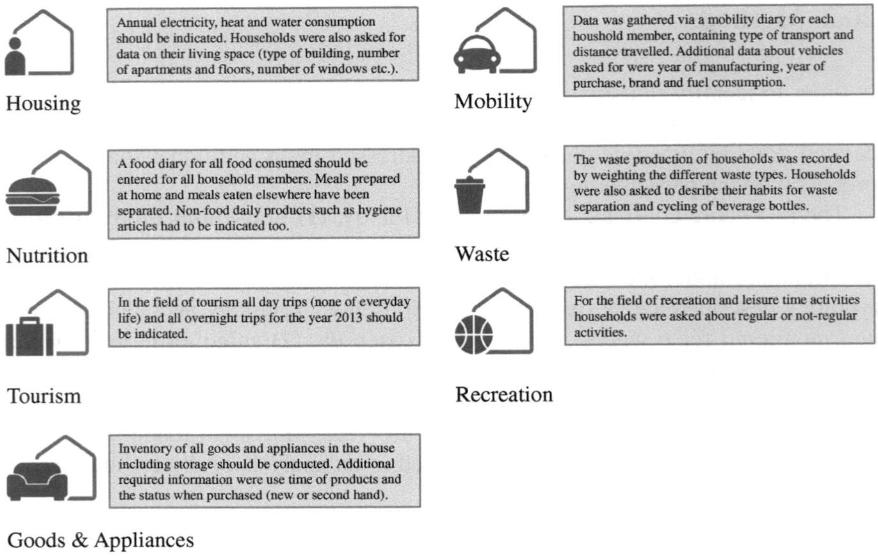

Fig. 20.1 Questionnaires and their content

20.2.2 Step 2: Calculation and Evaluation of the Material and Carbon Footprint

The calculations of resource consumption and greenhouse gas (GHG) emissions of households are based on Life Cycle Assessment (LCA) and Material Flow Accounting (MFA) methods (Fischer-Kowalski et al. 2011; ISO 1997). Both, LCA and MFA, usually cover all material and energy flows of a system from cradle to grave, from extraction of materials to the final treatment for end of Life (EoL). The Carbon Footprint indicator is drawn from characterisation factors, while the Material Footprint sums up all abiotic and biotic material flows in the system.

The Material Footprint (MF) indicator is based on the MFA method MIPS, which stands for Material Input per Service (Liedtke et al. 2014). In contrast to other methods for resource accounting, MIPS includes all material flows of a system and also accounts for resources which have not been put to a direct economic use, the so-called unused extraction. The MF sums up the categories abiotic and biotic raw materials and topsoil erosion in agriculture. In this study the categories abiotic and biotic raw materials were calculated for the MF.

The Carbon Footprint (CF) is defined as the total set of greenhouse gas emissions caused by a system. Since total greenhouse gas emissions would require a large amount of data, one usually refers to the global warming potential for 100 years (GWP 100a) of all relevant sources, sinks and storage options in a defined system. In order to calculate the CF in a LCA, characterisation factors of GWP 100a are used, which are provided by the Intergovernmental Panel on Climate Change (IPCC, Forster et al. 2007).

The calculation of Material Footprints (MFs) and Carbon Footprints (CFs) were based on a database for Life Cycle Inventories (Ecoinvent version 2.2, Hedemann and König 2007) which provides generic data for the production of basic materials, provision of food and feedstuff products as well as average data for transport and energy systems. The upstream and hidden flows had to be modelled to the purpose of the study. This so called functional unit (LCA) or service unit (MFA) can be considered the main metric to which every resource consumption and GHG emission is related. In SuslabNRW the unit for relation was 1 year/ person or household (per a*P or a*HH), as chosen in previous studies. While the mobility and housing questionnaires already included data on the per year consumption, other activities had to be scaled up to 1 year, such as the amount of waste and food per week or frequent recreation activities.[4]

Modelling was conducted with help of a software for Life Cycle Assessment (OpenLCA, Ciroth 2007). For both MF and CF calculations the inventory data was the same.

In the course of the project some simplifications and appraisements were required, due to the overall extent and complexity of the study, the limited information from the questionnaire and the need for consistent calculation rules. Simplifications and appraisements were also necessary wherever data availability was low or data for more comprehensive calculations was missing in the questionnaires. Due to this, the results reflect differences between households based on comparable inputs, while some individual choices would result in different absolute values, if processed in a household-specific LCA or MFA. Therefore, the calculations do not represent the original households of the study, but households with certain comparable activity metrics such as: distance to work, type of vehicle, number of goods and appliances like telephones, size of living space, amount of biotic garbage etc.

In addition, there are differences regarding the calculation of each field of household activity. A comprehensive household system would be very complex und cannot be accounted completely (at least within reason), especially considering the amount of inputs it would require by the participating households. Some activities also intervene and interrelate to a certain degree. In order to distinguish the fields of activity from each other and to prevent items from being counted twice, system boundaries were drawn for each field (removing any overlaps). The general system boundaries of each field and their most important capturing rules are listed in Table 20.1.

To the knowledge of the authors, this capture of household's natural resource consumption is one of the most comprehensive in literature. With exception of work, it covers all important ways a household can directly influence its resource

[4]Whenever the input data was inconsistent or questionable, households were asked to verify certain inputs in the questionnaire.

Table 20.1 System boundaries and capturing rules

Activity	Sub-categories	System border	Specific calculation rules
Goods and appliances	• Electrical and electronic appliances • Books, newspapers and magazines • Clothing • Home textiles • Furniture • Large domestic appliances • Tools • Toys and leisure time equipment • Kitchenware	Cradle-to-gate: from material extraction to production of goods; no use phase with exception of use time in years; no transports and no EoL treatment	Database and literature based where ever possible; average goods where ever possible; one product type each; assembly and material losses often based on average industrial data
Mobility	• Car driving and production • Bicycle driving and production • Bus driving • Train driving • Driving of subways and similar vehicles • Walking	Full service life cycle including infrastructure; no inclusion of rides with other persons or in taxis (if not explicitly stated by household)	Average vehicles, scaled according to the given weight and fuel consumption; production of every vehicle is attributed to every household member accordingly; no impacts for walking
Waste	• No sub-categories	Production of accumulated waste and transports to nearest collection point with an average distance; consideration of returnable bottles, but no consideration of non-frequent waste (e.g. due to building renovations)	Extraction of secondary materials (recycling) according to German average only if household separates the specific type of waste; returnable bottles based on average number of returns of bottle types
Nutrition	• Meat and fish • Fruit and vegetables • Dairy products and eggs • Sweets, sugar and oil • Cereal and potato products • Water, soft drinks and juice • Coffee and tea • Alcoholic beverages • Instant meals and non-domestic meals • Disposable goods and sanitary products • Transports	Cradle-to-gate: from extraction to production, wherever possible, as well as transports from point of origin to Germany; no inclusion of cooking	Database and literature based wherever possible

(continued)

Table 20.1 (continued)

Activity	Sub-categories	System border	Specific calculation rules
Tourism	• Day trips • Overnight stays • Mobility to and at holiday location	Full service life cycle including infrastructure for all journeys; rough estimation of overnight stays in 6 categories (stays in private homes included)	Travels cannot be completely separated from mobility; business trips are included in some cases
Recreation	• Entertainment • Activities • Others	Building and energy use for entertainment and activities; no production of equipment	Average buildings and their energy use; some buildings have been aggregated to accommodate different activities
Housing and Energy	• Building • Estate • Electricity • Heating • Water	Full service life cycle	Average buildings in up to three classes; two different electricity mixes (conventional and eco mix)
Pet food	• No subcategories	Cradle-to-gate: from extraction to production as well as transports from point of origin to Germany	Database and literature based wherever possible

consumption and GHG emissions. However, there are clearly uncertainties in the results, which are not easily quantifiable. For one, every included database process is, for some or even most parts, based on secondary literature with varying data quality and researcher assumptions. In addition, matching these processes to activities results in matching-errors: Every real product or service a household uses differs from a hypothetical average to some unknown degree, weather on a scale of time (year of production), space (production of parts in different countries) or the level of aggregation (one model of one product can be considered unique in comparison to an average product type). Even within simplified system boundaries for some cases, the researchers of this study had to make assumptions. Against this background, one should be careful to compare the results to other studies. System boundaries would have to be matched and LCI data to be drawn from the same sources (in this case the database ecoinvent). It would be possible though, to calculate comparable Material and Carbon Footprints based on questionnaire inputs by other studies.

However, when comparing the households to each other, these uncertainties are less relevant with exception of the researcher's interpretation of participant inputs. Given the limited relevance of the uncertainties, the authors claim to have calculated a best available estimation of Material and Carbon Footprints of households,

Table 20.2 Data availability, data quality and uncertainty of calculations

Activity	Data availability and quality	Uncertainty (absolute scale)
Goods and appliances	Low	High
Mobility	High	Low
Waste	High	Mediocre
Nutrition	Low	High
Tourism	Mediocre	Low
Recreation	Low	Mediocre
Housing and energy	High	Low
Pet food	Low	High

suitable for comparison between these households. Since the uncertainties are not quantifiable, the authors judged the quality of each activity calculation on a three level scale, using mobility as reference: low, medium and high. According to Table 20.2 high quality and low uncertainty is attributed to mobility and housing, while goods and appliances, nutrition (including pet food) and recreation are subject to low data availability and high uncertainties in the results. The latter in particular was difficult to capture, as there is a high variety of possible recreation activities and low data availability in general. Some are part of a manageable system such as reading or walking, while others would require the consideration of many subsystems in order to calculate the complete Footprints (golfing for example).

20.2.3 Step 3: Road Mapping Process

In order to reflect the results with the households, a workshop was conducted. Seven of the 16 households attended this workshop. After discussing the results, possible lifestyle changes for reducing the Material Footprint were discussed. Assisted by the project team, the households were asked to choose from given options to reduce their Material and Carbon Footprint or to develop new options. While most of the given options were behavioural changes, some addressed consumption patterns. This basis enabled the households to create individual roadmaps towards lower environmental pressure in the fields of housing, mobility, nutrition and goods and appliances. The roadmaps were structured into short (6 month), mid (6–18 month), and long time (over 18 month) periods.

Another workshop took place in September 2015. Just three households participated. Its aim was be to elicit and understand the households' experiences with efforts to reduce the Carbon and Material Footprint. This workshop took the form of a focus groups discussion that sought to identify drivers and barriers with regard to the fields of activity. We define drivers as specific factors leading to reduced resource consumption and/or GHG emissions. Barriers are defined as factors leading to counterproductive behavioural change and thus to no reduction or even increase of resource consumption and/or GHG emissions.

20.3 Results and Discussion

Figures 20.2 and 20.3 show the results of the study per year and person. The resource consumption ranges from 16.2 to 59.2 t of resources (factor 3.7). The lowest calculated Carbon Footprint is calculated at 5.2 t and the highest at 17.2 t (factor 3.3). On average of the sample, one person per year consumes 29.8 t of resources and emits 9.0 t of GHG emissions.

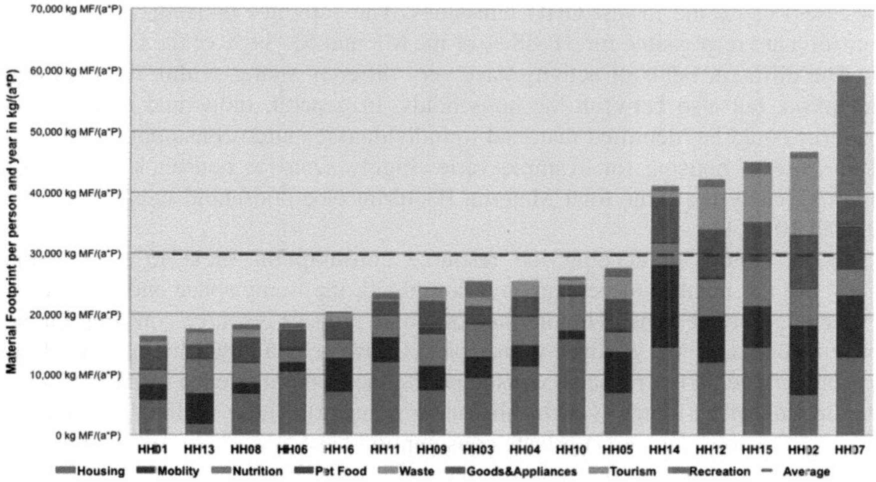

Fig. 20.2 Material Footprint of analysed households in increasing order

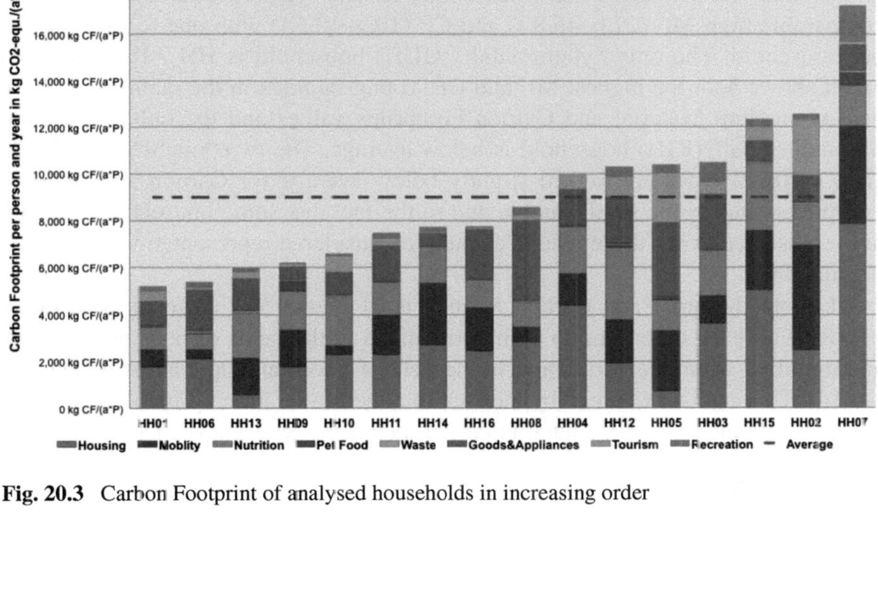

Fig. 20.3 Carbon Footprint of analysed households in increasing order

The activities of household 7 have the highest and those of household 1 the lowest impact. In a ranking of the fields of activity, housing and energy dominates the Material Footprint in 13 of 16 cases, while in three households nutrition (HH 13), tourism (HH 2) or recreation (HH 7) dominate. The two remaining households (4 and 9) have no or a very low resource consumption for recreation. Waste and pet food have the lowest impact in 14 cases. Regarding the Carbon Footprint, housing and energy dominate in 11 households, nutrition in two (12 and 13) and goods and appliance also in two (5 and 8). Pet food (12 cases) and waste (4 cases) cause the lowest GHG emissions. The activities housing, mobility and nutrition are responsible for 51–88 % of the MF and 53–89 % of the CF.

The different fields of activity have very different shares within the total consumption, but also between the households. In general, individual consumption patterns could be identified that lead to individual resource consumption patterns. The share of housing for example varies highly even for households that show approximately the same total Material Footprint (see household numbers 03, 04, 05, 09, 10 an 11).

Other factors influencing the resource consumption (next to social practises) are the number of persons in a household, the living space and income. As expected, more persons living in one household reduces resource consumption per person compared to e.g. single households. Conversely, a larger living space leads to comparable higher resource consumption, due to the influence of buildings and the heating of buildings (which falls under housing). These findings are not surprising. Previous studies on private consumption have also shown these influences for example for energy consumption (see e.g. Longhi 2015).

The households were further classified into different household types (see Chap. 19. Household Typology and social milieus), listed in Table 20.3. The factor of greatest influence—number of persons in the household—is also indicated in Table 20.3. Most households belong to the categories "young families in the mainstream" (JFM) and "established empty nesters" (ETH). The latter exhibits a comparable high MF (27.6–46.8 t) and CF (10.3–12.5 t) with one to two persons per household. The only "young adult" (JUH) household is HH 7 (two persons) and it shows both the highest MF and CF. Young families in the mainstream have low or medium Material and Carbon Footprints values and the only "cosmopolite-intellectual" (KIH) household is below average. The two "mainstream/middle class" (HBM) types are ranked slightly below average for Carbon and Material Footprints. Due to the small sample and to the fact that some household types are not represented at all, these results cannot be considered representative of German population.

The total Material and Carbon Footprints of households correlate to a factor of 0.84 with each other, due to high correlations in the fields of housing, nutrition and mobility, while the correlation in the field of goods and appliances is negative (–0.14). This can be explained by the fact, that the Material Footprint is more sensitive to small amounts of resource intensive metals in electronics (such as gold and silver), while Carbon Footprint mainly accounts for emissions during processing, which do not increase (per g in the product) for precious metals on the same scale.

Table 20.3 Household types and total material Footprint (increasing MF order)

Household number	Household type abbreviation	Household Type	Total MF per person and year (kg/(a*P))	Total CF per person and year (kg/(a*P))
HH01	JFM	Young families in the mainstream	16,219	5,173
HH13	JFM	Young families in the mainstream	17,500	5,967
HH08	JFM	Young families in the mainstream	18,193	8,543
HH06	JFM	Young families in the mainstream	18,379	5,382
HH16	KIH	cosmopolitan-intellectual	20,303	7,731
HH11	N.N.	N.N.	23,424	7,452
HH09	HBM	Main stream/ middle Class	24,303	6,173
HH03	JFM	Young families in the mainstream	25,402	10,470
HH04	JFM	Young families in the mainstream	25,458	9,977
HH10	N.N.	N.N.	26,145	6,576
HH05	ETH	Established empty nesters	27,595	10,372
HH14	HBM	Main stream/ middle Class	41,148	7,690
HH12	ETH	Established empty nesters	42,221	10,299
HH15	ETH	Established empty nesters	45,114	12,316
HH02	ETH	Established empty nesters	46,770	12,540
HH07	JUH	Young adults	59,206	17,242

20.3.1 Focus Heating: Comparison Within All Fields of Activity

A significant share of resource consumption and GHG emissions can be attributed to housing. This field of activity consists of the construction of buildings, water use and electricity for goods as well as heating. Heating in the sample amounts to 7 % of the resource consumption on average, but to 20 % of the GHG emissions in the sample. High heating demand, as well as the use of heating sources with high specific GHG emissions (for example oil) result in higher Carbon Footprints.

Most of the households in the study use gas for heating and live in detached houses or apartments. Although factors like living space, number of persons and the type of energy source clearly affect the resulting Material and Carbon

Footprint of heating, it can be assumed that the households also differ highly in their heating behaviour. To illustrate this point, five households with a similar overall Material Footprint are compared to each other (see Figs. 20.4 and 20.5). Four of these households use gas for heating (HH 11, HH 9, HH 3 and HH 4) and one (in addition to electricity) wood pellets (HH 5). While housing and energy contributes to the overall Material Footprint with 26–52 % (the black line), heating is responsible for 3–13 % of the overall resource consumption. The resource consumption for heating of HH 4, consisting of three persons and living in an apartment from 2002, is 4.6 times higher than the Material Footprint of HH 11 with five persons living in a detached house from 1919. This considerable difference cannot be attributed solely to the economy of scale and the quality of heating systems and insulation, but are an indication for differences in heating behaviour as well.

In contrast to the Material Footprint, the Carbon Footprints for heating vary more and in a slightly changed order. While in HH 4 heating has a share of 25 %, this is only 2 % in HH 5. HH 5, using wood pellets for heating, benefits from the fact that the Carbon Footprint discounts CO_2 emissions bound in plants, while the Material Footprint accounts for the long time removal (extraction) of biotic resources from nature.

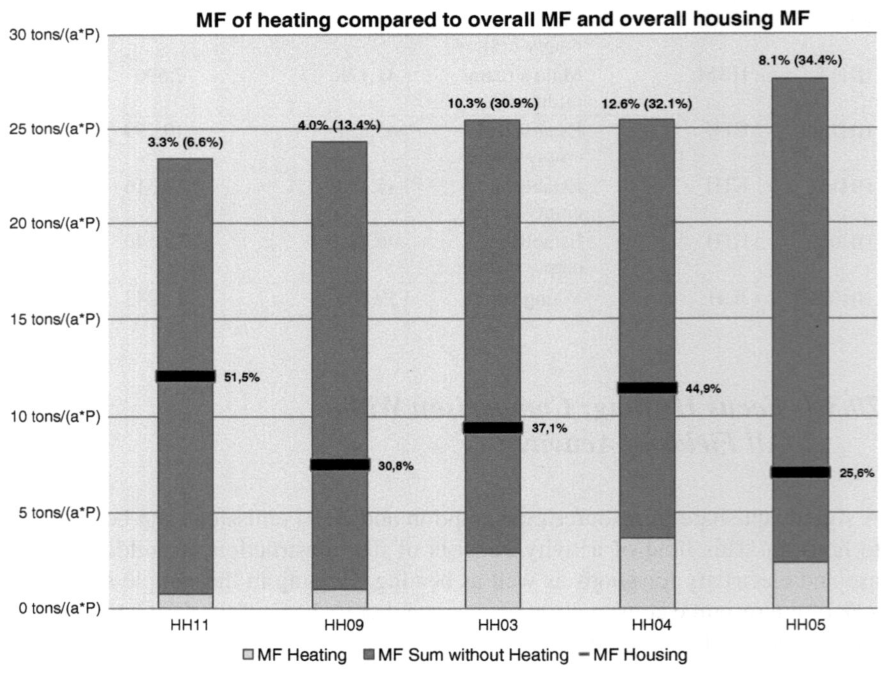

Fig. 20.4 MF of heating compared to overall MF. Share of heating in overall MF and in field of housing stated above column

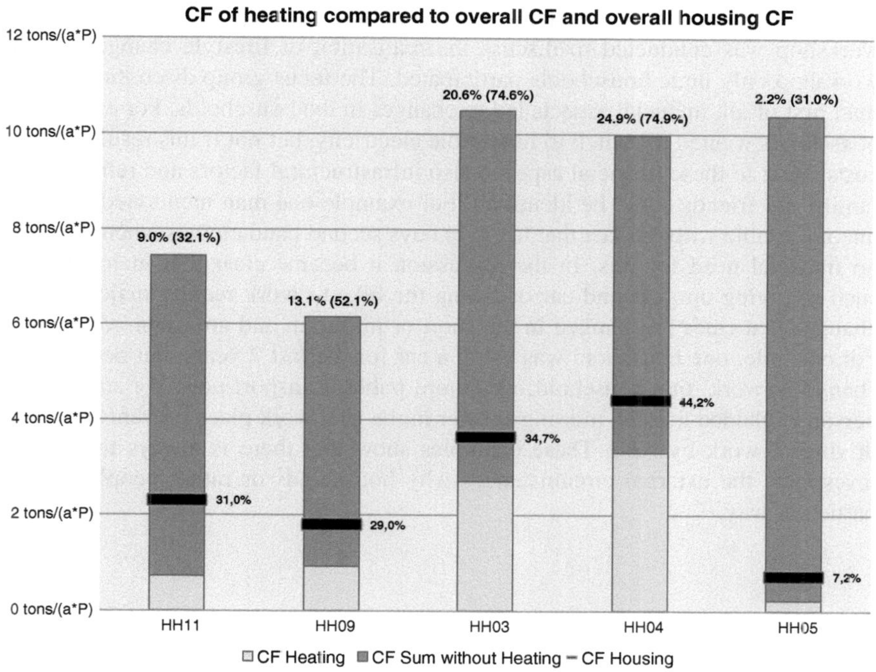

Fig. 20.5 CF of heating compared to overall CF. Share of heating in overall CF and in field of housing stated above column

20.3.2 Roadmaps, Drivers and Barriers for Behavioural Changes

The results of the Road Mapping Process revealed a high affinity of the volunteers for lifestyle changes. Some households told the researchers about changes they had already made during the analysis period e.g. going by bike instead of by car, combining trips by car and offering others a lift. In the case of mobility nearly all households wanted to try to use bicycles in the summer or public transport instead of a car and try to combine trips by car (like shopping and trip to work) in the short term. Two households wanted to try car sharing. In the long term two households wanted make do with just one car instead of two. In the case of housing, most households wanted to change their airing (ventilation) behaviour in the short term. In the mid term three households wanted to switch to renewable electricity. In the long term two households wanted to change their living space and two wanted to buy devices/appliances that are more energy efficient. In the case of nutrition nearly all households wanted to try to buy more organic, regional, seasonal or fair-trade food in the short term. Two households wanted to try a vegetarian and vegan lifestyle and three wanted to drink more tap water. Two households wanted to try the last two changes in the long term. In September 2015 a second workshop was conducted to discuss the realisation of lifestyle changes. In this workshop only three households participated. The focus group discussion showed that, first of all, financial aspects led to changes in the households. For example all households wanted to switch to renewable electricity, but not if this results in extra costs. Next to these financial aspects, also infrastructural factors and influences by family and friends could be identified. For example one man mentioned that he is uncomfortable with the fact that his wife buys second hand clothes when they have no financial need for this. In the discussion it became clear that major changes such as giving up a second car or taking the bike to work require major lifestyle changes that can't be changed in the short or mid term and are often not planned. For example, one household was using a car for the last 2 years, but because of a change in work, this household uses more public transport now. By contrast one person explained that the missing shower in the new work place prevents him from driving to work by bike. These examples show that there is always the need to investigate the external circumstances why households or rather people act in a particular way.

20.4 Conclusions

Households' natural resource consumption (n = 16) and GHG emissions were calculated for seven fields of activity to analyse influencing factors, especially the impact of social practises. From this, the Material Footprint (MF) and Carbon Footprint (MF) have been calculated. Results of resource and GHG profiles show a large range of 3.7 (MF) and 3.3 (CF) factor differences within the volunteering households. In comparison, the results of a previous finish household study with 27 volunteering households of all kinds of household types showed a range of factor 9. A similarly large diversity of lifestyles was identified by Groezinger et al. (2013) between the participants and their material footprints, ranging from 8.5 to 69 t/person/year (factor 8). Due to the high similarity of households in this study, it can be assumed that a comparative analysis of the complete German society would result in higher ranges than 3.7 and 3.3. The recruiting method (newsletter by Innovation City Bottrop) focused on one single area (the city Bottrop) and presumably attracted households with high interest in environmental issues.

Further results show that next to technical options, there is a high potential for structural changes and social innovations not only in the fields of housing, mobility and nutrition, but also in the field of goods and appliances. The shares of fields of activity are very different between the households. This findings lead to the assumption that behaviour by individual persons in households is one deciding factor for different resource consumption and GHG emissions. Regarding household types no significant correlation with resource consumption could be found. There is, however, a tendency of higher resource consumption in 'empty nester' households (senior citizens with children having already moved out) in comparison to young families belonging to the mainstream.

The road mapping process showed that the possibilities for households to change their resource use are limited within the options of the urban environment. As described previously in a number of studies it is difficult for households to act alone (Geels et al. 2015; Hasselkuß 2013; Walker 2015). In addition to changes at the consumer level, the political and business levels must also be integrated into considerations of lifestyle changes.

With this holistic assessment of household consumption, an analysis was attempted of the factors which most influenced the resource consumption and GHG emissions of households. It can be shown that individual behaviour and social practises in households have a high impact. The individual data shows whether there is a greater impact for example in food consumption or from mobility. For further evaluation it would be necessary to question volunteers about the circumstances that lead to such social practises, which was not done in this study. Although an analysis was attempted of these circumstances and routines within road mapping workshops with a subset of households, for further investigation such analysis should be integrated into the questionnaire. This is also tried within the resource calculator that has been developed on this resource analysis approach (www.ressourcen-rechner.de). Up to now over 4,000 data sets ("resource profiles") have been gathered. Currently the calculator is available in German only. An English version is in planning stage.

With this resource analysis a possible holistic approach has been tested to be included in the toolkit of sustainable Living Labs. It was possible to describe influencing factors in the different fields of activity. Due to this analysis, the field of heating can be placed within the whole household's resource consumption, and it can be measured what impact a reduced demand in one field might have on different fields of activity. It has been shown that, while some activities dominate the environmental impact, all activities should be taken under consideration. This is of special interest for developers of sustainable Product-Service-Systems (PSS), which could become an important part of LivingLab research. Since this study also revealed that the described methodology is limited in capturing the practises behind household consumption, one could complement it with other methods such as time use analysis. This would allow for a facilitated and more informed development of transformation supporting solutions within the user-integrated LivingLab approach.

References

Bringezu, S., & Bleischwitz, R. (2009). *Sustainable resource management: global trends, visions and policies*. Greenleaf Publishing.

Brookes, L. (1990). The greenhouse effect: The fallacies in the energy efficiency solution. *Energy Policy, 18*(2), 199–201. doi:10.1016/0301-4215(90)90145-T.

Buhl, J. (2014). Revisiting rebound effects from material resource use. indications for Germany considering social heterogeneity. *Resources, 3*(1), 106–122. doi:10.3390/resources3010106.

Buhl, J., & Acosta, J. (2015). Work less, do less?: Working time reductions and rebound effects. *Sustainability Science*,. doi:10.1007/s11625-015-0322-8.
Ciroth, A. (2007). ICT for environment in life cycle applications openLCA—A new open source software for life cycle assessment. *The International Journal of Life Cycle Assessment, 12*(4), 209–210.
Fischer-Kowalski, M., Krausmann, F., Giljum, S., Lutter, S., Mayer, A., Bringezu, S., et al. (2011). Methodology and indicators of economy-wide material flow accounting: state of the art and reliability across sources. *Journal of Industrial Ecology, 15*(6), 855–876. doi:10.1111/j.1530-9290.2011.00366.x.
Forster, P., Ramaswamy, V., Artaxo, P., Berntsen, T., Betts, R., & Fahey, D. W., et al., (2007). Changes in atmospheric constituents and in radiative forcing. Chapter 2. *Climate Change 2007. The Physical Science Basis*.
Geels, F. W., McMeekin, A., Mylan, J., & Southerton, D. (2015). A critical appraisal of Sustainable Consumption and Production research: The reformist, revolutionary and reconfiguration positions. *Global Environmental Change, 34*, 1–12. doi:10.1016/j.gloenvcha.2015.04.013.
Giljum, S., Dittrich, M., Lieber, M., & Lutter, S. (2014). Global patterns of material flows and their socio-economic and environmental implications: A MFA study on All countries world-wide from 1980 to 2009. *Resources, 3*(1), 319–339. doi:10.3390/resources3010319.
Groezinger, R., Verbree, R., Hicks, C., & Ritola, M. (2013). *Scenarios for sustainable lifestyles* (pp. 41–46). Pathways toward a positive future, Ökom: Ökologisches Wirtschaften, Munich.
Hasselkuß, M. (2013). *Transformation towards sustainable consumption: Changing consumption patterns through meaning in social practices*. Paper for Presentation. Zurich.
Hedemann, J., & König, U. (2007). Technical documentation of the ecoinvent database. Final report ecoinvent data v2. 0, No. 4. *Swiss Centre for Life Cycle Inventories, Dübendorf, Switzerland*.
Huysman, S., Sala, S., Mancini, L., Ardente, F., Alvarenga, R. A. F., De Meester, S., et al. (2015). Toward a systematized framework for resource efficiency indicators. *Resources, Conservation and Recycling, 95*, 68–76. doi:10.1016/j.resconrec.2014.10.014.
ISO. (1997). *Environmental Management: Life Cycle Assessment: Principles and Framework* (Vol. 14040). ISO.
Kotakorpi, E., Lähteenoja, S., & Lettenmeier, M. (2008). *Household MIPS - Natural resource consumption of Finnish households and its reduction*. http://hdl.handle.net/10138/38369.
Kuijer, L., & De Jong, A. M. (2011). Practice theory and human-centered design: A sustainable bathing example. *Nordes*, (4).
Lettenmeier, M., & Wuppertal Institut für Klima, Umwelt, Energie (Eds.). (2009). *Resource productivity in 7 steps: how to develop eco-innovative products and services and improve their material footprint*. Wuppertal: Wuppertal Inst. for Climate, Environment and Energy.
Lettenmeier, M., Hirvilammi, T., Laakso, S., Lähteenoja, S., & Aalto, K. (2012). Material footprint of low-income households in Finland—Consequences for the sustainability Debate. *Sustainability, 4*(12), 1426–1447. doi:10.3390/su4071426.
Lettenmeier, M., Liedtke, C., & Rohn, H. (2014). Eight tons of material footprint—Suggestion for a resource cap for household consumption in Finland. *Resources, 3*(3), 488–515. doi:10.3390/resources3030488.
Liedtke, C., Jolanta Welfens, M., Rohn, H., & Nordmann, J. (2012). LIVING LAB: User-driven innovation for sustainability. *International Journal of Sustainability in Higher Education, 13*(2), 106–118. doi:10.1108/14676371211211809.
Liedtke, C., Bienge, K., Wiesen, K., Teubler, J., Greiff, K., Lettenmeier, M., & Rohn, H. (2014). Resource Use in the production and consumption system—The MIPS approach. *Resources, 3*(3), 544–574. doi:10.3390/resources3030544.
Liedtke, C., Baedeker, C., Hasselkuß, M., Rohn, H., & Grinewitschus, V. (2015). User-integrated innovation in Sustainable LivingLabs: An experimental infrastructure for researching and developing sustainable product service systems. *Journal of Cleaner Production, 97*, 106–116. doi:10.1016/j.jclepro.2014.04.070.

Longhi, S. (2015). Residential energy expenditures and the relevance of changes in household circumstances. *Energy Economics, 49*, 440–450. doi:10.1016/j.eneco.2015.03.018.

Reckwitz, A. (2002). Toward a theory of social practices: A development in culturalist theorizing. *European Journal of Social Theory, 5*(2), 243–263. doi:10.1177/13684310222225432.

Rockström, J., Steffen, W., Noone, K., Persson, Å., Chapin, F. S., Lambin, E. F., et al. (2009). A safe operating space for humanity. *Nature, 461*(7263), 472–475. doi:10.1038/461472a.

Schmidt-Bleek, F. (2009). *The earth: Natural resources and human intervention*. Haus Pub.

Schmidt-Bleek, F. (2013). *Wieviel Umwelt braucht der Mensch?: MIPS—das Maß für ökologisches Wirtschaften*. Springer.

Walker, G. (2015). Beyond individual responsibility: Social practice, capabilities and the right to environmentally sustainable ways of living. In Y. Strengers & C. Maller (Eds.), *Social practices*. Beyond behaviour change. Routledge: Intervention and Sustainability.

Ward, K., & Neumann, F. (2012). *Consumer in 2050: The Rise of the EM Middle Class*. Global Economics, Oct: HSBC Global Research.

Watson, D., Acosta Fernández, J., Wittmer, D., Gravgård Pedersen, O., & European Environment Agency. (2012). *Environmental pressures from European consumption and production a study in integrated environmental and economic analysis*. Luxembourg: Publications Office. Accessed November 9, 2015 from http://dx.publications.europa.eu/10.2800/70634.

Chapter 21
FoodWatch and Food Resource Flows

Jesper Knutsson

Abstract Food consumption represents a significant environmental impact, and in terms of climate impact, food consumption ranks among the top three contributing sectors. By changing dietary choices and reducing food waste, significant reductions in food-derived GHG emissions can be achieved. However, food consumption behavior depends on several interrelated factors, some of which have proved difficult to influence. Further research and new approaches in data collection and intervention design are needed to identify effective strategies. Here, a web-based tool for collecting highly disaggregated data on household food consumption and food waste behavior, called Food Watch, is presented. I present the results of an early version field trial, with detailed analysis of food waste categories and a discussion about intervention and feedback design. A roadmap for future research and development of the Food Watch application is also presented.

Keywords Sustainable food consumption · Consumer behavior · Ecological footprint · Food waste reduction

21.1 Introduction

Addressing food consumption and management behavior is of great interest for several reasons, one being food security and resource conservation; to be able to feed a growing global population in a secure and sustainable way, adjustments to wasteful practices has to be pursued (Finn 2014; Kahiluoto et al. 2014). Furthermore, the reduction of the climate impact of food consumption in the industrialized countries will be invaluable to meet the policy goals set by the European Union.

J. Knutsson (✉)
Chalmers University of Technology, Civil and Environmental Engineering,
Sven Hultins Gata 8, 41296 Gothenburg, Sweden
e-mail: jesper.knutsson@chalmers.se

D.V. Keyson et al. (eds.), *Living Labs*, DOI 10.1007/978-3-319-33527-8_21

Consumption of food and beverages contribute significantly to the overall CO_2 footprint of private households. It has been shown that there a considerable potential for reduction of the CO_2 intensity by addressing changes in dietary choices, reduction of overall caloric intake and reducing the amount of edible waste generated. Earlier studies have shown that the process of planning, purchasing, cooking and consuming meals all influence consumption patterns of food. It is a challenge to influence attitudes, knowledge and awareness in a way so that food consumption habits are persistently geared towards less CO_2 intensive alternatives (Quested et al. 2013).

21.2 Ecological Impact of Food Consumption

Household demand is the primary driver for overall consumption, and it is being argued that in order to reduce the resource use intensity in society, efforts are best directed towards household behavior. Household food consumption contributes contributed roughly 12 % to the overall carbon dioxide emissions from private consumption in Sweden 2002 (Carlsson et al. 2006), in UK the corresponding number is 19 % in 2009 (Berners-Lee et al. 2012). This makes household food consumption one of the three most important economy sectors together with petroleum products and energy in terms of climate impact.

It has been estimated that by conservative changes in dietary choices a 22–47 % reduction of GHG emissions derived from food consumption is possible (Berners-Lee et al. 2012), with other options, like reducing overall caloric intake and reducing the amount of food waste, not being considered (Scarborough et al. 2014). Such a reduction would correspond to a 5 % reduction of overall CO_2 emissions from private consumption. While this may seem a modest reduction, it is suggested that target CO_2 reductions are unlikely to be achieved by any single measure, or in other words, there no 'silver bullet' when it comes to GHG mitigation, but rather we expect success to come from many incremental reductions in different areas.

Food waste reduction potential in western economies are significant, especially when compared to benchmark number in developing economies and regions with non-western food culture. Average food waste per person in western type economies ranges from 138 kg per capita and year (UK) and 78 kg in EU 27, with substantial national differences (EUSTAT).

21.2.1 Pathways to Sustainable Food Consumption

Recent research shows that the food consumption of the developed countries, and especially western countries, is both excessive in amount and sub-optimal in composition (diet) (Carlsson-Kanyama 1998; Carlsson-Kanyama and González 2009; Reynolds et al. 2014; Scarborough et al. 2014). Furthermore, there is a lot of waste in both supply and retail sectors, as well as on the consumer side (Berners-Lee et al. 2012). Thus there is a clear potential to decrease the environmental impact, as well as the detrimental health effects coupled with western style diets. This is important also for secondary effects as living standards rise and consumption patterns spread to developing economies as well.

21.2.2 Sustainable Diets and Waste Reduction

There has been some research performed on food waste behavior (Finn 2014; Graham-Rowe et al. 2014; Quested et al. 2013), but the main body of this work has focused on what food is being wasted and attitudes to food waste, rather than the reason that the food waste is generated. Food is an integrated part of our cultural identity, and as such food is not just another commodity. People generally attach a great deal of emotional value to food consumption, apart from being a mere nutritional provision it is also contributing to overall experience of well-being and personal identity. Especially behavior and actions leading to the generation of food waste and its outcome is characterized by separation in time and a conceptual disconnect, making it difficult to predict food waste by using traditional behavioral models, such as Triandis' theory of interpersonal behavior (Triandis 1977) and the Theory of planned behavior (Ajzen 1991).

However, using components from such models can still give useful insight into factors that should be considered with regards to food waste and its prevention. Suggested key factors influencing consumer choice include *habits, knowledge, facilities, attitudes and socio-cultural pressure* (Quested et al. 2013).

The implementations of interventions in the current version of the Food Watch application was not designed to target specific factors separately, but it was assumed that the feedback presented to the participants could affect at least their knowledge and attitude on diet and waste.

Food product flows through homes have been studied over an extended time period. The products are scanned through their bar codes and the input-output considered in terms of ecological and water footprints as well as land use. An application, Food Watch, has been developed from the database and is currently being adjusted for use in the HSB Living Lab. The Food Watch application allow different modes of feedback interventions to be presented to the user, and their influence on behavior and choices to be assessed.

21.3 Method

A novel approach has recently been suggested which combines a high level of disaggregation and automation with a data management system, called Food Watch (Harder et al. 2014). The Food Watch web application stores data about household purchases and solid waste flow through the scanning of product barcodes which are present on the vast majority of the food item packages consumed in industrial economies. In Food Watch products are linked to life-cycle analysis (LCA) data

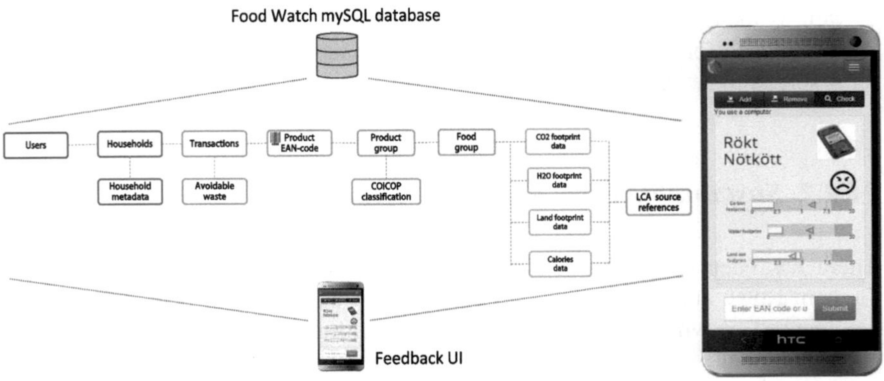

Fig. 21.1 Schematic description of the main database tables and their relations, and a illustration of the user interface shown on a handheld device

on carbon, water and land use footprint, and through the unambiguous identification of products it is also possible to present an immediate and detailed feedback on the ecological impact of the food consumption of the household.

However, there are several challenges in creating a database to meet the requirements of the Food Watch application. Importantly, the database needs to be comprehensive enough to cover a majority (for example >90 %) of regular purchases by any given household. Considering the number of available food products on the market this is no trivial requirement. Secondly there need to be LCA data connected to each product in the database, and the data have to be relevant to the context where it is presented. Third, when associating data with tangible items, some degree of abstraction will be necessary, and in order to be able to assess data the level of abstraction and simplification need to accompany the data.

The overall design of the Food Watch database is shown in Fig. 21.1. Products are linked to LCA and ecological foot print data through a product group, which is based on the UN's Classification of Individual Consumption According to Purpose (COICOP), and then through a food group, which is the final link to the LCA data. A certain food group may be linked to several sources of LCA data, which enable composite data to be used in the application, i.e. multiple data sources associated with a single product and parameter.

This design approach requires products to be classified and assigned to a food group, and subsequently LCA data associated with that food group. Both the association of LCA data and the products with the food group infer an abstraction, or a certain level of mismatch between the product and the original LCA data. As such the footprint data generated by the application should be regarded as indicative. There are two important comments to be made here, one being that from a research perspective footprint data should not be considered at face value, and secondly that for the purpose of presenting eco feedback and intervention design, the footprint data relevance is 'good enough'.

{"ean":"7310500026039","name":"Babymor\u00f6tter","amount":"0.500","unit":"kg","carbon_footprint_per_product":"0.2775000"}

Fig. 21.2 Sample output of the Food Watch API, detailing the carbon footprint of the product 0.500 kg "Baby carrots". The output is returned in json format and can be fetched programmatically through HTTP-calls to the API endpoint

21.3.1 Open Access and Innovation

To promote innovation and co-creation around the Food Watch platform, an API (advanced programming interface) was developed, to provide an access point to the product and LCA data for use in third party applications. Access to the API is free, but at the moment of writing, an application key is required. A sample output of the API can be seen in Fig. 21.2.

Further information about the API can be obtained from the author, or by visiting the Food Watch website (www.food-watch.se), looking for the API menu option.

21.3.2 Intervention and Eco Feedback

Basic methodology calls for the possibility to gather baseline data, and the subsequent selective introduction of intervention and feedback. The interventions and feedback of the design iteration described here were inspired by the decision model described by Quested et al. (2013), including the decision process in the planning and purchase stages, but also addressing the practices in storing, cooking and discarding (or not) food items. The intervention targeting the decision in the purchasing situation (a) is shown in Fig. 21.3. This function is intended to be

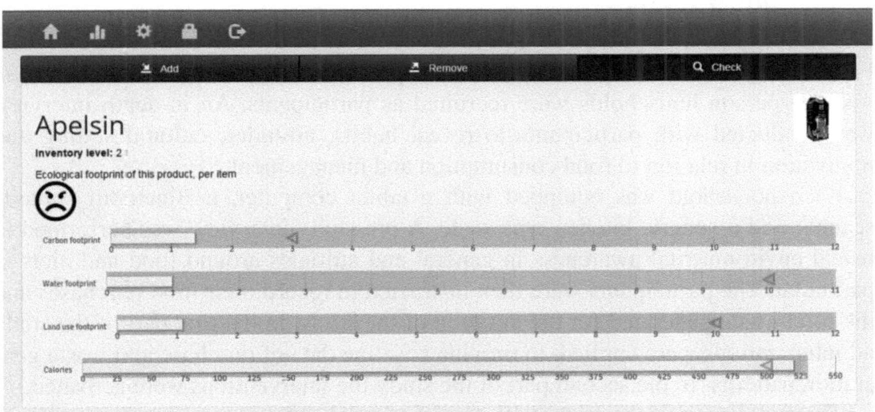

Fig. 21.3 Excerpt from the FoodWatch application showing the visualisation of the LCA footprint data associated with a sample product of orange juice

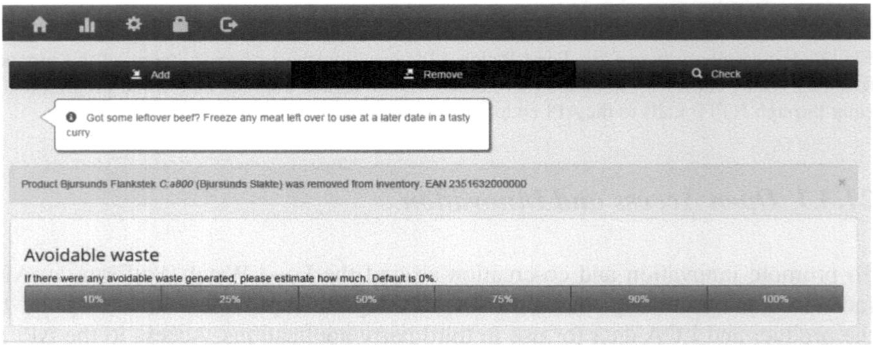

Fig. 21.4 Excerpt from the FoodWatch application showing an intervention in the form of a tip of what can be done with leftover beef

accessed using a hand held device, such a mobile phone, to quickly get information on inventory level (how much do I have at home?), ecological impacts in the form of CO_2 footprint, water footprint and land use footprint. The climate impact of the product is abstracted and presented in the form of an "emoticon", showing an increasingly discontent face the higher the CO_2 footprint/calories ratio of the product.

The other main type of intervention consist of advice on practices, like how to store or food for increased longevity, how to best use leftovers, or facts and advice on dietary choices, i.e. how carbon intensive foods could be exchanged for climate smart substitute products, see example in Fig. 21.4.

21.3.3 Study Design

For the purpose of testing the Food Watch application and the intervention design, 4 single-person households were recruited as participants. An in-depth interview was conducted with participants to reveal habits, attitudes, cultural setting and motivation in relation to food consumption and management.

Each household was equipped with a tablet computer, a Bluetooth barcode scanner and a generic kitchen type scale. A pre-study interview was performed to reveal environmental awareness in general and attitudes around food and diets in particular. The participants were then instructed to record their food purchases and the food waste generated for the duration of the study. In the first part of the study no interventions were applied, to provide baseline data of purchase and waste generation patterns. In the second part of the study the interventions were activated.

The conclusion of the study will consist of a debriefing type interview to assess if there were any changes to household practices, but also a co-creation session with the intention to generate improvements to the Food Watch application itself.

21.4 Results

At the time of the writing the field study has not yet been concluded. Nevertheless, 264 purchase and 67 waste transactions have so far been registered in the database. The transactions were registered over a period of 3 months, with a three week interruption. There were significant drop-out of the participating households, and during the intervention phase only two participants remained active.

Due to the small sample, some odd artefacts were present in the collected data, as can be seen in Fig. 21.5, where cucumber waste was the food category most

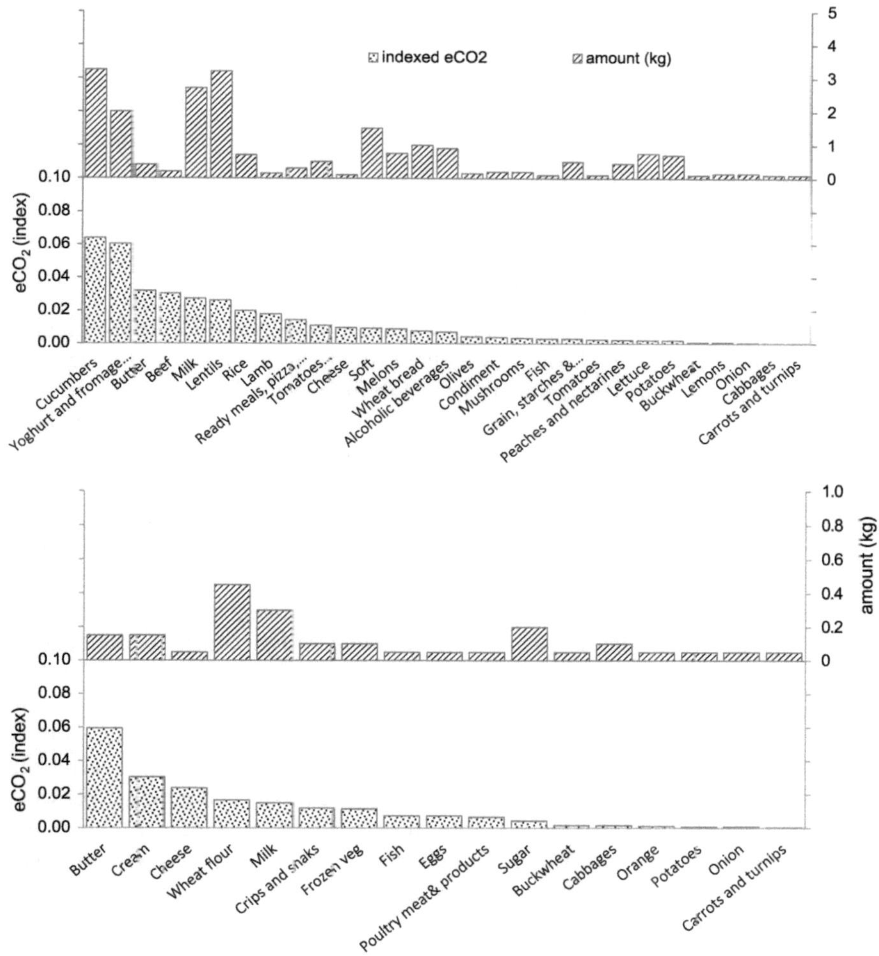

Fig. 21.5 Graphs showing waste derived carbon dioxide footprint index and total waste amount per food category, before (*top panel*) and during (*bottom panel*) the intervention period

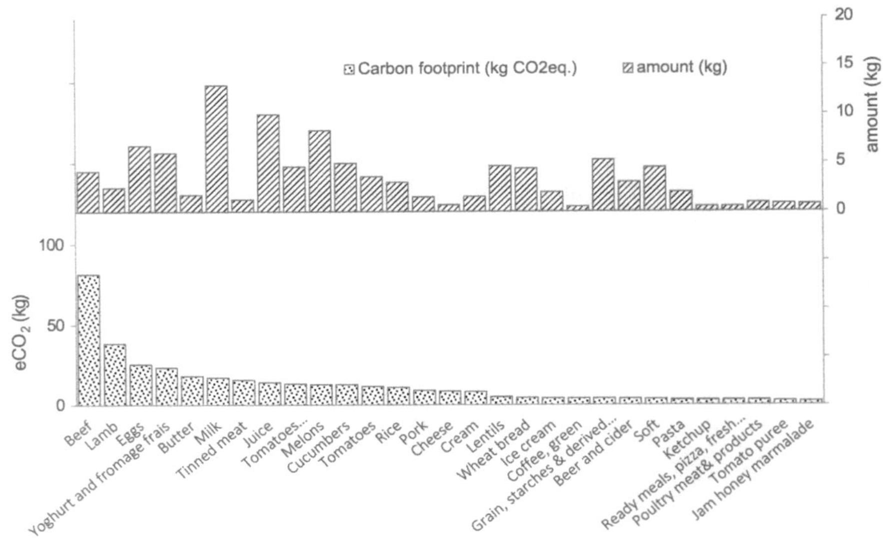

Fig. 21.6 A graph showing the carbon dioxide footprint and total amounts per food category associated with household purchases

significantly contributing to the carbon dioxide footprint, simply because one participant cleaned out the cupboard during the testing period and in the process discarded a large amount of cucumber. The high-impact food beef is not the main contributor to waste related carbon dioxide footprint—in neither the pre-intervention nor the intervention phase. Interestingly, butter (10.30 kg eCO_2 per kg) is appearing as an important product category with an indexed eCO_2 (kg eCO_2/total food purchases in kg) of about 0.06 during both phases. It can also be observed that other dairy category products like cream, milk and cheese (indexed eCO_2 0.03–0.015) are found among the top 10 contributing food waste categories, highlighting the potential for carbon dioxide savings in this category.

Looking at the purchase side of the registered food consumption data the results show that beef is a significant contributor to the carbon dioxide footprint, with 81.6 kg eCO_2 (0.21 indexed eCO_2), and also dairy products such as yoghurt, butter and milk showed important contributions to total footprint (see Fig. 21.6).

Another interesting outcome of the field trial was the categorization of the generated food waste. It was evident that participants experienced a "large package" as an important reason (21.6 kg eCO_2) for the occurrence of food waste, both edible and non-edible (see Fig. 21.7). "Wrong storage" was another important reason stated (6.9 kg eCO_2). The issue with large packages is an often cited issue in the literature (Graham-Rowe et al. 2014; Quested et al. 2013). Optimal packing size from waste often conflicts with marketing considerations and marketplace offers, where large packages often offer better value for money. Package size and pricing strategies are part of ongoing discussions between producers, retailers and policy makers.

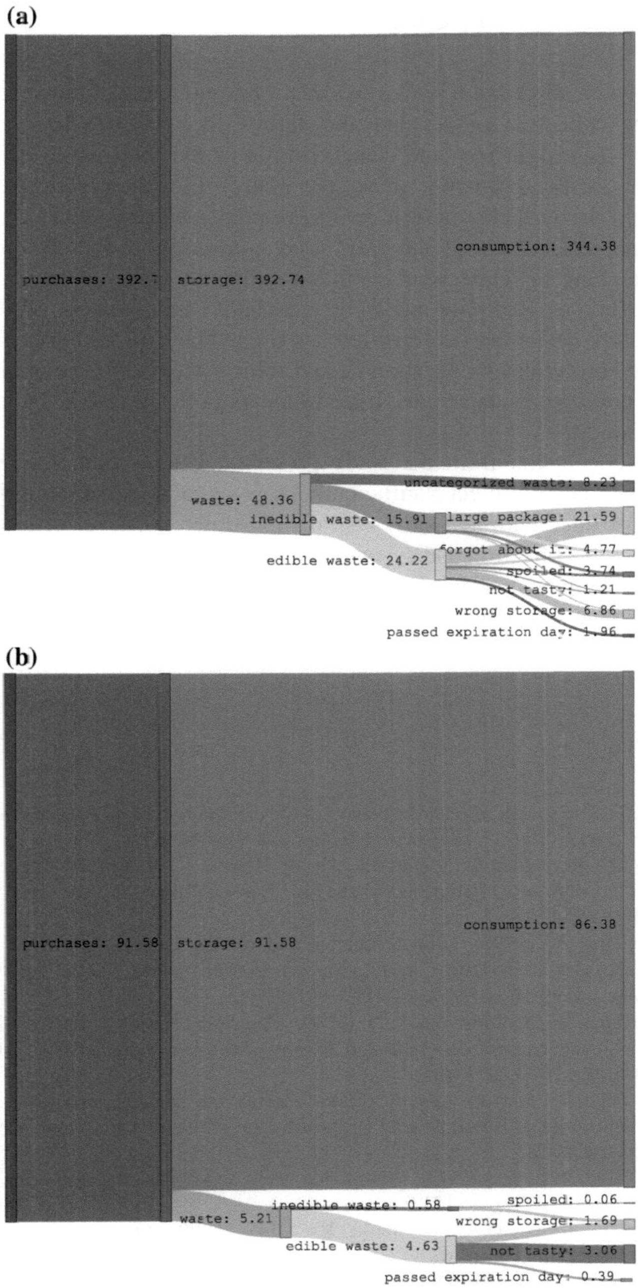

Fig. 21.7 Sankey diagram of output data from the Food Watch application used by four test households during the field trial. The diagram visualize the categorized flow of food represented as carbon dioxide footprint (kg CO_2e), including characterization of generated food waste, before (*top panel*) and after (*bottom panel*) the introduction of interventions

21.5 Conclusions and Outlook

The Food Watch application serves two main purposes, one being a research tool, allowing the collection of detailed and highly disaggregated food consumption data, as well as insight into and categorization of the food waste generated. The other purpose of Food Watch is to act as a platform for intervention and feedback on household performance, which could essentially be considered an educational and motivational tool about the own food consumption and its environmental impact. Designing such interventions in a way that allows targeting specific factors influencing food consumption behavior is a significant challenge, and will be the focus for future design and co-creation activities. One conclusion of the field trial was that the current design iteration lacked some components for the intervention, in terms of consistency and focus. Improvements in this area will be important for subsequent studies.

Another conclusion from the study presented here is that it is necessary to gather a larger data set to get a sufficiently large signal to noise ratio when evaluating impact of interventions, but is also demonstrated the level of disaggregation, detail in the data collection and systematization of data that is achievable.

Future development of the Food Watch application will include specific interventions designed to address socio-cultural pressure, motivation and habits, and their evaluation.

References

Ajzen, I. (1991). The theory of planned behavior. *Organizational Behavior and Human Decision Processes, 50*(2), 179–211. doi:10.1016/0749-5978(91)90020-T.

Berners-Lee, M., Hoolohan, C., Cammack, H., & Hewitt, C. N. (2012). The relative greenhouse gas impacts of realistic dietary choices. *Energy Policy, 43*, 184–190. doi:10.1016/j.enpol.2011.12.054.

Carlsson-Kanyama, A. (1998). Climate change and dietary choices—How can emissions of greenhouse gases from food consumption be reduced? *Food Policy, 23*(3–4), 277–293, doi:http://dx.doi.org/10.1016/S0306-9192(98)00037-2.

Carlsson-Kanyama, A., & González, A. D. (2009). Potential contributions of food consumption patterns to climate change. *The American Journal of Clinical Nutrition, 89*(5), 1704S–1709S. doi:10.3945/ajcn.2009.26736AA.

Carlsson, A., Palm, V., & Wadeskog, A. (2006). Energy use and CO_2-emissions for consumed products and services. IPP-indicators for private and public consumption based on environmental accounts SCB.

Finn, S. M. (2014). Valuing our food: Minimizing waste and optimizing resources. *Zygon, 49*(4), 992–1008. doi:10.1111/zygo.12131.

Graham-Rowe, E., Jessop, D. C., & Sparks, P. (2014). Identifying motivations and barriers to minimising household food waste. *Resources, Conservation and Recycling, 84*(0), 15–23, doi:http://dx.doi.org/10.1016/j.resconrec.2013.12.005.

Harder, R., Kalmykova, Y., Morrison, G. M., Feng, F., Mangold, M., & Dahlén, L. (2014). Quantification of goods purchases and waste generation at the level of individual households. *Journal of Industrial Ecology, 18*(2), 227–241. doi:10.1111/jiec.12111.

Kahiluoto, H., Kuisma, M., Kuokkanen, A., Mikkilä, M., & Linnanen, L. (2014). Taking planetary nutrient boundaries seriously: Can we feed the people? *Global Food Security, 3*(1), 16–21. doi:10.1016/j.gfs.2013.11.002.

Quested, T. E., Marsh, E., Stunell, D., & Parry, A. D. (2013). Spaghetti soup: The complex world of food waste behaviours. *Resources, Conservation and Recycling, 79*, 43–51, doi:http://dx.doi.org/10.1016/j.resconrec.2013.04.011.

Reynolds, C., Buckley, J., Weinstein, P., & Boland, J. (2014). Are the dietary guidelines for meat, fat, fruit and vegetable consumption appropriate for environmental sustainability? *A Review of the Literature. Nutrients, 6*(6), 2251–2265.

Scarborough, P., Appleby, P., Mizdrak, A., Briggs, A. M., Travis, R., & Bradbury, K., et al. (2014). Dietary greenhouse gas emissions of meat-eaters, fish-eaters, vegetarians and vegans in the UK. *Climatic Change, 125*(2), 179–192. doi:10.1007/s10584-014-1169-1.

Triandis, H. C. (1977). *Interpersonal behavior.* Brooks/Cole Pub. Co.

Part VI
Case Studies on Exploring Energy Feedback and Visualisation with Users

Chapter 22
Making Energy Feedback Understandable

David V. Keyson and Natalia Romero Herrera

Abstract In this chapter several aspects in the design of home energy feedback are discussed along with the presentation of energy feedback design examples. The examples include the Ampul interface that was tested in the SusLab Concept House. Ampul was designed to enable users to maximize the use of available solar energy. A second example referenced is an aquarium metaphor in which plant, water and fish qualities are coupled to energy consumption and climate sensor data, which was developed as a demonstration and thirdly Powercord is described as an interface able to translate energy consumption into auditory sounds.

Keywords Energy feedback · Energy use

22.1 Introduction

Several factors may influence energy consumption including, user behavior, climate, the physical characteristics of the built environment, equipment and devices in the household, and energy related control technology. A range of variables such as cultural background and income levels may shape user behavior.

Home automation could significantly influence electricity consumption. For example, electricity consumption through technology could be reduced by about 50 % and by about 30 %, given a minimum investment (Eichhammer et al. 2009; Ellis and Jolland 2009). At the same time, differences in behaviors among end users contributes to the variability in household's energy consumption levels

D.V. Keyson (✉) · N. Romero Herrera
Delft University of Technology, Delft, The Netherlands
e-mail: d.v.keyson@tudelft.n

N.R. Herrera
e-mail: n.a.romero@tudelft.nl

D.V. Keyson et al. (eds.), *Living Labs*, DOI 10.1007/978-3-319-33527-8_22

(Lutzenhiser 1993; Sanquist et al. 2012). Sanquist et al. (2012) found that more than 40 % of electricity consumption in households can be attributed to lifestyle factors. Guerra-Santin and Itard (2010) found that approximately 12 % of the variation in energy use for space heating could be explained by occupant behavior. Similarly, Dietz et al. (2009) estimated that approximately 20 % reduction in household carbon emissions could be achieved through behavioral changes.

In cases in which household members do not understand how to efficiently utilize a technology or how to adjust their behavior accordingly, the potential impact on energy reduction may be lost through 'wrong' usage of equipment As reported (Geelen et al. 2013). For example, using an air-conditioner on a hot day while the windows are open would result in higher electricity consumption. Rebound effects may also occur, meaning that potential energy savings as a result of technical improvements are not achieved due to behavioral changes that counteract the energy saving potential (Sorrell 2007). An example of a Direct rebound effect is when end users replace light bulbs for more efficient ones and then leave the lights on longer. Another example here would be the case of end users who take advantage of increased comfort afforded by a new more energy-efficient heating system by heating more rooms. Studies have shown that households with programmable non-learning thermostats and balanced ventilation, i.e. a ventilation system where air supply and exhaust consist of approximately equal quantities of fresh outside air and polluted inside air respectively, tended to heat more rooms and use the heating system for more hours than households with manual thermostats and mechanical exhaust ventilation (Guerra-Santin and Itard 2010).

Current approaches to providing the user with information on the amount of electricity consumed or locally produced and information pertaining to data collected from indoor climate sensors typically entail data like views of patterns using line charts and histograms on a timeline. Given data-centric views the user is required to interpret the data in order to discover a trend. By portraying electricity or other real-time sensor measurements in terms of digits and graphs, real-time flows remains invisible and are given no tangible meaning.

Ambient displays have been considered as a means to translate energy consumption into a real-world representation such as color light or the physical motion of a mechanism. For example, the Wattson display uses colors to show relative energy consumption level, while cost and kilowatt-hours can also be displayed as an overlay on the device panel. Other examples of ambient displays include the Power Cord and Plant digital display developed at the Interactive Institute in Sweden (http://www.tii.se/), Energy Puppet (Do and Abdelmohsen 2008) and the Ambient Orb configured as an energy ORB (Chapa 2007). The idea of the Plant suggests an element of care as part of an ecosystem. If home electricity is continually not well managed, the plant will wilt over time. Similarly, the direct result of one's actions is reflected in a monthly energy bill, but many factors over time influence cost. Novel designs for interaction with energy information, such as the use of gestures, have also been developed at the Ubicomp Lab at the University of Washington.

Despite the wealth of creative ideas for ambient displays, they often do not provide the user with a level of feedback that can assist in changing behavior, as only total consumption levels are provided rather than appliance level feedback. Secondly, limited or no information is provided on the degree to which available renewable energy is being utilized.

In terms of insights into energy usage and planning one can simply think of three phases in time, namely past, present and future. Secondly, information should be provided at a level of detail such that the user can take action. Often the term goal-based or actionable feedback is used here (Do and Abdelmohsen 2008). The design challenge then is to provide actionable feedback without creating a highly complex display. Home energy displays are typically limited to providing only total energy consumption levels, rather than including feedback at the appliance level. The user may also want to receive a concrete suggestion or tips on saving energy. An additional key design challenge of energy feedback systems is the need to support a broad range of potential users in a single household. At any given moment, a user may just want to casually glance at a display to gain an overview of relative energy consumption levels, while still being able to easily gain deeper insights into energy usage and production patterns. Via progressive disclosure, deeper insights can be given without sacrificing usability. Furthermore a mixed initiative user interface design can reduce the load on attention resources (Horvitz 1999).

User experience issues in relation are highly inter-twined in terms of shaping the overall user experience. Three key factors that make up the landscape of user experience are: (a) understanding and sense of being in control, (b) emotionally appealing and engaging, and (c) expected and perceived functional performance (Keyson 2008). In demonstrating the role of mixed initiative and collaborative dialog to reach energy savings and comfort goal and intelligent thermostat was developed (Keyson 2008). The interface was built with a hierarchical task model, such that the user could see things to say at any given point of interaction. For example stating "help me save energy" could trigger the response, you only use your bathroom 1 hour a day, shall I heat it during this hour and add the event to your agenda? The user could also manually access the agenda as a touch screen (Fig. 22.1).

As a more abstract metaphor, a 3D graphical aquarium interface was developed during the SusLab project and linked to the SusLab sensor boxes for demonstration purposes. Though not tested in the field the concept was designed to equate comfort and energy parameters with a real world like aquarium environment. For example color was equated to temperature, fish movement speed to energy, plant growth to CO_2, and water clarity to air quality. Rich character graphics and background element common to game design formed an integral part of the design approach (Keyson et al. 2013).

Through self-reporting apps and devices building containing multiple occupants, such as a shared office place can input their preferences and at the same time view actual energy consumption. For example in student work (Savvkai 2015) a physical room display was developed in the form of a clock to indicated average reported levels for air quality, temperature, light and observed energy use (Fig. 22.2).

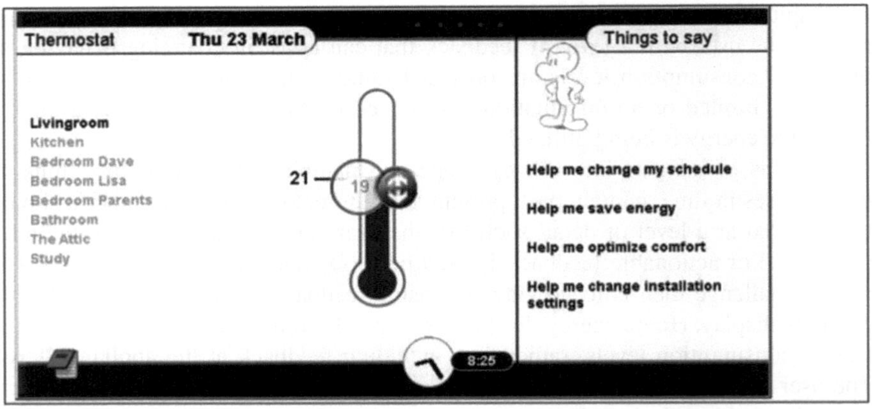

Fig. 22.1 Screen capture of the intelligent thermostat designed to support speech input with things to say based upon a hierarchical task model, and combined touch screen

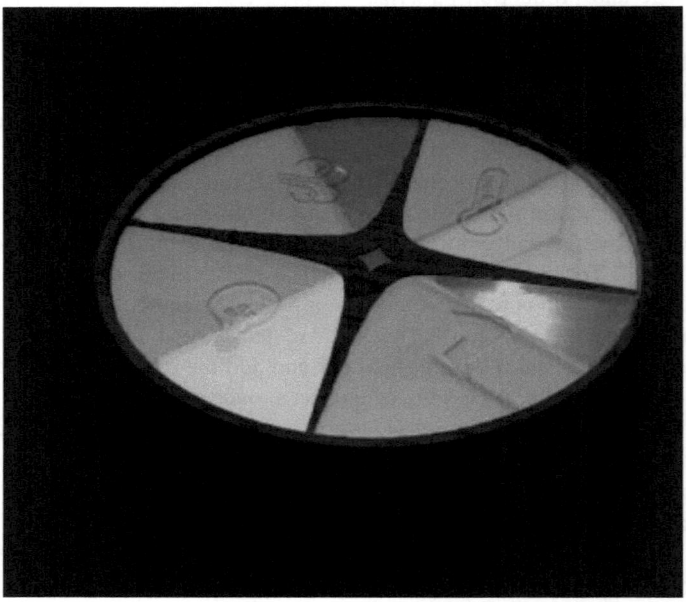

Fig. 22.2 Illustration of an energy feedback display for offices, which shows observed energy usage and average, preferred temperature, light, and perceived air quality

Summarizing, the introduction of energy efficient technology and home automation into the household may theoretically lead to changes in energy consumption, however when behavior in the household is not aligned, potential energy savings may not be realized. This does not imply that end users should always have to adjust their behavior to technology. Technology should also fit end user needs, wishes and abilities. Technology and behavior thus have to complement each other (Geelen et al. 2013).

References

Chapa, J. (2007). THE ENERGY ORB: Visualize Electricity Consumption! Retrieved September 1 2007, from inhabitat.com.

Dietz, T, Gardner, G. T., Gilligan, J., Stern, P.C., & Vandenbergh M. P. (2009). *Proceedings of the National Academy of Sciences USA*, *106*(44), 18452–18456 (2009). doi:10.1073/pnas.0908738106.

Do, E. Y., & Abdelmohsen, S. (2008). Energy puppet: An ambient awareness interface for home, energy consumption. In *7th International Workshop on Social Intelligence Design, Designing socially aware interactions, SID 08*, December 3–5, 2008 Universidad de Puerto Rico, San Juan, PR

Eichhammer, W., Fleiter, T., Schlomann, B., Faberi, S., Fioretto, M., Piccioni, N., et al. (2009). Study on the energy savings potentials in EU member states, candidate countries and EEA countries.

Ellis, M., & Jolland, N. (2009). *Gadgets and gigawatts, policies for energy efficient electronics*. Paris: International Energy Agency.

Geelen, D., Reinders, A., & Keyson, D. (2013). Empowering the end-user in smart grids: Recommendations for the design of products and services. *Energy Policy, 61*, 151–161.

Guerra Santin, O., & Itlard, L. (2010). Occupants' behavior determinants and effects on residential heating consumption. *Building Research and Information, 38*, 318–338.

Horvitz, E. (1999). Principles of mixed-initiative user interfaces. In *Proceedings of the SIGCHI conference on human factors in computing systems* (pp. 159–166).

Keyson, D. V. (2008). The experience of intelligent products in product experience. In Schifferstein & Hekkert (Eds.), Elsevier Science.

Keyson, D. V., Al Mahmud, A. Romero, N., & Augusto J. C., et al. (Eds.): Living Lab and research on sustainability: Practical approaches on sustainable interaction design. AmI 2013, LNCS 8309, 229–234, 2013. Springer International Publishing Switzerland 2013.

Lutzenhiser, L. (1993). Social and behavioral aspects of energy use. *Annual Review of Energy and the Environment, 18*, 247–289.

Sanquist, T. F., Orr, H., Shui, B., & Bittner, A. C. (2012). Lifestyle factors in U.S. residential electricity consumption. *Energy Policy, 42*, 354–364.

Savvaki, K. (2015). Sustainability and comfort in the workplace. Master Thesis, Delft University of Technology.

Sorrell, S. (2007). The rebound effect: An assessment of the evidence for economy-wide energy savings from improved energy efficiency. UK Energy Research Centre.

Wattson display. (http://www.treehugger.com/clean-technology/wattson-monitor-your-homes-energy-usage.html).

Chapter 23
Powerchord: Exploring Ambient Audio Feedback on Energy Use

Dan Lockton, Flora Bowden and Claire Matthews

Abstract Influencing energy use is a major research topic. However, many approaches lump 'energy demand' together, disconnected from everyday artefacts, the realities of household life, and people's diverse understandings of the systems around them. There is an opportunity for research through design which addresses relationships with the invisible concept of energy through new kinds of feedback. *Powerchord* is an ongoing (2014—) exploration of sonifying energy use in near-real time. The prototypes developed so far monitor multiple household electrical appliances in parallel, turning readings of the instantaneous power being drawn into various kinds of sounds. *Powerchord* provides a form of ambient experiential feedback intended to fit with the soundscapes of everyday domestic life, while (perhaps) enabling a deeper understanding of the characteristics of energy use. The concept was developed from ideas suggested by householders during co-creation sessions as part of the European SusLabNWE project, funded by INTERREG IVB, as part of our wider exploration of the invisibility of energy which also led to 'Drawing Energy' (see Chap. 14 'Participatory Drawing in Ethnographic Research').

Keywords Energy use · Sonification · Ubiquitous computing

D. Lockton (✉) · F. Bowden
Royal College of Art, London, UK
e-mail: dan@danlockton.co.uk

F. Bowden
e-mail: flora.bowden@gmail.com

C. Matthews
Independent, London, UK
e-mail: claire.dyb@gmail.com

D.V. Keyson et al. (eds.), *Living Labs*, DOI 10.1007/978-3-319-33527-8_23

23.1 Background

Design for sustainable behaviour has grown significantly as a field of research in recent years (see Chap. 7 'Design with Intent and the field of design for sustainable behaviour' of this volume). It aims to reduce the 'undesirable' environmental and social impacts of products and services, or increase the desired impacts, through design (in a broad sense) concentrating on understanding and influencing people's interactions with technology. It is inherently multidisciplinary, drawing on knowledge, perspectives and models from a number of fields relating to human behaviour and social practices.

Mainstream 'interventions' largely take the form of redesign of products and services themselves, or the design of interfaces, usually digital, and usually visual, which give users information and feedback (and sometimes *feedforward*) on use or the impacts of their actions. The digital approach builds on significant work in HCI on persuasive technology (Fogg 2009) and on the effectiveness of behavioural feedback from other disciplinary perspectives.

Energy use is one of the major issues on which design for sustainable behaviour has concentrated. Reducing humanity's energy demand, and in turn carbon dioxide (CO_2) impact, through influencing public behaviour, is a significant research topic across multiple intersecting technological and social science disciplines. Aside from design work on technology or infrastructure change, the majority of work on influencing energy use through behaviour change concentrates on numerical, visual feedback displays for electricity or gas use, in both domestic and commercial environments. There are numerous studies and meta-analyses looking at the effectiveness of different kinds of feedback (real-time, summary, normative, and so on) in this context, and the adoption of these kinds of displays within household life. As technology develops, the opportunities afforded by networked smart meters, which enable adaptive pricing changes (as well as providing energy utilities with much more detailed usage data) are also being explored, driven by legislation.

While some influence on behaviour, leading to changes in energy use, has been found from feedback displays (e.g. Kobus et al. 2012), the situation is complex: simple numerical feedback may not take account of the realities of household life (Brynjarsdóttir et al. 2012; van Dam et al. 2009; Hargreaves et al. 2013) or people's *understanding* of units and quantities (Strengers 2011), nor link people to wider comprehension of the energy system (Boucher et al. 2012). Most visual displays require the householder to look at the display—often a small LCD (Fig. 23.1a/b), or a web dashboard—regularly, and actually be able to act on it, for it to have any effect, assuming a model of individual householders as "micro-resource managers" (Strengers 2011). While there have been some more ambient coloured light-based systems for displaying electricity use, such as Gustafsson and Gyllenswärd's (2005) *Power Aware Cord*, DIY Kyoto's *Wattson* and Ambient Devices' *Orb*, and clever use of thermal imaging (Goodhew et al. 2015), these are exceptional.

(a)

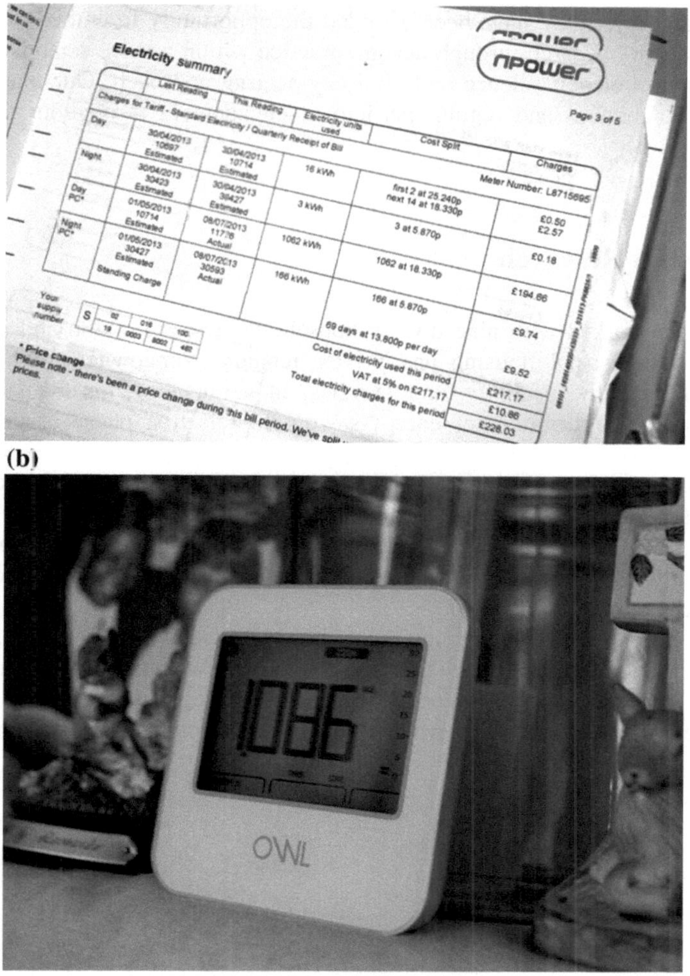

(b)

Fig. 23.1 a/b Bills and simple numerical displays are common interfaces for energy use

Many approaches lump ‘energy demand’ together as a number, disconnected from everyday artefacts, the realities of household life, and people’s diverse understandings of the systems around them. There is an opportunity for design-led research placing artefacts and the narratives of practices more centrally, because, on the most basic of levels, it is through both everyday appliances, and artefacts such as these in-home displays that people actually experience energy use. Within HCI and design research, Pierce and Paulos (2010) call for more work on *materialising* energy, while novel approaches such as Mazé and Redström (2008) and Boucher et al. (2012) bring an artefact-driven perspective to the field.

As part of SusLabNWE, a large multidisciplinary collaborative European Living Lab project, we have been afforded the opportunity to embed this kind of artefact-driven 'research through design' practice within a more established quantitative environmental science and civil engineering approach. Our methodology employs co-creation and regular participant engagement throughout the design process.

23.2 Initial Research

Initially, we worked with nine diverse households across London and the south-east, including social housing and private tenants, homeowners, older people, people affected by disability, and families, in a range of housing stock, both unmodified and retrofitted for energy-saving. In this first phase of the project, we visited people at home, investigating stories of daily interactions with heating, lighting, appliances, and electricity monitors, and people's understanding of energy use. This was followed with a 'logbook' probe study including activities exploring themes such as metaphors for energy, social influences on energy use, and narrating everyday energy-related routines and frustrations through annotation (Fig. 23.2; Lockton et al. 2014).

Based around themes emerging from these interviews and logbooks, we then ran a co-creation workshop with our householders, in which, working with designers, they created concepts for new kinds of interface or device which they felt would help them reduce their energy use. This was followed by a 'Home Energy

Fig. 23.2 A householder's annotation of her gas meter as 'difficult to use', because of its position in an exterior cupboard

Hackday' at the Science Museum's Dana Centre, bringing together designers, energy experts, and developers from the Internet of Things community to explore new ways of understanding and engaging with energy, building on the ideas from householders (Fig. 23.3a/b).

One of the main themes that emerged was the general invisibility of energy in modern life, confirming Pierce and Paulos's (2010) argument, and the consequences of this for behaviour and everyday practices. Householders' mental models of energy itself, and energy-using systems such as heating (Revell and Stanton 2014; Lockton et al. 2013), together with the relative importance

Fig. 23.3 a/b Scenes from the Home Energy Hackday at the Dana Centre, London

of different energy-using systems in the home, were partly determined by what was most salient, such as lighting, rather than 'hidden' uses such as heating and cooling (this aligns with other research, e.g. Attari et al. (2010) and Kempton and Montgomery (1982)).

By people's own admission, much of the energy 'wasted' at home through particular behaviours, such as leaving heating on when going out, or leaving lights on elsewhere in the house, was partly due to its invisibility from the perspective of where they were at the time. People questioned how they could change how they use energy when they can't easily see or feel it, or get a sense of the changing rate at which it is being used. We found confusion with the different characteristics of energy use by different appliances (e.g. the 'spike' of a kettle compared with the continuous power drawn by lighting), and units, for example between kilowatts as a measure of power and kilowatt-hours as a measure of energy. As also recounted by Bowden et al. (2015), one householder told us:

> "I worked out that through gas and electricity every year, the average house gets the equivalent of a bit over three tons of coal delivered completely silently and without any mess. And go back a hundred years ago and everyone would have a really good quantitative understanding of how much energy they used because they had to physically shovel the stuff."

We decided to explore this theme of energy's invisibility through both visualisation beyond numbers (see Chap. 14 'Participatory Drawing in Ethnographic Research' of this volume), and transitioning to another sense: *sound*.

23.3 Sonification

The issue of energy's invisibility suggested opportunities for visualisation beyond numbers, but also non-visually, for example *sonification* (Walker and Nees 2011) of energy use.

In our co-creation workshop, one person suggested that being able to 'listen' to whether appliances were switched on or not, and what state they were in (e.g. listening to a washing machine will give a good idea as to where it is in its cycle), was potentially more useful for understanding, experientially, the links between practices and energy demand, than a visual display. Another householder suggested, in response to discussion of smart metering and demand-based pricing changes, that being able to 'hear' the load on the grid (for example, a pleasant background hum could become discordant as the grid's frequency changes due to high demand, or the tick of a clock could become temporarily faster) would be less intrusive than, for example, a text message or a flashing light. There was discussion around the quality of the sound, e.g. whether a lower-pitched 'rumbling', like thunder, would be more appropriate for greater rate of energy use (i.e. power) than a higher pitch, and whether there could be a music system that somehow 'distorted' what it played when the house's energy use was higher than normal.

There are echoes of early work in calm technology and ubiquitous computing, such as Natalie Jeremijenko's *Live Wire (Dangling String)* (Weiser and Brown 1995), or Ernevi et al's (2007) *Erratic Radio*, in which the 'display' fits with the existing daily visual landscape and *soundscapes* (Schafer 1977) of the environment. Sonification of energy use along these lines could enable ambient comprehension of energy use with multiple appliances, including pattern recognition and state changes (Serafin et al. 2011). Relating sound to energy use is not unknown. In explicit data sonification work, Opower's *Chicago in the Wintertime* (Tinjum and Ben-Meshulam 2013) turns the city's residential electricity use over winter 2012–13 into piece of music; less directly, Foster et al's (2011) *Power Ballads* made use of aversive feedback based around UK chart music, automatically posted to the user's Facebook wall, based on high levels of electricity use.

To explore near-real-time energy sonification, we chose CurrentCost electricity monitors, as supplied to many utility customers in the UK, including some of our participating households. The CurrentCost 'ecosystem' includes a bridge connecting to a router and posting data to a website, and individual appliance monitors (IAMs) wirelessly connected to the base unit, enabling disaggregated data. CurrentCost has been used in a number of Internet of Things (IoT) academic studies. The system can also monitor gas use, if a household has a compatible meter.

Building on others' code for extracting CurrentCost data (e.g. by Colin R Williams: http://crwilliams.co.ukprojects/arduino-currentcost-lcd/), we developed *Powerchord*, an Arduino-based system which parses the CurrentCost's XML output every 6s, extracting the IAM power figures for individual appliances, and mapping these figures to ranges defined in code. Three IAMs are used, although the system could support up to nine. Initially we worked with a GinSing synthesizer shield for the Arduino, producing different tones (with various effects) mapped to power ranges, but, lacking experience in sound design, the results we were able to produce were aesthetically unattractive to say the least.

We decided instead to build on the idea from our co-creation work with householders around fitting into the existing daily soundscapes of the home—something more like the tick of a clock, or the sound of distant church bells, 'repurposing' them with extra energy information rather than being part of the "increasing clutter of beeps and bleeps" (Serafin et al. 2011) of feedback. This 'blended sonification' (Tünnermann et al. 2013) meant that recordings of these sounds, suitably modified, could be used; we linked the Arduino to a Robertsonics WAV Trigger, enabling polyphonic playback for multiple audio files simultaneously.

Power ranges were defined to match the typical ranges found in household appliances, from <10 W for trickle charging, to >2 kW for electric heaters. For each power range, for each appliance, the WAV Trigger plays a particular audio track, looped until the power range changes (Table 23.1). Any audio tracks can be used, including tones, sound effects or music—on a user-replaceable micro SD card. Powerchord can thus act as a platform for different kinds of ambient energy sonification research (Figs. 23.4, 23.5 and 23.6).

For the initial demonstration prototypes, aside from trying out actual power chords from 90s rock, we have primarily used *birdsong* (from http://xeno-canto.

Table 23.1 Mappings of power ranges to sound files—per appliance, in parallel

Power range (W)	Sound files used
0–5	Nothing played
6–30	Track A (lowest intensity)
31–150	Track B
151–390	Track C
391–500	Track D
501–900	Track E
901–1700	Track F
≥1701	Track G (highest intensity)

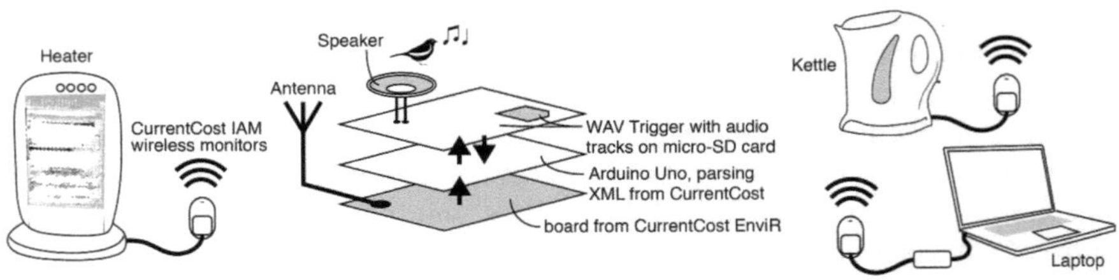

Fig. 23.4 How the Powerchord system works

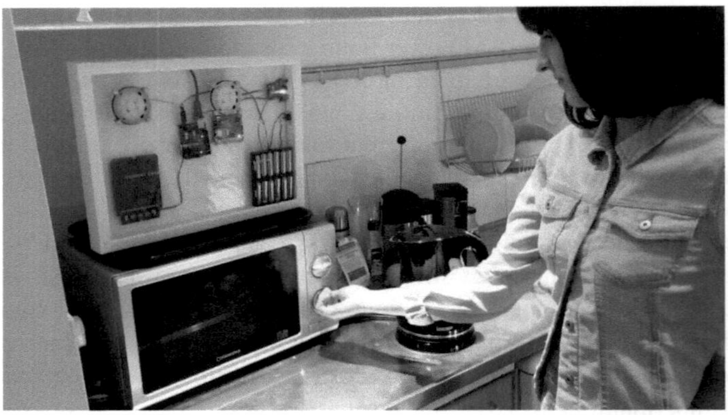

Fig. 23.5 A Powerchord prototype being tested in a kitchen environment, monitoring a microwave oven and an electric kettle

org)—blackbirds, house sparrows and herring gulls—edited so that different intensities of song (number of birds, agitation level) map to power ranges. The fact that we hear birds calling and singing every day, and notice when they are abnormally loud or agitated, yet are usually unable to understand what the sounds 'mean', suggests that birdsong represents an opportunity for this 'unused auditory bandwidth' to be exploited as a channel for information. The subjective beauty of most bird vocalisation, such that even birds' alarm calls are usually considered pleasant

Fig. 23.6 A Powerchord prototype being tested monitoring a hairdryer

(and very unlike the alarm sounds generated by most electronic devices), offers a different quality of experience to direct tone-mapped sonification.

Connecting people better to the wider, complex systems around them, in which their behaviour plays a part, such as energy use, suggests that 'natural' metaphors may be particularly relevant. One might equally imagine the sound of a river, waterfall, or the sea, or wind in the trees, as being appropriate in this context, or indeed other weather-based sound collections. We selected common garden birds whose calls and song would likely already be familiar and potentially part of householders' daily soundscapes—blackbirds and house sparrows—and also, as a counterpoint, the distinctive calls of herring gulls, which potentially evoke seaside memories, but which are not 'tranquil' in the same way. In some demos, we have mapped the herring gulls to an electric kettle, since it is used intermittently rather than continuously, and the startling sound of the gulls is aligned with the sudden change in household energy use that switching on a kettle normally entails.

Demonstrating a Powerchord prototype alongside the Drawing Energy activity at the Victoria and Albert Museum's 'Digital Design Weekend' 2014 (Fig. 23.7a/b), around 40 members of the public tried it out, switching on a fan heater at different power levels to hear the changes in birdsong. We used this activity to engage people in completing quick response sheets suggesting the kinds of sounds they believed they would find useful for understanding the energy use characteristics of different appliances and activities. This has enabled us, working with sonic interaction designer Claire Matthews (http://claire-matthews.com), to create and explore the possibilities of a range of 'sound packages' for Powerchord, including ever-more-complex jazz and acoustic guitar schemes, a log fire progressing from crackling to roaring, the sound of increasing numbers of coins being dropped, other natural sounds such as rain intensity, different animal sounds from mice to elephants, and even dog whistles so that the family dog perhaps learns to howl at high energy use, becoming the 'household expert' on it.

Fig. 23.7 **a/b** Members of the public trying out Powerchord monitoring a fan heater, at the V&A Digital Design Weekend 2014, London

23.4 Discussion and Further Work

While Powerchord is still work in progress (the project has a home on the web at http://powerchord.me), insights from public engagement together with our work with householders, have enabled us as design researchers, working in a largely quantitative multidisciplinary project, to broach questions around public understanding of energy, novel forms of feedback, and the affordances and value of research through design methodologies. Using 'things' (prototypes, artefacts, drawings) in this way enables ways of knowing which are qualitatively different to those enacted in other disciplines addressing the same broad questions, in this case around design for sustainable behaviour and energy use. Artefacts enable the materiality of energy (Pierce and Paulos 2010) to be manifested, discussed, and explored practically.

In the next stage of our work with Powerchord we are aiming to explore this value further, through field studies with householders, incorporating a variety of sound schemes as outlined above. The research questions centre on how people's *understanding* of the energy use characteristics of appliances change (or not) with this kind of ambient feedback, and how people make use of the system in practice. Can knowledge derived from 'longer term use' of such a design intervention, for instance, help inform us about people's evolving understandings and interactions between technology, meanings and practices? How, over time, do people construct meaning (Dourish 2001; Fantini van Ditmar and Lockton 2016) for their data? There are also many specific energy-related avenues to explore, including linking sound to the load on the electricity grid (particularly where this may lead to different pricing per unit) and applications in local or community microgrids where generation as well as consumption (and the balance between them) comes into consideration.

References

Attari, S. Z., DeKay, M. L., Davidson, C. I., & Bruine de Bruin, W. (2010). Public perceptions of energy consumption and savings. *Proceedings of the National Academy of Sciences, 107*(37), 16054–16059.

Boucher, A., Cameron, D., & Jarvis, N. (2012). Power to the people: Dynamic energy management through communal cooperation. In *Proceedings of DIS 2012*, Newcastle.

Bowden, F., Lockton, D., Gheerawo, R., & Brass, C. (2015). *Drawing energy: Exploring perceptions of the invisible*. London: Royal College of Art.

Brynjarsdóttir, H., Håkansson, M., Pierce, J., Baumer, E. P. S., DiSalvo, C., & Sengers, P. (2012). Sustainably unpersuaded: How persuasion narrows our vision of sustainability. In *Proceedings of CHI 2012*, Austin, Texas.

Dourish, P. (2001). What we talk about when we talk about context. *Personal and Ubiquitous Computing, 8*(1), 19–30.

Ernevi, A., Palm, S., & Redström, J. (2007). Erratic appliances and energy awareness. *Knowledge, Technology and Policy, 20*(1), 71–78.

Fantini van Ditmar, D., & Lockton, D. (2016). Taking the code for a walk. *Interactions, 23*(1), 68–71.

Fogg, B. J. (2009). A behavior model for persuasive design, Persuasive 2009. In *4th International Conference on Persuasive Technology*. New York: ACM.

Foster, D., Linehan, C., Lawson, S., & Kirman, B. (2011). Power ballads: Deploying aversive energy feedback in social media. In *CHI '11 Extended Abstracts on Human Factors in Computing Systems*. New York: ACM.

Goodhew, J., Pahl, S., Auburn, T., & Goodhew, S. (2015). Making heat visible: Promoting energy conservation behaviours through thermal imaging. *Environment and Behaviour, 47*(10), 1059–1088.

Gustafsson, A., & Gyllenswärd, M. (2005). The power-aware cord: Energy awareness through ambient information display. In *Poster Proceedings of CHI 2005*. New York: ACM.

Hargreaves, T., Nye, M., & Burgess, J. (2013). Keeping energy visible? Exploring how householders interact with feedback from smart energy monitors in the longer term. *Energy Policy, 52*, 126–134.

Kempton, W., & Montgomery, L. (1982). Folk quantification of energy. *Energy Policy, 7*(10), 817–827.

Kobus, C. B. A., Mugge, R., & Schoormans, J. P. L. (2012). Washing when the sun is shining! How householders interact with a household energy management system. *Ergonomics, 56*(3), 451–462.

Lockton, D., Harrison, D., Cain, R., Stanton, N. A., & Jennings, P. (2013). Exploring problem-framing through behavioural heuristics. *International Journal of Design, 7*(1), 37–53.

Lockton, D., Renström, S., Bowden, F., Rahe, U., Brass, C., & Gheerawo, R. (2014) Energy storytelling through annotating everyday life. In *BEHAVE 2014: 3rd European Conference on Behaviour and Energy Efficiency*, 3–4 Sept 2014, Oxford.

Mazé, R., & Redström, J. (2008). Switch! Energy ecologies in everyday life. *International Journal of Design, 2*(3), 55–70.

Pierce, J., & Paulos, E. (2010). Materializing energy. In *Proceedings of DIS Conference on Designing Interactive Systems*. Arhus, Denmark. DIS '10. New York, NY: ACM Press.

Revell, K. M. A., & Stanton, N. A. (2014). Case studies of mental models in home heat control: Searching for feedback, valve, timer and switch theories. *Applied Ergonomics, 45*(3), 363–378.

Schafer, R. M. (1977). *The soundscape: Our sonic environment and the tuning of the world.* Rochester, VT: Destiny Books.

Serafin, S., Franinovic, K., Hermann, T., Lemaitre, G., Rinott, M., & Rocchesso, D. (2011). Sonic interaction design. In T. Hermann, A. Hunt, & J. G. Neuhoff (Eds.), *The sonification handbook* (pp. 87–110). Berlin: Logos.

Strengers, Y. (2011). Designing eco-feedback systems for everyday life. In *Proceedings of CHI 2011*, Vancouver.

Tinjum, A., & Ben-Meshulam, Y. (2013). This is the sound of Chicago's energy consumption. Opower blog, 20 Nov 2013. http://blog.opower.com/2013/11/power-rock-this-is-the-sound-of-chicagos-energy-consumption/.

Tünnermann, R., Hammerschmidt, J., & Hermann, T. (2013). Blended sonification: Sonification for casual information interaction. In *Proceedings of International Conference on Auditory Display* (ICAD2013), Łódź, Poland.

van Dam, S., Bakker, C. A., & van Hal, J. D. M. (2009). The mediating role of home energy management systems. In *Proceedings of First European Conference on Energy Efficiency and Behaviour*, Maastricht.

Walker, B. N., & Nees, M. A. (2011). Theory of sonification. In T. Hermann, A. Hunt, & J. G. Neuhoff (Eds.), *The sonification handbook* (pp. 9–39). Berlin: Logos.

Weiser, M., & Brown, J. S. (1995). Designing calm technology. http://www.ubiq.com/hypertext/weiser/calmtech/calmtech.htm.

Chapter 24
Designing Ampul: Empowerment to Home Energy Prosumers

Natalia Romero Herrera, Jaap Rutten and David V. Keyson

Abstract Ampul is a home energy management system (HEMS) designed for the emergent group of home energy prosumers (producers and consumers of energy at home). The innovative visualizations and interactions provided by Ampul offer prosumers timed, situated and appliance specific information on the energy that is produced and consumed. Ampul has been developed by applying in-situ interventions in real homes and Concept House Prototype 1, a home living lab facility located in a neighbourhood in Rotterdam, The Netherlands. Residents of the same neighbourhoods were invited to actively collaborate in the design process. Ampul is presented in this section as an inspiring example of how Living Labs methods facilitate the role of future users as active collaborators in gathering insights and experimenting earlier concepts in homes and home living labs.

Keywords Home energy management systems · Prosumers · Living lab · In-situ interventions

24.1 Introduction

The increasing market of rooftop solar panels over the last decades has resulted in households and businesses being not only energy consumers but also energy producers, which in one word is referred to as 'prosumers' of energy. Energy

N. Romero Herrera (✉) · J. Rutten · D.V. Keyson
Delft University of Technology, Delft, Netherlands
e-mail: n.a.romero@tudelft.nl

J. Rutten
e-mail: jaaprutten@me.com

D.V. Keyson
e-mail: d.v.keyson@tudelft.nl

D.V. Keyson et al. (eds.), *Living Labs*, DOI 10.1007/978-3-319-33527-8_24

prosumers are able to produce, consume, manage and trade energy under their own decisions. This trend is shifting also the role of home and office owners and residents to move from passive energy users to active energy managers. In most cases the energy produced from rooftop solar does not always match the demand profile of a home or office building's occupancy. When the demand is lower than the production, three main solutions are to consider: sell unused excess of energy back to the grid, store energy on site for later use and/or change behaviour to use more energy whilst it is available. Governmental and regulatory barriers currently prevent the implementation of flexible scenarios to manage energy (produce, buy and sell) at optimal prices. On-site storage solutions are yet to be market accessible. Therefore the latter, behavioural change becomes a relevant scenario to study and design for.

Eco-feedback (Froelich et al. 2010; Holmes 2007) combines advances in energy monitoring technologies and human computer interfaces to provide home residents with relevant and user-friendly information about energy consumption for different appliances as well as for the total household. They are promising in providing effective information to influence sustainable behaviour, however private energy production has not been targeted sufficiently yet. Design needs to address issues such as how to enhance the experience of producing energy and how to inform the relation of production and consumption in a less technical way than the existing commercial visualizations (Katseff and Wangel 2015).

According to Chetty et al. (2008), household members are mostly unaware of their in-moment energy consumption or power usage (kW) for different appliances as well as their total household energy consumption or energy usage (kWh). Utility systems are in the background, and therefore eco-feedback is needed to "encourage householders to reflect on and re-engage with these aspects of the home's infrastructure." (Chetty et al. 2008).

Despite the growing effort to develop effective home energy systems or eco-feedback systems no conclusive impact on reducing energy consumption has been observed (van Dam et al. 2010). On the one hand, data-centric visualizations are difficult to link to existing daily life practices. They are perceived as too technical and therefore losing meaning for the user (Heller and Borchers 2011; Fischer 2008). On the other hand, ambient displays may well integrate into the home context but are criticized by their limited usefulness with regard to the abstract type of information given. Understanding the different needs and usages of energy visualizations in homes reveals the need for systems that provide different layers of information, such as consumption over time (currently, weekly, etc.), comparisons between appliances, comparison per location and per activity; as well as different modes of interaction: ambient and localized (van Dam 2013), personalized (Schwartz et al. 2014), and social (Petersen et al. 2009).

Ampul is a product-system that is designed to empower home users of energy to become an active player in their household electricity consumption through positive stimulation (Rutten 2013). Unlike other eco-feedback products, Ampul supports users in optimizing the use of the micro-generated electricity instead of only reducing energy consumption.

Ampul was developed following a user-centred design research approach using a Living Lab environment. The Concept House Village was used, a Living Lab setting with the involvement of real homes residents and a living lab home. Residents participated in an in-situ intervention implemented in their homes, to capture their perception of energy consumption at home and attitudes towards energy production. The first Ampul prototype was implemented and used as a second intervention installed in Concept House (CH). Residents spent a weekend in CH to explore Ampul in context and embedded in daily life practices. Insights from this exercise resulted in a fully developed concept. In this chapter, both interventions are described together with the final Ampul concept.

24.2 Concept House Prototype I

Concept House, by the time of this project, consisted of the Concept House Prototype I, a fully equipped home living lab environment and the Concept House Village, which involves the neighbourhood of Heijplaat, Rotterdam. The home living lab, from now on referred simply as Concept House (CH), is fully equipped for future users to live there for a period of time and engage in the conceptualization, prototyping and experimentation of innovative sustainable technologies. It is also constantly adapted to fulfil requirements of different research studies on energy consumption, with a focus to investigate and explore the users appropriation and adoption of newly introduced sustainable technologies.

For the design of Ampul, the Concept House Village was involved in gathering insights through an in-situ intervention. An extended invitation to residents in and outside the village was sent to test the first prototype in CH by spending 2 days living there.

24.3 Gathering Insights—In-situ Intervention in Real Homes

From an early survey with 90 participants, users of energy systems reported the lack of knowledge of the energy consumption of different electrical appliances and would like to use existing feedback possibilities like price and percentages to effectively understand the impact of their activities. Therefore, a proposal to design granular views of energy use may help users build up understanding of their total energy use by comparing the impact of different appliances. Still this leaves open the question of how users can link the impact of energy appliances to daily behaviour. The presented in-situ intervention aims to unveil users' motivations and obstacles related to the decisions made on energy consumption by confronting them with provoking energy visualizations.

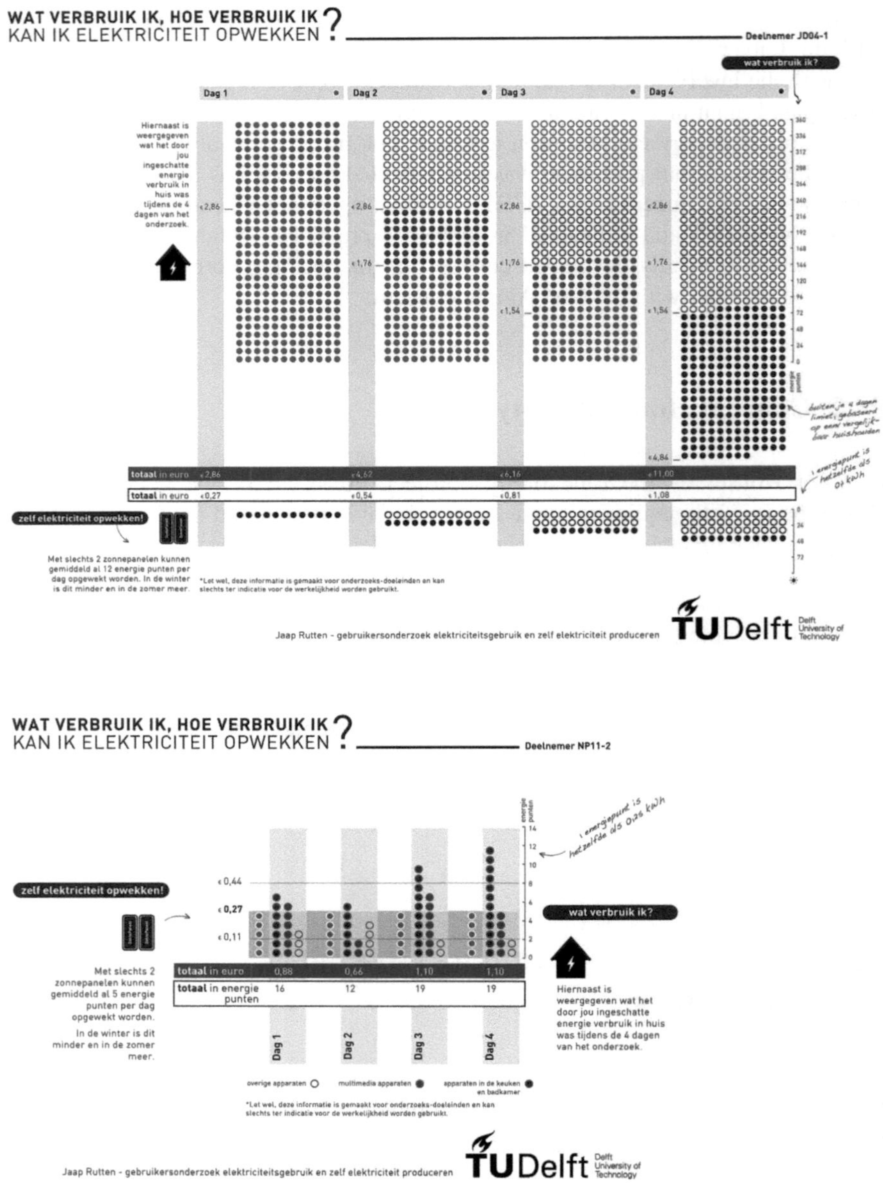

Fig. 24.1 *Top*—energy overview of the 'Limited Energy' condition; each day is colour-coded; below the *grey bar* information of how much would be covered by two photovoltaic panels is presented. *Bottom*—energy view of the 'Energy clustered' condition; each column represents one appliance category; the *green dots* represent the energy provided by 2 photovoltaic panels

What are the drivers to manage your energy consumption at home and to become self-sufficient by producing most of the energy you consume? To answer this question, two visualizations, the limited energy and the clustered energy, were implemented (Fig. 24.1). Both use a defined unit of electricity consumption to visualize electricity used. The unit is represented by a coloured dot that equals to 0.02 kilowatt hour (kWh). This number is based on a comparison between a full washing cycle and an energy-saving bulb of 11 watts used for 2 h. With this unit value, the comparison that users can make is of one washing machine cycle equals to 98 times the energy used by an energy-saving bulb.

The visualizations aim to trigger different modes of reflection about energy consumption. The 'Limited visualization' represents electricity as a limited resource, where a maximum number of units to be consumed are given per week. The 'Clustered visualization', represents electricity clustered by appliances, thus the visualization of electricity consumed on that day is distributed per type of appliance.

Participants are asked to provide the input needed for the visualizations via an electronic diary. In the evening, they annotated consumption per appliance by indicating the amount of time or times an appliance was used that day (Fig. 24.2). The list of appliances and characteristics (energy label or consumption load) was customized for each participant, based on an intake interview. The times provided by participants were translated into one of the two visualizations. Users were asked to reflect on it by answering two questions: which activities they think had influenced the consumption of that day and whether they would change anything for the next day, and why. The study period closed with an in depth interview using the experience and outcome of the intervention as sensitizing material and probes to trigger reporting.

Six households participated for four days excluding the final interview. From the overall discussion, the following points were captured as insights and requirements for the development of a prosumer interface.

The diary served as a sensitizing activity in the final interview, as users were triggered to think about their electricity consumption on a daily basis. The visual feedback triggered reflection and discussion regarding the information that is needed for an effective energy interface. For instance, fine-grained information was reported as necessary to compare energy costs per appliance. The visual overviews also triggered users to reflect on whether and how to act differently ("more sustainable"). Sustainable practices discussed by users were limited to address reduction of ligh use. Other strategies than reducing light use were not imaginable, as other activities than lighting were considered as non-negotiable in reducing its frequency. This opens the opportunity, when considering the introduction of photovoltaic panels, to support other sustainable strategies such as shaving or shifting energy consumption peaks based on energy production.

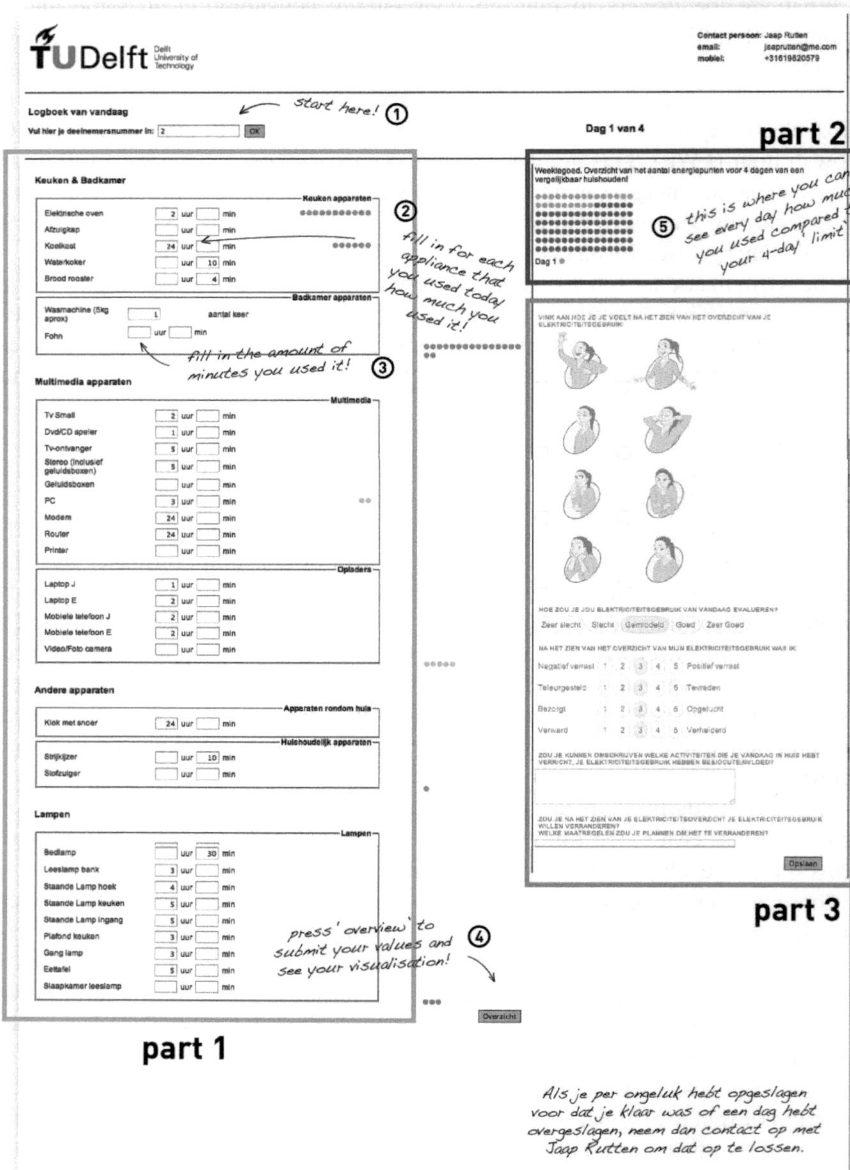

Fig. 24.2 web-based input interface for both conditions. It consists of three parts. Part 1—contains an adapted list of electronic appliances of a household and the correspondent input, clustered in four categories: kitchen and bathroom appliances; multimedia artefacts; others; and lamps. Part 2—represents the visualization of the energy consumption in *coloured dots*. Part 3—contains the self-report form

24.4 Ampul Concept

The concept is named 'Ampul' which is derived from the Turkish word *ampul*, which translates as light bulb and is in turn related to the French *ampoule* which means 'glass capsule'. Ampul consists of two visible parts: a wall-mounted display and an application on a smart phone that complement each other in their working principle, see Fig. 24.3.

The Ampul concept aims to bridge the gap between the distant technological solutions of micro-generation of energy and energy-related daily practices. A three-layer model of information and interactions is used to represent this goal:

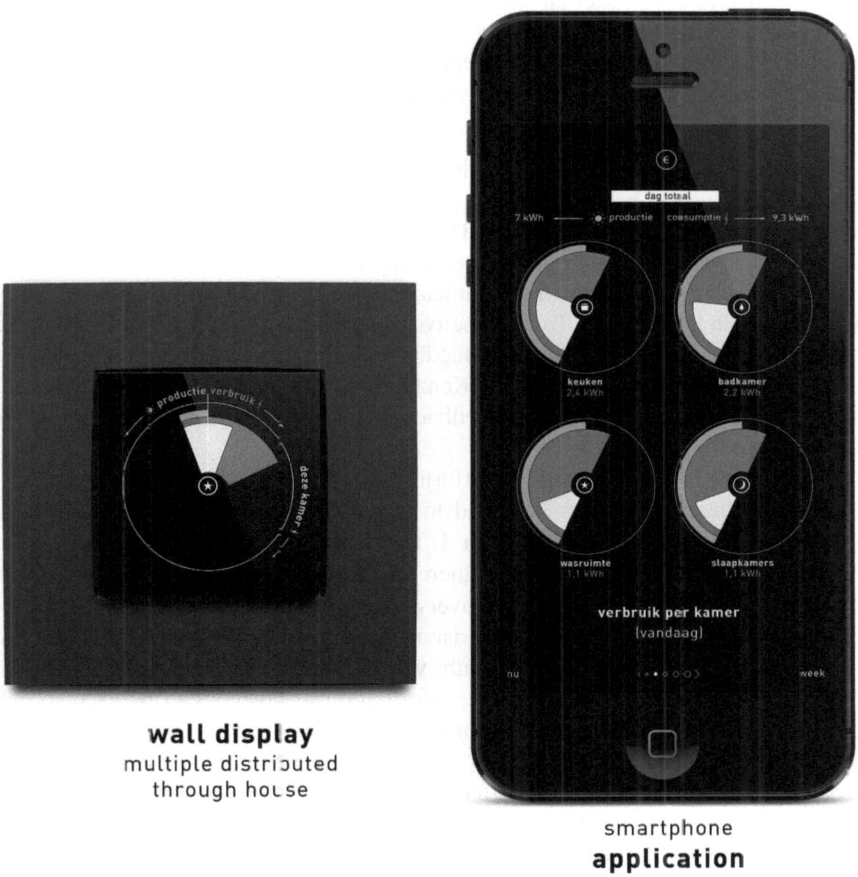

Fig. 24.3 Ampul concept, two layers of information in a wheel-like visualization. *Left*—Layer 1, the wall display; *Right*—Layer 2, the mobile application

- The first layer, the integration layer, connects to the daily life activities of users. It refers to the information and interactions that should take place close in space and time to the activity in question. This layer aims to encourage the use of renewable energy technologies and discourages the use of energy from conventional parties.
- The second layer, the learning layer, supports users' own energy literacy development in an interactive information display with overviews of their electricity consumption and production (prosumption) activities. It shows historical data and device specific consumption. Device specific data projects whether devices can run on electricity from the solar cells, or may need additional electricity from the grid.
- The third layer, the external layer, projects the impact of the current situation and compares it with the players of the market of micro-generation technologies and prosumption practices: from personal PV panels to PV parks in the neighbourhood, PV parks versus wind energy parks, etc.

In this chapter, the first two layers of Ampul are described (Fig. 24.3). The first layer, where integration takes place with the daily life practices of users, is implemented by the use of prompts, as they have been considered an effective strategy for feedback (Froehlich et al. 2010). The action of prompting is implemented for the appropriation of the technology and emergence of sustainable behaviours. The prompting visually integrates the technical impact of prosumption with the daily life activities in households. To be effective the prompts target specific behaviours, are always present to provide direct feedback, and are unobtrusive to blend into the household existing practices (McKenzie-Mohr 2011). Ampul implements in-situ prompts (timed and placed) so feedback is provided as close in time and place as possible to the targeted behaviour.

Layer 2 serves as a connecting platform between integration-layer 1 and external-layer 3, therefore it does not need to have the characteristic of blending in, to the same extent as integration-layer 1. It is however expected to work in close collaboration with the first layer and therefore has to share some of its characteristics. Layer 2 aims to provide detail overviews of the different localized information from Layer 1, allowing a comparison of the impact of energy behaviours at different time frames: day, week, month, year. It shares the same visual metaphor of layer 1 to support comparison. The information in layer 2 can be used to discover trends in both production and consumption of electricity. Users can compare e.g. how much electricity did the activity of cooking consume on different days, or discover how their consumption patterns are related to the patterns of production. Users can use the visualisation to stimulate themselves to balance their consumption of electricity with the production.

The concept uses a wheel-like metaphor to present production of electricity as a counter clockwise movement and consumption as a clockwise movement. For validation, several visualizations were generated and tested with users. The final visualization contains four layers of 'animated pie-charts' growing clockwise or counter-clockwise in relation to the data collected on consumption and production respectively. The base layer is the production (yellow) placed in the bottom, topped by the total consumption layer (red). On top of that the room consumption (white) and finally the appliance consumption (blue). The chart that represents production starts from a "12 o'clock" position and grows counter-clockwise. The chart for consumption starts from the line demarking current production value and grows clockwise. Both charts integrated explain the net consumption. A line at "12 o'clock" represents the point of parity: when the consumption chart passes that line the electricity from the grid will be used; if it stays on the right side of the line, solar energy will be delivered back to the grid. This emphasises that the production of electricity can be used for consumption and reveals an empty surface of yellow when less electricity is consumed than what is produced, thereby hinting to users that they have a surplus of electricity, which can be consumed.

24.5 Ampul Prototype

A working prototype of Ampul was implemented in Concept House with the aim to let users experience and explore the value of the prompts in supporting awareness, reflection and action towards sustainable behavior. Therefore, visualizations were implemented up to the layer of room consumption for the current consumption (history and device specific consumption were not implemented).

A wall-mounted-display hosts the 'integration layer' at room level (Fig. 24.4 top). In Concept House two main rooms were equipped with such displays: the utility room (washing machine) and the living room/kitchen room. These 'room-displays' show the instant changes of total production and consumption of the household and the electricity consumption of the designated room. These three streams of information are visualised in a single 'prosumption' visualisation. This single visualization aims at providing a 'sixth sense' to users to enable a 'sensory' understanding of prosumption behaviour per room. No 'direct' active interaction takes place between the user and the 'room-displays'.

A mobile device application implements the 'learning layer' (Fig. 24.4 bottom). The application gives the user a range of possibilities and allows 'direct' active interaction. The user can interact with the application to learn more about the details of each room in the house at different times of the day and remotely from the room itself. The app aims to provide a feeling of control and confidence by accessing different clusters of information and learning from them. Clusters of information emerge from the physical separation unit 'room-displays'. As clusters are connected to a tangible concept, rooms, they implicitly connect to a set of activities that are related to each other. For instance, a kitchen is connected to a certain set of activities, like cooking a meal, boiling water for tea, running a dishwasher, a refrigerator, and so on. Those activities carry a high contextual meaning and are therefore highly suitable to become a unit of visualization to navigate from. Therefore, from an overview of the total household, information becomes more detailed by navigating through overviews of the 'prosumption' situation

Fig. 24.4 Ampul prototype. *Top*—room display. *Bottom*—mobile app

in different rooms and an overview of the situation inside the room (appliance level feedback is not implemented in this prototype). When the application is turned on, the first layer on the screen presents the total household 'prosumption' (Fig. 24.5—2nd layer). After that the user can navigate downwards to see more details regarding rooms and devices in those rooms (see Fig. 24.5 3rd and 4th layer). When navigating upwards, a specific overview of the electricity generated by the solar cells is presented (first layer) to provide the user with information regarding return on investment, maximum production hours and days, etc.

The prototype ran independently from other systems installed in Concept House. In total, three 7-inch tablets were used to implement both room-displays and the app. For the room-displays a frame was designed so the visual area on the screen was adjusted to focus the attention on the wheel-like metaphor. All appliances except for the hairdryer, chargers and lights, were located in these two rooms in the house: the utility room and the combined kitchen and living room. The prototype measured all electrical outlets in these rooms, while the fixed

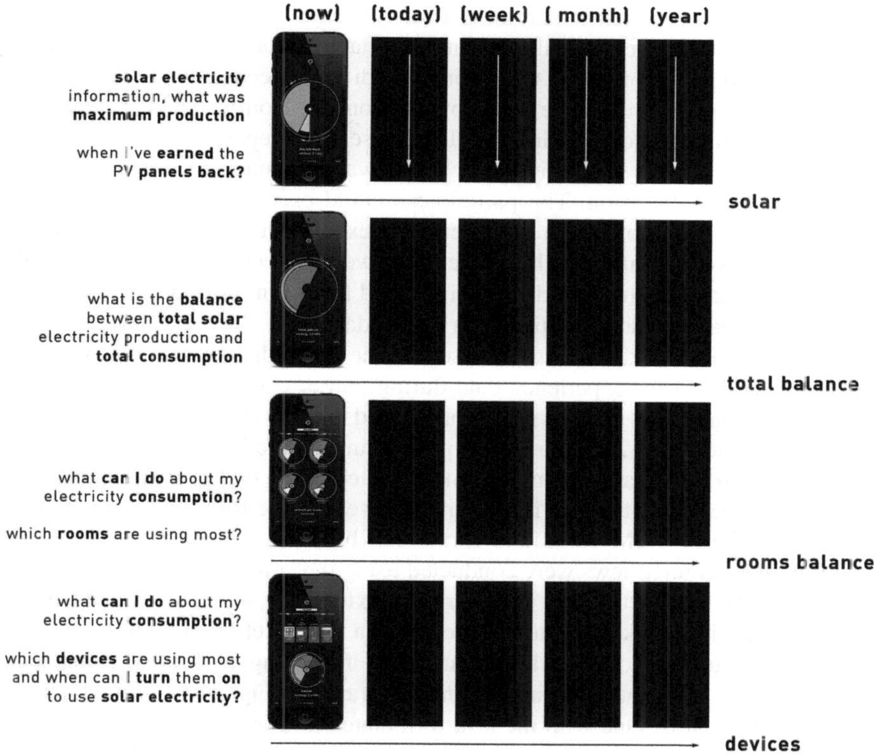

Fig. 24.5 Ampul prototype. Overview navigation of mobile app (4th layer was not implemented)

appliances were attached to a fixed outlet that was measured separately. The available outlets throughout the combined kitchen and living room were counted as separate appliances. As explained earlier, only the part of the 'now' visualization of Ampul was implemented in the prototype.

24.6 User Exploration—In-situ Intervention in Concept House

Local residents of the Concept House Village and outsiders were invited to participate in this intervention. Four families confirmed their participation: a single male participant of age 57, a couple between the age of 25 and 30, a family with a single child around the age of 15 and parents between 50 and 60 and a single female participant of age 56.

Participants were asked to stay in CH for two days and two nights doing their normal activities as they usually do on such days. In addition, the participants were requested to perform four tasks corresponding to electricity related activities that fit the use of Ampul in the setup of the Concept House: cooking, tea/coffee preparation, laundry and watching TV. The house was prepared with all the appliances necessary for these activities, such as a washing machine, electric cooking equipment and a television. The participants could execute each activity on any given point in time, though the activities were expected to take place at a certain time of day. Cooking and watching television were expected to take place in the evening, whereas washing laundry was indicated to be run on solar energy, requiring the participants to do the activity during the day.

Three in-situ methods were used based on the methods introduced in Chaps. 2 and 12, to collect user experience data during and after the intervention. During the intervention, ESM cards were implemented based on Experience Sampling Method (Hektner et al. 2007) to collect participants' reflection on a task right after it was done; an ESM cards form was implemented at the end of participants' stay, asking participants to stick the ESM cards answers in the form and evaluate their overall experience (ESM cards and forms can be seen in context in Fig. 24.6); finally individual interviews were conducted using the completed forms as probes.

Three ESM card sets were placed right next to the appliance related to the requested activity: cooking plate for preparing a meal, kettle and coffee machine for preparing tea or coffee, washing machine for doing a laundry using solar energy and TV for watching an evening programme (Fig. 24.6—top). Each set contained four cards. One with the task formulated as a question on one side, and space to write down an answer on the other side. The other three cards ask participants to assess the interaction with Ampul after doing the task on the basis of three scales: feeling of satisfaction; difficulty of understanding and useful information. The scale-cards contained a question of one of the measuring scales on one side, and the scale to indicate an answer on the other side.

The specific questions to describe each task in the ESM cards were:

Activity	Task
Cooking	How much electricity did you use during the process of cooking? Indicate how you found the answer
Tea/coffee preparation	What is the contribution of preparing tea or coffee to your total consumption of electricity? How did you find out?
Laundry	Can you indicate if the washing machine ran using solar energy? How did you find out?
TV	While watching TV at night, could you indicate when you use more electricity than what you have generated over the entire day?

In the evaluation form (Fig. 24.6—bottom) participants were asked to glue each card on a designated place, which resulted in a clustering of four themes to facilitate reflection. The themes originated from the ESM cards aim to compare the four tasks on the basis of: energy consumption and the user experience with relation

Fig. 24.6 *Top*—The four sets of ESM cards; *Bottom*—the evaluation form

to satisfaction, understanding and usefulness. Participants were then asked to rate their overall experience based on four semantic scales: surprise, satisfaction, peace of mind and information. A final open question addressed whether they have changed their attitude towards home micro-generation of electricity.

During the closing interview participants were asked to elaborate on their answers given in the individual reflection. Important qualities to address in the interview were:

1. Feelings of reward and victory as result of being informed about your 'home grown' electricity.
2. Annoyance as a result of constantly being reminded of your energy consumption.
3. Guidance and self-confidence as result of the prompts that give constant live feedback.

The room-displays were perceived as useful feedback to understand how own production may cover fully or partly the electricity needed by an activity. The ambient characteristic of the room-displays built up users' awareness of the impact of solar production on the electricity consumption and consequently their (in) dependence of the grid. A feeling of connectedness with nature was also reported as the visualization showed dynamically the degree by which solar energy, depending on the time of the day and weather, can meet or not the household's electricity demand.

On the other hand, sudden peaks of energy use caused by certain electrical appliances (e.g. kettle) were linked to a feeling of frustration, as they would in most cases surpass the production even though the total consumption of such device would be less than what was produced at that moment. The absence of accumulated and historical data aggravated this feeling.

All participants agreed that the wheel-like metaphor was the most valuable part of the interface in particular in the room-display, being reported as "simple and understandable at a glance". In contrast, only few participants preferred the mobile app to the room-display; for two participants, with little experience with mobile apps, it was not used at all. Overall, the combination of both interfaces was perceived as pleasant and useful.

24.7 Conclusions

The presented work illustrates a design research project in the area of sustainability using the facilities of a Living Lab setting. Insights were gathered from in-situ interventions in real homes, and Ampul was developed as a quick and functional prototype in Concept House. As a design intervention, Ampul was deployed to bring existing and potential users of solar panels to explore and provide feedback to an interactive and visual concept for home prosumers. This setup allows for quick iterations in the design and development of a final concept, involving users as collaborators in the design and testing of quick prototypes. In a future stage, a full-developed service/product will be installed in real homes for an evaluation in context and for long periods of time. The outcomes are expected to provide quantitative and qualitative results on the impact of such concept on sustainability (e.g. optimization of energy production and consumption) as well as on people's lifestyle (e.g. increase of comfort, quality of life, personal values).

Acknowledgment The development of Ampul is part of the graduation project of Jaap Rutten (Rutten 2013) at the Faculty of Industrial Design Engineering in Delft University of Technology, funded by SusLabNWE Interreg (www.suslab.eu). All visual materials in this document have been provided by Jaap Rutten. Special thanks to Marcus de Hoogh for developing the prototype, and to all participants involved in the design process of Ampul.

References

Chetty, M., Tran, D., & Grinter, R. E. (2008). Getting to green: Understanding resource consumption in the home. In *Proceedings of the 10th International Conference on Ubiquitous Computing* (pp. 242–251). ACM.

Fischer, C. (2008). Feedback on household electricity consumption: A tool for saving energy? *Energy Efficiency, 1*(1), 79–104.

Froehlich, J., Findlater, L., & Landay, J. (2010). The design of eco-feedback technology. In *Proceedings of the SIGCHI Conference on Human Factors in Computing Systems* (pp. 1999–2008). ACM.

Heller, F., & Borchers, J. (2011) PowerSocket: Towards on-outlet power consumption visualization. In *Proceedings CHI'11 Extended Abstracts on Human Factors in Computing Systems* (pp. 1981–1986). ACM.

Hektner, J. M., Schmidt, J. A., & Czikszentmihalyi, M. (2007). *Experience sampling method: Measuring the quality of everyday life*. Sage Publications.

Holmes, T. G. (2007). Eco-visualization: Combining art and technology to reduce energy consumption. In *Proceedings of the 6th ACM SIGCHI Conference on Creativity and Cognition* (pp. 153–162). ACM.

Katzeff, C., & Wangel, J. (2015). Social practices, households, and design in the smart grid. In *ICT innovations for sustainability* (pp. 351–365). Springer International Publishing.

McKenzie-Mohr, D. (2011). *Fostering sustainable behavior: An introduction to community-based social marketing*. New Society Publishers.

Petersen, D., Steele, J., & Wilkerson, J. (2009). WattBot: A residential electricity monitoring and feedback system. In *CHI'09 Extended Abstracts on Human Factors in Computing Systems* (pp. 2847–2852). ACM.

Rutten, J. (2013). Designing Ampul: Exploring future prosumer interactions. Master thesis in Industrial Design Engineering, Delft University of Technology.

Schwartz, T., Stevens, G., & Jakobi, T. (2014). What people do with consumption feedback: A long-term living lab study of a home energy management system. *Interacting with Computers, 27*(6), 551–576.

van Dam, S. S. (2013). *Smart energy management for households*. Doctorate thesis, Delft University of Technology.

van Dam, S. S., Bakker, C. A., & van Hal, J. (2010). Home energy monitors: Impact over the medium-term. *Building Research and Information, 38*(5), 458–469.

Chapter 25
Energy Feedback Objects

Tomasz Jaskiewicz, Aadjan van der Helm and David V. Keyson

Abstract It is often assumed that providing occupants with feedback about their energy consumption will encourage them to understand their own contribution to energy consumption and stimulate them to save energy as a result. However, providing such feedback in the form of raw data is known to be too difficult for occupants to interpret. There are many examples where raw data has been replaced by easy to read data visualisations, communicated through metaphors, translated to specific tips, or even turned into playful interfaces and games. However, even such approaches often have short-lived impact on occupant behaviour, as they are often not embedded into complex social practices taking place in building environments, and providing individual feedback to occupants proves insufficient. The challenge of developing energy-feedback designs which may trigger lasting behaviour change by engaging social practices of building occupants was taken up by students following the "Interactive Technology Design" (ITD) course at the IDE faculty of TU Delft.

Keywords Energy feedback · Interactive technology design

T. Jaskiewicz (✉) · A. van der Helm · D.V. Keyson
Delft University of Technology, Delft, The Netherlands
e-mail: t.j.jaskiewicz@tudelft.nl

A. van der Helm
e-mail: a.j.c.vanderhelm@tudelft.nl

D.V. Keyson
e-mail: d.v.keyson@tudelft.nl

D.V. Keyson et al. (eds.), *Living Labs*, DOI 10.1007/978-3-319-33527-8_25

25.1 Energy Feedback Objects

Non-residential buildings, such as offices or universities, contribute a large share of the energy consumption in Europe's building sector. Various estimates indicate that even over 30 % of that energy could be saved by changing the behaviours of building occupants (Lopes et al. 2012). To achieve such change, occupants would only need to adjust their daily practices to avoid unnecessary waste of energy, which typically doesn't require much additional effort or time, or diminish occupants' comfort. Some examples of such actions can be turning off lights and other equipment when not in use, turning down heating instead of opening windows in the winter when too warm, or ventilating by opening windows instead of using air conditioning on a mild summer day. However, it is difficult for building occupants to change their daily routines, and, once changed, to maintain them without falling back into old habits.

It is often assumed that providing occupants with feedback about their energy consumption will encourage them to understand their own contribution to energy consumption and stimulate them to save energy as a result. However, providing such feedback in the form of raw data is known to be too difficult for occupants to interpret. There are many examples where raw data has been replaced by easy to read data visualisations, communicated through metaphors, translated to specific tips, or even turned into playful interfaces and games (Spagnolli et al. 2011). However, even such approaches often have short-lived impact on occupant behaviour, as they are often not embedded into complex social practices taking place in building environments, and providing individual feedback to occupants proves insufficient. The challenge of developing energy-feedback designs which may trigger lasting behaviour change by engaging social practices of building occupants was taken up by students following the "Interactive Technology Design" (ITD) course at the IDE faculty of TU Delft (Aprile and van der Helm, 2011).

Four groups of students received a design brief to design new kinds of energy feedback solutions that would be naturally integrated into daily practices of non-residential building occupants. The building of the IDE faculty was used as a test case, and each group was assigned to one of the rooms of this building as a design context. These included a two-person office, an open office, a flexible design studio space and a large lecture room. Students iteratively conceptualised, prototyped and tested the solutions. Throughout this process, which consisted of 20 workdays during a full semester, they sought to gain a deep understanding of the intricacy of the studied situations, leading to revisions and improvements of their concepts. Ultimately, four concepts and working prototypes were realised.

The first concept is called "Volt" (Fig. 25.1). It is a desktop power hub with three sockets, designed for a flexible workspace. When a person working around the table connects her laptop or phone charger, the device indicates the amount of energy flowing into the socket. When another person connects to the same hub, the energy starts to flow to her socket, and stops flowing to the socket of the first person using the Volt device. By tilting the device, the flow of energy can be changed

Fig. 25.1 The Volt power socket hub provides light feedback to indicate electric energy consumption and enables users to share energy with each other by physically tilting the device

back, or a balanced output can be reached. The device is filled with granular material and tilting it additionally provides tactile feedback, giving users the feeling of electrical energy physically flowing from one outlet to another. Because the interaction with the device is very playful, the annoyance of losing power has been diminished. Instead, the use of the device triggers people to think about the energy they consume, to develop curiosity about exact energy consumption and the origin of the energy, and provides an opportunity to discuss these issues with others around the work desk.

The next concept is a solution for providing energy feedback in the studio space. Students found that inefficient control of lights was the main contribution to wasting energy by occupants in the studio. "Akiko" (Fig. 25.2) is a device designed to replace the light switches, and to become a "host" of the room. Akiko monitors and controls the lights in the room. It also uses light to express "emotion" corresponding to its understanding of the efficiency of the use of light in the room. Users wishing to work in the room pick up Akiko at the entrance. The room lights up sufficiently for users to move around it, and when placed on a desk, Akiko turns on the stronger lights above. The longer the lights are on, the more "upset" Akiko gets, and the more erratic the emitted light pattern becomes. When Akiko is left alone at the table, the intensity of this pattern reaches its maximum, reminding users to bring it back to the dock at the entrance and by this to switch the lights off. However, the aim of Akiko is not to punish the users of the space by sending annoying reminders, but to create an emotional bond with them. By tapping Akiko, its users can "assure" it that the energy is put by them to good use, and Akiko calms down and its light pattern returns to its mild state.

Fig. 25.2 The Akiko prototype uses energy monitoring and control to create an emotional bond between the user and the studio lighting

The concept presents an alternative to traditional light switches, where occupants of a room build a social relationship with the room based on the room's energy consumption. In short, the Akiko device helps to assure users that energy is used for a good purpose while thinking about energy as a limited resource. It makes them feel appreciated when saving energy and enforces a habit which is likely to be carried on also in other rooms not equipped with Akiko.

The "Neo" concept addresses the situation of a two-person office. Unlike the studio or flexible office context, the two-person office is assumed to be fully occupied by the same two persons. In such a case, more detailed feedback on energy can help the users to analyse their daily practices. The specific dynamics of two persons occupying a room are considered so as to stimulate them to mutually encourage one another to save energy. Neo was designed to provide office users with an ability to set a daily goal for using energy in the office. It is an aesthetically attractive device hanging on the wall, which uses a diaphragm mechanism to show the status of the daily energy use (Fig. 25.3). When approached, it displays an exact value of energy that has been used during the day. In this way Neo provides office occupants with an attractive energy display, integrating an abstract and explicit feedback and bringing the energy use into the social sphere of the office context.

The fourth concept called "Hotspot" addresses the situation of energy consumption in the lecture room (Fig. 25.4). A typical lecture room is continuously heated or cooled, regardless whether it's unoccupied, occupied by a small audience, or used to full capacity. In response to this problem the students designed a new chair for the lecture room. The chair is individually heated or cooled by rubbing the armrest. In addition, the chair promotes the clustering of the occupants to optimise the use of energy since the local warmth around a single chair is

Fig. 25.3 Neo is an aesthetically attractive wall-mounted device that allows office occupants to set an energy goal for the day and provides both ambient, and detailed feedback on current energy consumption at the office

Fig. 25.4 The prototype of the Hotspot project involved four lecture room chairs that would warm up when rubbed, but only if people sat next to each other

increased by adjacent warmed occupied chairs. In this way, the concept also supports people to cluster in groups with particular indoor climate preferences. In the Hotspot concept, feedback on energy consumption is formed through requiring a regularly repeated effort of the occupant to trigger heating or cooling, and the effect of energy used can be instantly felt and appreciated, not only by one user, but the entire group of users in her direct proximity.

For occupants of nonresidential buildings, energy is a resource that is very difficult to perceive, and, as a result, also difficult to conserve. The four concepts described in this section present a spectrum of alternative approaches for not only increasing occupants' awareness of energy use, but also for triggering emergence of new kinds of social practices around the goal of saving energy. The initial user tests performed by the students during the development of the projects and during their public exhibition received a positive reception. This encourages further investigation of energy feedback solutions that, instead of only passively providing users with energy-use information, actively engage them in rich, interactive experiences.

Acknowledgments We would like to thank the following students for their design work in ITD course as presented in this chapter.
Volt: Nina van Adrichem, Joric Koghee, Bob van Iersel, Kathrin Höfer, Lianne Siemensma.
Akiko: Nadege Heyligar, Zoé Martial, Daan Oldenhof, Jip Warendorf, Rochelle Simons.
Neo: Heleen Bouma, Daan Krijnen, Jip Eilbracht, Lisa Vork, Maibrit Andersen.
Hotspot: Josephine Vlaanderen, Tin An Wang, Dioni Schoop, Mert Tosun, Jesse Beem.
Coaches: Ianus Keller, David Keyson, Tomasz Jaskiewicz, Aadjan van der Helm.

References

Aprile, W. A., & van der Helm, A. (2011). Interactive technology design at the Delft University of Technology-a course about how to design interactive products. In *DS 69: Proceedings of E&PDE 2011, the 13th International Conference on Engineering and Product Design Education*, London, UK, 08. 09.09. 2011.

Lopes, M. A. R., Antunes, C. H., & Martins, N. (2012). Energy behaviours as promoters of energy efficiency: A 21st century review. *Renewable and Sustainable Energy Reviews, 16*(6), 4095–4104.

Spagnolli, A., Corradi, N., Gamberini, L., Hoggan, E., Jacucci, G., Katzeff, C., et al. (2011). Eco-feedback on the go: Motivating energy awareness. *Computer*, *44*(5), 38–45.

Part VII
Understanding Comfort and Energy Practices in Residential Buildings

Chapter 26
Relationship Between Building Technologies, Energy Performance and Occupancy in Domestic Buildings

Olivia Guerra-Santin

Abstract Building regulations have been updated to improve the energy performance of buildings. However, research has shown large differences between expected and actual energy performance of buildings. The differences have been attributed partially to occupant behaviour. Occupants have a large influence on the actual performance of buildings, creating uncertainties related to the actual energy savings, payback periods for low carbon technologies, and actual comfort in the buildings. This section explores the influence that building occupants have on the actual performance of domestic buildings and the consequences in the development of new and renovated low and zero energy housing. Monitoring building performance before and after renovation for retrofit projects, and monitoring building performance in experimental Living Labs and after the occupancy of buildings are discussed as potential solutions for occupancy uncertainties.

Keywords Energy consumption · Occupancy · User behaviour · Building performance · Performance gap

26.1 Introduction

Worldwide, building regulations have been updated to improve the energy performance of buildings. New buildings and buildings that undergo a large renovation mostly comply with building envelope and systems requirements or a minimum energy performance. As an outcome, buildings recently constructed have airtight envelopes and efficient heating and ventilation installations, which in theory should reduce their energy consumption. However, research has shown

O. Guerra-Santin (✉)
Delft University of Technology, Delft, Netherlands
e-mail: o.guerrasantin@tudelft.nl

D.V. Keyson et al. (eds.), *Living Labs*, DOI 10.1007/978-3-319-33527-8_26

large differences between expected and actual energy performance of buildings (Sunikka-Blank and Galvin 2012; Guerra-Santin and Itard 2009). The differences have been attributed to the actual building characteristics and occupant behaviour.

Faults in building envelope and systems can be identified and reduced by conducting proper commissioning of systems and building envelope tests during the construction process and before occupancy, such as air permeability tests and infrared thermography surveys (Guerra-Santin et al. 2013). As a contrast, the effects of occupancy are more difficult to predict and avoid. Occupants have a large influence on the actual performance of buildings, creating uncertainties related to the actual energy savings, payback periods for low carbon technologies, and actual comfort in the buildings.

This section explores the influence that building occupants have on the actual performance of buildings and the consequences in the development of new and renovated low and zero energy buildings.

26.2 User-Related and Building-Related Energy Consumption

The effect of the occupancy, and the way to minimize such effect largely depends on the type of energy targeted. In residential buildings, energy consumption can be divided into energy for space heating, space cooling, ventilation, artificial light, domestic hot water, cooking, auxiliary energy and power for appliances and electronics. These can be classified into building-related and user-related energy consumption.

Building-related energy consumption is the energy used for services related to the building itself, such as space heating and cooling, ventilation, lighting and domestic hot water. These energy services can be directly influenced through design both in new and renovated buildings. These energy requirements can be reduced by delivering a better design (e.g. passive design) that allows the building to retain heat gains in winter, avoid heat gains in summer, and maximizes the use of natural light. For example, space heating can be reduced by improving the building envelope characteristics such as increasing insulation level or using double or triple glazing. Space cooling can be reduced by providing with the right amount of shading and ventilation during the summer. Artificial light demand can be reduced by using as much as possible natural light.

User-related energy consumption is considered to be mostly influenced by the building's occupants. Within user-related consumption we can find the energy used for cooking, use of electronics and appliances. Although the use of energy efficient appliances and electronics could reduce the energy consumption, the purchase of such products are mostly in the hands of the occupants. Designers and building regulators have almost no influence on these choices. In order to influence user-related energy consumption, we could make use of feedback mechanisms and interventions for behavioural change.

Building regulations often only focus on building-related energy consumption, which is also called 'regulated energy consumption', while few solutions are offered to influence the user-related (or unregulated) energy consumption. However, building performance is often assessed as a whole. The energy consumption of a building is metered and compared against a benchmark, design expectations or building simulations. As a consequence, results of these investigations show the above-mentioned differences between the expected and actual energy consumption.

To carry out building performance assessment, it is necessary to understand the different effects that the occupants have on building-related and user-related energy consumption in order to be able to determine the instances in which actual energy reductions can be achieved. We have identified four ways in which occupancy affects the energy performance of buildings. These are: rebound effect, pre-bound effect, differences between households, and user-building technology interaction and building control. These factors have two main consequences in the building industry and on building research: (1) they create uncertainty related to the actual payback periods for low carbon technology, and (2) they influence the actual energy consumption (performance of buildings). Following sections discuss these factors.

26.3 Occupancy Impact on Uncertainties During Design and Financing

Building renovation projects, and very low or zero energy new buildings are associated with high costs and long payback periods. The actual performance of these buildings is often unpredictable due to the uncertainty provided by occupant behaviour (Virote and Neves-Silva 2012).

Building simulations are used within new and renovation projects to predict the building-related energy performance of the buildings. However, simulated occupant behaviour is often based on assumptions rather than measured observations. Heating set-points, occupancy and schedules, as well as use of lighting and appliances to calculate internal heat gains are based on an 'average' household with an 'average' occupancy. Therefore, current simulation tools are not a totally reliable instrument to predict the performance of energy efficiency measures in a building (Virote and Neves-Silva 2012).

Expected user-related energy consumption is usually based on benchmarks or on energy calculations based on assumed number and type appliances owned in a household, and the hours of use per appliance. Although calculating the electricity consumption with an appliance audit could generate more accurate results in a building with known occupancy, when the assumptions are based on average occupancy, the results can be as uncertain as with building simulations.

As a consequence of these uncertainties, when the expected energy consumption is compared to the actual energy consumption, the outcome is the so-called 'performance gap'. The performance gap is actually caused by two different effects: the pre-bound effect, and the rebound effect. These are explained in following sections.

26.3.1 Pre-bound Effect

The pre-bound effect has been defined as the situation in which, before the renovation, less energy is consumed than expected. According to Sunikka-Blank and Galvin (2012), as renovations cannot reduce energy that is not actually consumed, this has implications for the economic viability of thermal retrofits. The expected energy consumption in renovation projects can be higher than in reality given than, as mentioned before, in building simulations and energy calculations, an 'average household' and 'average building occupancy' are often employed.

There is a large diversity in household characteristics, preferences and lifestyles of buildings' occupants, and therefore, large differences have been found between standardized occupancy patterns and actual occupancy patterns (Guerra-Santin and Itard 2012). For example, a single young adult with full-time employment would spend only few hours at home everyday, therefore requiring less energy for heating than a household with two retired seniors that spend most of their time at home and who, because of their age, require higher thermostat settings.

In renovation projects in which someone has to finance the low carbon technologies, payback periods and return of investments are very important for the success of a project and to keep implementing these technologies in new projects. More complexity is added in multi-family properties, and rental properties, in which the incentives for saving energy are split between tenants and landlords.

Building research focused on developing occupancy profiles, occupancy patterns, and diversity profiles can help to diminish the uncertainties on payback periods and investments in low carbon technologies.

26.3.2 Rebound Effect

The rebound effect has been widely studied in recent years. The rebound effect can be defined as the increase on energy consumption in services for which improvements in energy efficiency reduce the energy costs (Herring and Sorrell 2009). Rebound occurs when people compensate for efficiency improvements by increasing their spending (Hens et al. 2010). The rebound effect can be direct or indirect. A direct effect is seen when energy consumption increases on the same service in which the efficiency was improved (e.g. space heating). An indirect effect is seen when energy consumption is increased on a different service or activity (e.g. savings on energy for heating are 'spent' on an intercontinental flight). Within the context of this research, we only focus on the direct effect.

The rebound effect is found mostly on energy consumption for heating. For example, after renovation, a household can expect lower energy bills and thus decide to set the thermostat to a higher temperature level to increase their comfort. However, it is important to investigate the origin of the extra expenditure, since the increase on temperature could be a response to an actual need, for example fuel poverty: a household in the past might have set the thermostat to a lower temperature level in order to keep energy bills low even thought they always needed a higher temperature.

In addition, it is important to consider that the rebound is in some cases, not a consequence of the user's choices or behaviour, but a consequence of new technologies. For example, changing a manual thermostat for a programmable thermostat during renovation might increase energy consumption. The programmable thermostat will switch on the heating system before the time stated in the program, as the system requires some time to heat the spaces at the desired temperature.

The rebound effect can also be a matter of switching energy demand from one service to another. Old houses often only have natural ventilation or mechanical exhaust ventilation, which implies a null or very low energy consumption related to ventilation. New and renovated buildings usually have heat recovery ventilation, which employs more energy than mechanical exhaust or natural ventilation. However, a heat recovery ventilation system would imply less heating losses, and thus, lower energy demand for space heating. A similar effect can be found in new and renovated buildings due to the use of auxiliary energy to run systems. This type of energy is often not included in benchmarks, since there is no comparison point before the renovation or in old buildings. Because of this matters, it is important to carry out energy performance assessments with sub-metered data.

26.3.3 Performance Gap

The rebound and pre-bound effects, as well as the performance gap, could be considered an 'artificial' or research-made problem, since the gap is only produced when an 'expected' energy consumption is defined (source: results from workshop UserTec, Copenhagen 2015). Therefore it is important to assess well how the expected performance was calculated, what assumptions were considered, what type of energy services were included, and so on. Assumptions are particularly important when the performance baseline is a building simulation calculation, while knowing the services included and the type of reference building are particularly important when the baseline is a benchmark.

For example, during building simulations and energy calculations, it might be assumed that in the building, the clothes will be air-dried, that a building is empty during weekends (e.g. schools), or that the systems will be switch off during non-working hours. On the other hand, a benchmark might not include energy consumed by elevators, or security cameras, as well as the use of auxiliary energy for new technologies.

Rebound and pre-bound effects can be minimised by knowing better the context of the users, their actual requirements and their actual capacity for changing behaviour. This information can be obtained through monitoring campaigns and research in Living Labs.

26.4 Occupancy Impact on Actual Energy Consumption

The effect of occupancy on actual energy consumption refers to the comparison of a dwelling's energy consumption with similar dwellings or relative to the dwelling itself. The consequences of this influence are not related to financing or uncertainties, but to the design and use of the buildings.

It is common to read within building research that the actual energy consumption of similar dwellings varies up to a factor of 2 (Fokaides et al. 2011; Doran 2005; Danielski 2012; Bell et al. 2010). These differences can be caused by two main factors: by differences between households, and by the interaction between user and building technology. These topics are reviewed in this section.

26.4.1 Differences Between Households

A third way in which the user affects the actual energy consumption in buildings is due to the differences between households. As mentioned before, the pre-bound effect is caused by considering an average household as a basis for calculations. However, differences between households can be quite large. The factors identified as influencing energy consumption are:

(1) Demographics can greatly influence the energy requirements in residential buildings. Two of the main factors affecting energy consumption in dwellings are household composition (Schipper et al. 1982; McLoughlin et al. 2012), and age (Kane et al. 2015; Wei et al. 2014) especially in regard to the presence of children and elderly people (Verhallen and Raaij 1981; Guerra-Santin and Itard 2010; Guerra-Santin et al. 2015; Yohanis et al. 2008). For example, older occupants might require more energy for heating than younger occupants given their low activity level and (sometimes) poor health condition. However, younger households might require more energy for electronics than older households. Large households with children will consume more energy than smaller households since they occupy more spaces, require

more hot water, and consume more meals. Demographics of a household are factors that cannot be changed, and they would always have an effect on energy use. It is important to add, however, that these factors will change over time within the household. The use of household profiles in building simulation is therefore important in building research.

(2) The background of the household could also have an effect on energy demand. Occupants from warmer countries might require higher heating settings than occupants from cooler countries. In addition, culture could also have an effect on airing patterns and cooking habits. This aspect is especially important when we address renovation projects in segregated areas or in mixed neighbourhoods.

(3) Lifestyle and schedules of households also affect energy consumption. Households spending more time at home would consume more energy than those with busy schedules away from home. In the same way more active households could be using less energy than more sedentary households. These factors are closely related to household composition, which are relatively permanent, but also to socio-economical factors, which could change over time.

4) Socio-economical factors also affect energy consumption, for example employment status (Verhallen and Raaij 1981; Yohanis et al. 2008; Kane et al. 2015), income (Meyers et al. 2010; Wei et al. 2014), tenure (Kane et al. 2015; Guerra-Santin et al. 2009; Wei et al.2014); and educational level (Wei et al. 2014). However, the effect of these factors might depend on the country or region of study.

(5) Other issues such as health related habits (i.e. smoking), attitudes towards energy saving or environmental awareness, and preferences for comfort can also have a great influence on energy use. These factors are more difficult to determine, since they are more subjective and can very across time. However, these could have a large effect on the success of energy reduction campaigns and interventions for behavioural change.

Some occupant-related effect on energy consumption can be minimised by providing people with the right feedback and information regarding the impact of their choices: for example, by changing airing routines (use of a higher setting on the heat-recovery ventilation instead of opening windows on cold days) or not heating unoccupied spaces. Changing behaviour towards more sustainable practices can significantly reduce the energy consumption in buildings (Palmborg 1986; Ryckaert et al. 2010; Yun et al. 2010; Green Building Council 2011; Nguyen and Aiello 2013). However, there are conditions that cannot be changed. For example, elderly people spend more time and home and heat to a higher degree because of their low activity level. However, these practices follow an actual need and cannot be changed. As a contrast, a household that heats their home when is not occupied could change these practices.

26.4.2 User-Building Interaction and Control

With the incorporation of new technologies in renovated and new buildings, occupants are faced with complex systems that are difficult to operate, which can lead to an increase on energy consumption and a decrease in overall satisfaction (Cole et al. 2008; Stevenson et al. 2012; Stevenson and Leaman 2010; Maier et al. 2009; Thomsen et al. 2005).

According to Meyers et al. (2010) the barriers to the adoption of monitoring and control technologies (programmable thermostats, smart meters, zone heating, automated sensors and wireless communication infrastructures) are lack of consumer awareness of the technologies, high costs due to lack of economies of scale, and difficult user interfaces. Complex systems will not be adopted if consumers cannot invest the time to understand and use the system. Such systems should be designed for usability with intuitive interfaces.

In order to increase the probability that a building performs as expected, it is also necessary to provide occupants with clear instructions to understand and operate the building. Several authors have stressed the importance of providing clear operation manuals and inductions to new homes (Stevenson and Rijal 2010; Leaman et al. 2010). Within building research, we can make use of re-enactment and home walkthroughs to investigate the way in which occupants operate their homes.

26.5 LivingLabs to Reduce User-Related Uncertainties

In recent years, research has focused on occupancy monitoring studies to (1) provide data for building simulation to improve building models and energy calculations (Heo et al. 2012; Sun et al. 2014; Raftery et al. 2011; Kashif et al. 2013; Lee and Malkawi 2014), (2) using real time monitoring data to control building systems (Dominguez et al. 2013; Oldewurtel et al. 2012; Privara et al. 2012; Alvarez et al. 2013; Yang and Becerik-Gerber 2014; Aria and Akbari 2014), and (3) to provide interfaces to change behaviour (Kanga et al. 2012; Jain et al. 2013; Huebner et al. 2013). Monitoring campaigns are however, time consuming given the complexity of working in real-life settings. Issues such as privacy and safety often rise in this type of studies. Living Labs can provide the necessary infrastructure to implement this type of building research.

The investigation of occupancy patterns (presence at home), indoor conditions (temperature, indoor air quality), occupant behaviour (use of heating and ventilation systems) and occupancy practices, can help to determine and reduce the effect of occupants on building-related energy consumption. These studies can be carried out in pre-renovation monitoring campaigns, or in post-occupancy monitoring campaigns. These are discussed in the following sections.

26.5.1 *Pre-renovation Monitoring to Diminish User-Related Uncertainties*

Pre-renovation monitoring campaigns have been suggested as a mean to increase the energy performance of renovated buildings (Gupta and Chandiwala 2010). Pre-renovation campaigns can be carried out to provide actual user information for building simulation programs and to determine the actual user requirements in renovation projects. This type of occupancy investigation can help to reduce uncertainties related to pre-bound effects and can also foresee issues related to energy performance that rise from differences between households.

Monitoring campaigns can provide with detail information to be used as input on building simulation software, or can be used to calculate electricity consumption when combined with the power of appliances. According to various authors, more accurate occupancy patterns could improve the results of building simulations (Lee and Malkawi 2014; Johansson et al. 2011; Virote and Neves-Silva 2012). The disadvantage of these studies is related to the fact that the results cannot be generalized to a wider population; at best, the patterns can be considered as characteristics of the type of household and building in which the monitoring campaign was carried out (D'Oca and Hond 2015).

26.5.2 *Post-renovation/Post-occupancy Monitoring Campaigns*

Post-renovation or post-occupancy (new buildings) monitoring campaigns are mostly carried out on experimental or prototyping projects. One of the main objectives of these campaigns is to inform designers to improve future projects. In these projects it is possible to test the impact of the user on the low carbon technologies, and the level of control and understanding of the occupants.

Post-renovation or post-occupancy (new buildings) monitoring campaigns are less often carried out in real-life projects, with the actual occupants. Often, these sorts of studies are commissioned by the owner of the building. Through these types of monitoring campaigns, it is possible to provide a building diagnosis to improve the performance of the building. These occupancy studies can focus in three different aspects: energy consumption, building operation, and occupants' comfort (Guerra-Santin and Tweed 2015a, b). Ideally, these campaigns would be for a minimum of one year in order to capture the performance of the building in all seasons (especially when comfort is investigated) and in order to allow for enough time of building fine-tuning.

26.6 Closing Remark

Living Labs offer new possibilities to study buildings' occupants in their context, a factor of particular importance to establish occupancy practices and personal comfort. Further research should be directed towards methods to integrate the results from these studies into building design, perhaps through the implementation of a feedback mechanism embedded into BIM systems.

References

Álvarez, J. D., Redondo, J. L., Camponogara, E., Normey-Rico, J., Berenguel, M., & Ortigosa, M. P. (2013). Optimizing building comfort temperature regulation via model predictive control. *Energy and Buildings, 57*, 361–372.

Aria, H., & Akbari, H. (2014). Integrated and multi-hour optimization of office building energy consumption and expenditure. *Energy and Buildings, 82*, 391–398.

Bell, M., Wingfield, J., Miles-Shenton, D., & Seavers, J. (2010). *Low carbon housing: Lessons from Elm Tree Mews*. York: Joseph Rowntree Foundation.

Cole, R. J., Robinson, J., Brown, Z., & O'shea, M. (2008). Re-contextualizing the notion of comfort, *Building Research and Information*, 36-4, 323–336.

D'Oca, S., & Hong, T. (2015). Occupancy schedules learning process through a data mining framework. *Energy and Buildings, 88*, 395–408.

Danielski, I. (2012). Large variations in specific final energy use in Swedish apartment buildings: Causes and solutions. *Energy and Buildings, 49*, 276–285.

Domínguez, M., Fuertes, J. J., Alonso, S., Prada, M. A., Morán, A., & Barrientos, P. (2013). Power monitoring system for university buildings: Architecture and advanced analysis tools. *Energy and Buildings, 59*, 152–160.

Doran, S. (2005). Improving the thermal performance of buildings in practice. BRE client report no. 78132, for the office of the deputy Prime Minister, Building Research Establishment, East Kilbride, Glasgow.

Fokaides, P. A., Maxoulis, C. N., Panayiotou, G. P., Neophytou, M. K. A., & Kalogirou, S. A. (2011). Comparison between measured and calculated energy performance for dwellings in a summer dominant environment. *Energy and Buildings, 43*, 3099–3105.

Green Building Council. (2011). *Carbon reductions in existing non-domestic buildings*. London: UK Green Building Council.

Guerra-Santin, O., & Itard, L. (2010). Occupants' behaviour: Determinants and effects on residential heating consumption. *Building Research and Information, 38(3)*, 318–338.

Guerra-Santin, O., & Itard, L. (2012). The effect of energy performance regulations on energy consumption. *Energy Efficiency, 5(3)*, 269–282.

Guerra-Santin, O., Itard, L., & Visscher, H. (2009). The effect of occupancy and building characteristics on energy use for space and water heating in Dutch residential stock. *Energy and Buildings, 41*, 1223–1232.

Guerra-Santin, O., Tweed, A. C., Jenkins, H., & Jiang, S. (2013). Monitoring the performance of low energy dwellings: Two UK case studies. *Energy and Buildings, 64*, 32–40.

Guerra-Santin, O., & Tweed, A. C. (2015a). In-use monitoring of buildings: An overview and classification of evaluation methods. *Energy and Buildings, 86*, 176–189.

Guerra-Santin, O., & Tweed, A. C. (2015b). In-use monitoring of buildings: An overview of data collection methods. *Energy and Buildings, 93*, 189–207.

Guerra-Santin, O., Silvester, S., & Konstantinou, T. (2015). 2nd Skin approach to zero energy rental properties: Occupancy patterns to improve energy simulation. In *Proceedings PLEA 2015*, Bologna 09–11 Sept.

Gupta, R., & Chandiwala, S. (2010). Understanding occupants: Feedback techniques for large scale low-carbon domestic refurbishments. *Building Research and Information, 38(5)*, 530–548.

Hens, H., Parijs, W., & Deurinck, M. (2010). Energy consumption for heating and rebound effects. *Energy and Buildings, 42*, 105–110.

Heo, Y., Choudhary, R., & Augenbroe, G. A. (2012). Calibration of building energy models for retrofit analysis under uncertainty. *Energy and Buildings, 47*, 550–560.

Herring, H., & Sorrell, S. (2009). *Energy efficiency and sustainable consumption. The rebound effect. Energy, climate and the environment series*. UK: Palgrave Macmillan.

Huebner, G. M., Cooper, J., & Jones, K. (2013). Domestic energy consumption—what role do comfort, habit, and knowledge about the heating system play? *Energy and Buildings, 66*, 626–636.

Jain, R. K., Gulbinas, R., Taylor, J. E., & Culligan, P. J. (2013). Can social influence drive energy savings? Detecting the impact of social influence on the energy consumption behavior of networked users exposed to normative eco-feedback. *Energy and Buildings, 66*, 119–127.

Johansson, D., Bagge, H., & Lindstrii, L. (2011). Measurements of occupancy levels in multifamily dwellings—application to demand controlled ventilation. *Energy and Building, 43*, 2449–2455.

Kane, T., Firth, S. K., & Lomas, K. J. (2015). How are UK homes heated? A city-wide, socio-technical survey and implications for energy modelling. *Energy and Buildings, 86*, 817–832.

Kanga, N. N., Choa, S. H., & Kimb, J. T. (2012). The energy-saving effects of apartment residents' awareness and behaviour. *Energy and Buildings, 46*, 112–122.

Kashif, A., Ploix, S., Dugdale, J., & Le, X. H. B. (2013). Simulating the dynamics of occupant behaviour for power management in residential buildings. *Energy and Buildings, 56*, 85–93.

Leaman, A., Stevenson, F., & Bordass, B. (2010). Building evaluation: Practice and principles. *Building Research and Information, 35–5*, 565–577.

Lee, Y. S., & Malkawi, A. M. (2014). Simulating multiple occupant behaviors in buildings: An agent-based modeling approach. *Energy and Buildings, 69*, 407–416.

Maier, T., Krzaczek, M., & Tejchman, J. (2009). Comparison of physical performances of the ventilation systems in low-energy residential houses. *Energy and Buildings, 41*, 337–353.

McLoughlin, F., Duffy, A., & Conlon, M. (2012). Characteristic domestic electricity consumption patterns by dwelling and occupant socio-economic variables: An Irish case study. *Energy and Buildings, 48*, 240–248.

Meyers, R. J., Williams, D., & Matthews, H. S. (2010). Scoping the potential of monitoring and control technologies to reduce energy use in homes. *Energy and Buildings, 42*, 563–569.

Nguyen, T. A., & Aiello, M. (2013). Energy intelligent buildings based on user activity: A survey. *Energy and Buildings, 56*, 244–257.

Oldewurtel, F., Parisio, A., Jones, C. N., Gyalistras, D., Gwerder, M., Stauch, V., et al. (2012). Use of model predictive control and weather forecasts for energy efficient building climate control. *Energy and Buildings, 45*, 15–27.

Palmborg, C. (1986). Social habits and energy consumption in single-family homes. *Energy, 11*(70), 643–650.

Prívara, S., Vána, Z., Záceková, E., & Cigler, J. (2012). Building modeling: Selection of the most appropriate model for predictive control. *Energy and Buildings, 55*, 341–350.

Raftery, P., Keane, M., & Costa, A. (2011). Calibrating whole building energy models: Detailed case study using hourly measured data. *Energy and Buildings, 43*, 3666–3679.

Ryckaert, W. R., Lootens, C., Geldof, L., & Hanselaer, P. (2010). Criteria for energy efficient lighting in buildings. *Energy and Buildings, 42*, 341–347.

Schipper, L., Ketoff, A., & Meyers, S. (1982). International residental energy demand use data: Analysis if historical and present day structure and dynamics. *Energy, 7*(2), 205–212.

Stevenson, F., & Leaman, A. (2010). Evaluating housing performance in relation to human behavior: New challenges (editorial). *Building Research and Information, 38(5)*, 437–441

Stevenson, F., & Rijal, H. B. (2010). Developing occupant feedback from a prototype to improve housing production. *Building Research and Information, 38(5)*, 549–563.

Stevenson, F., Carmona-Andreu, I., & Hancock, M. (2012). Designing for comfort-usability barriers in low carbon housing. In *Proceedings of 7th Windsor Conference: The Changing Context of Comfort in an Unpredictable World Cumberland Lodge*, Windsor, UK, 12–15 April 2012. London: Network for Comfort and Energy Use in Buildings.

Sun, K., Yan, D., Hong, T., & Guo, S. (2014). Stochastic modeling of overtime occupancy and its application in building energy simulation and calibration. *Building and Environment, 79*, 1–12.

Sunikka-Blank, M., & Galvin, R. (2012). Introducing the prebound effect: The gap between performance and actual energy consumption. *Building Research and Information, 40(3)*, 260–273.

Thomsen, K. E., Schultz, J. M., & Poel, B. (2005). Measured performance of 12 demonstration projects—IEA Task 13 advanced solar low energy buildings. *Energy and Buildings, 37*, 111–119.

Verhallen, T. M. M., & Raaij, W. F. V. (1981). Household behaviour and the use of natural gas for home heating. *Journal of Consumer Research, 8*, 253–257.

Virote, J., & Neves-Silva, R. (2012). Stochatic models for building energy prediction based on occupant behaviour assessment. *Energy and Buildings, 53*, 183–193.

Wei, S., Jones, R., & de Wilde, P. (2014). Driving factors for occupant-controlled space heating in residential buildings. *Energy and Buildings, 70*, 36–44.

Yang, Z., & Becerik-Gerber, B. (2014). The coupled effects of personalized occupancy profile based HVAC schedules and room reassignment on building energy use. *Energy and Buildings, 78*, 113–122.

Yohannis, Y. G., Mondol, J. D., Wright, A., & Norton, B. (2008). Real-life energy use in the UK: How occupancy and dwelling characteristics affect domestic electricity use. *Energy and Buildings, 40*, 1053–1059.

Yun, G. Y., Hwang, T., & Kim, J. T. (2010). Performance prediction by modeling of a light-pipe system used under the climate conditions of Korea. *Indoor and Built Environment, 19*, 137–144.

Chapter 27
Influence of User-Behavior on Energy Efficiency

Tanja Lovrić and Viktor Grinewitschus

Abstract How is the composition of the final energy consumption of private households for heating? Is there a portion which is influenced by user behavior, and can this portion be captured with suitable equipment? Is it possible to reduce this portion by applying appropriate auxiliary devices (assistance systems)? What is the potential in different types of buildings, which how much can be saved in this way?

27.1 Introduction

The demand of heating energy consumed as final energy in households for maintaining a comfortable indoor air temperature and temperature of the surrounding surfaces is composed by a share for heating and for ventilation. In this work we do not consider the energy used by the mechanical ventilation system because the households we monitored had no such system.

However, every human being has a different need for fresh air through the ventilation, depending on the personal preferences for air quality. Furthermore, each person performs this action in different ways (duration and type of opening of the window).

T. Lovrić (✉) · V. Grinewitschus
University of Applied Science Ruhr West, Mülheim, Bottrop, Germany
e-mail: Tanja.Lovric@hs-ruhrwest.de

V. Grinewitschus
e-mail: Viktor.Grinewitschus@hs-ruhrwest.de

D.V. Keyson et al. (eds.), *Living Labs*, DOI 10.1007/978-3-319-33527-8_27

Within the project period, three winter periods were monitored to answer the following questions:

- What is the current situation in the households concerning the heating and ventilation behavior of the residents?
- Does this behavior change and how does it change when we provide assistance devices to the user?
- Which assistance systems are suitable for this purpose?
- Which influence do the ventilation assistance systems have on the heating energy consumption in the households? Is an improvement of indoor air quality inevitably accompanied by an increased consumption of heating energy?
- What about the user acceptance of such assistance systems and is a so-called Honey-Moon effect observed when the systems are applied? We can observe the so-called Honey-Moon effect when assistance systems are applied for longer time period. The participants forget the former resolution concerning the change of the ventilation and heating behavior and goes back to the old habits.
- How does the heating demand change when we give feedback to the user about his daily heating energy demand?

27.1.1 Winter Pilot 1

Eighty households with different fuel kinds for heating system took part in the winter pilot 1. The year of construction of the houses was between 1895 and 2002 and the year of installation of the heating systems was between 1969 and 2012. Table 27.1 shows the structure of the heating systems in winter pilot 1.

27.1.2 Winter Pilot 2

In winter pilot 2, 40 households took part in the measurement. The houses were constructed between 1903 and 2000, while the year of installation of the heating system was from 1963 to 2013. Table 27.2 shows the structure of the heating systems in winter pilot 2.

Table 27.1 Heating systems of the registered households in Winter Pilot 1

Type of heating systems	N
Gas boiler	51
District heating	7
Oil boiler	13
Coal and wood-fired oven	2
Electrical night storage heater	6
Air-water heat pump	1

27.1.3 *Winter Pilot 3*

For winter pilot 3 an online survey was created with questions about the building and user behavior.

The year of construction of the houses was between 1900 and 1999. The year of installation of the heating systems was between 1992 and 2014 and 76 % of them were installed after 2000. Table 27.3 contains the main characteristics of our sample, information obtained from the online survey.

Our measurements could only be performed in the households with gas boiler, district heating or CHP. For other heating systems it was not possible to detect the consumption within the measurement period. In the Table 27.4 are listed different energy efficiency measures concerning the building, which participants carried out in last years.

A number of people living in the dwellings and their normal ventilation behavior are seen in the Tables 27.5 and 27.6.

Table 27.2 Heating systems of the registered households in Winter Pilot 2

Type of heating systems	N
Gas boiler	28
District heating	7
Oil boiler	1
Air-water heat pump	1

Table 27.3 Heating systems of the survey registered households in Winter Pilot 3

Type of heating systems	N
Gas boiler	41
District heating	3
Oil boiler	1
Wood pallet stove	1
CHP	1

Table 27.4 Energy efficiency measure on the buildings carried out since building construction?

Energy efficiency measures	N
Additional external insulation	14
Renewed window	25
Renewing heating system	29

Table 27.5 Number of people who live in households

Household size	N
Single-person household	3
Two-person household	15
Three-person household	12
Four-person household	6
Five-person household	5
Six-person household	2
Seven person household	1
Not specified	2

Table 27.6 Ventilation behavior of the participants

Window behaviour	N
The window is to ventilate sometimes even for more than 30 min on tilt ventilation	9
The window is fully opened for ventilation for a short period	24
The window tilted for ventilation for a period of 10–30 min	9
The window tilted for ventilation for a period of 5–10 min	2
The window tipped for fresh air only for a short time (less than 5 min)	1

27.2 Measuring Equipment

To compare measurements in different months, the energy consumption was corrected by using heating degree days and estimating the yearly energy consumption. We recorded the gas consumption with impulse measurements (0.1 m^3/impulse or 0.01 m^3/impulse). With the measured load response measurement we could analyze the gas consumption on daytime hours.

27.2.1 Measuring Equipment in Winter Pilot 1

Indoor air quality measurements were carried out using a mobile sensing device Testo 435 (Fig. 27.1) with a sensor probe containing a CO_2sensor, a temperature sensor and a relative humidity sensor. Data logging was possible on an internal

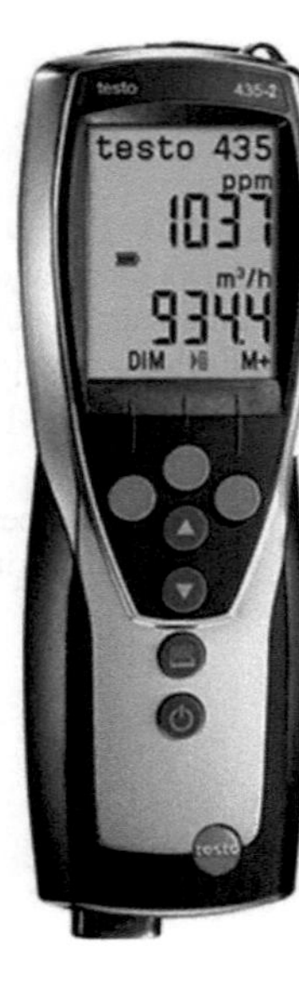

Fig. 27.1 Indoor air quality measurement by Testo 435

storage device. The measurement rate was every 3 min with infrared gas sensors (NDIR). A non-dispersive infrared sensor (or NDIR sensor) is a simple spectroscopic sensor often used as a gas detector. It is non-dispersive in the sense of optical dispersion since the infrared energy is allowed to pass through the atmospheric sampling chamber without deformation.

27.2.2 Measuring Equipment in Winter Pilot 2

Additionally to the IAQ Logger we used LUQA indoor air quality monitor in 2nd winter pilot. Besides the measurement of indoor air quality and displaying it LUQA monitor (Fig. 27.2) has a log function for the internal long-term recording and a USB interface including analysis software with charts (Fig. 27.3). It has been designed to monitor temperature, humidity, and CO2 levels as well as volatile organic compounds (VOC) in indoor air. LUQA IAQ monitor has an internal storage device for logging data for two weeks in one minute step. LUQA indoor airquality monitor has three light colours for signalling following three CO_2-concentration levels:

Green—CO2 level<1000ppm—good air quality and no action needed
Yellow—1000ppm<CO2 level<1500ppm—acceptable air quality
Red—CO2 level>1500ppm—bad air quality—airing recommended

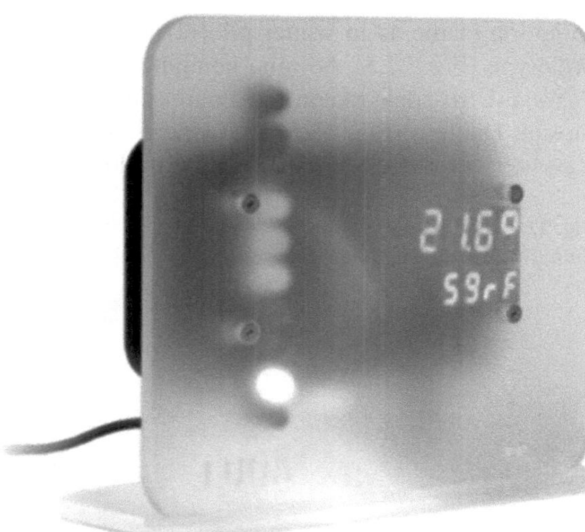

Fig. 27.2 LUQA indoor air quality monitor (*source* www.elk.de)

27.2.2.1 Home Automation Systems

Home Automation Systems consisting of convenience products such as RWE Smart Home (Fig. 27.3), were used in SusLabNWE Field Testing winter pilot 2014 in 19 households with the purpose of saving heat energy. One System consisting of door/window sensor, wireless radiator thermostat and gateway using radio connection for the communication with the database and the server was used. The door/window sensor consists of two elements, one of the parts is mounted on the moving part of the window or of the door, and the other part is placed on the frame.

If an action is detected, the window sensor sends a signal to the radio communication (846 MHz) network main controller (gateway). Wireless radiator thermostat is mounted on radiator heating valves and controls them by a motor. If the window is opened the radiator thermostat valve will reduce the heating for the time of the window opening. There is GUI (graphical user interface)for heating profile (Fig. 27.4) settings which allows setting different time-temperature-profile for each day and each room.

27.2.3 Measuring Equipment in Winter Pilot 3

27.2.3.1 HRW Prototype

In the winter 2014/15 (WP3), the RWE Smart Home System, and the HRW (Hochschule Ruhr West—University of Applied Sciences) prototype home automation system with GUI feedback of the indoor air quality and heat consumption (same as in winter pilot 2) were installed in 20 houses. For the HRW prototype (Fig. 27.5) the mini computer Odroid was used in combination with a Z-Wave-USB-stick (Aeon) as main controller (gateway) wireless radiator thermostat, Danfoss (Fig. 27.6) and window/door sensor. The Odroid is, just like the Raspberry Pi, a deck-of-cards sized ultra-low cost Linux computer.

Fig. 27.3 RWE Smart Home (http://www.rwe-smarthome.de)

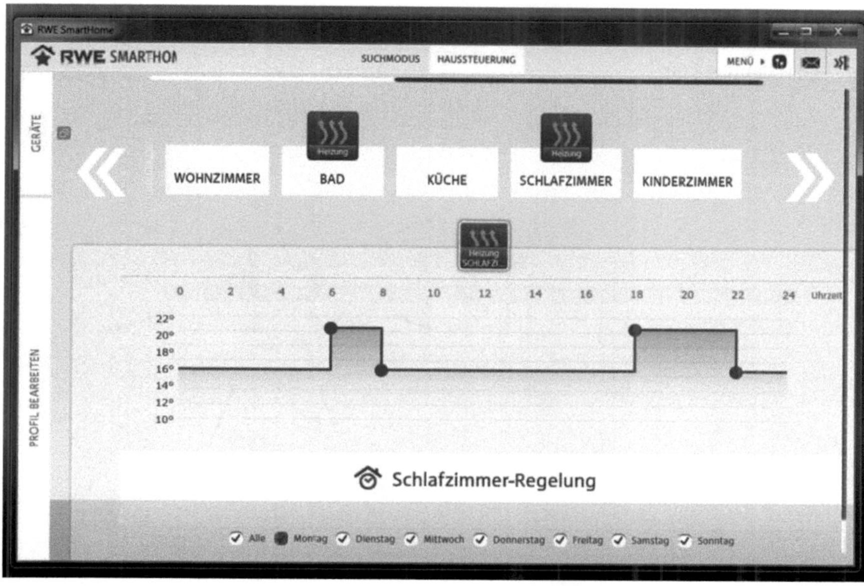

Fig. 27.4 Graphical User Interface—RWE Smart Home (*source* www.rwe-smarthome.de)

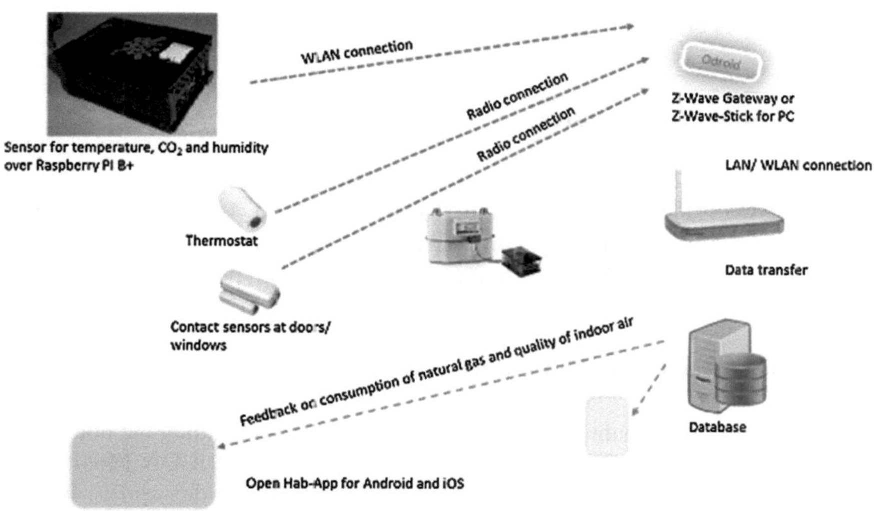

Fig. 27.5 HRW home automation prototype with Z-Wave radio communication, sensor box and gas consumption logging device

Fig. 27.6 Wireless radiator thermostat, Danfoss *Source* http://www.danfoss.com

The window/door sensor consists of two elements. One part is mounted on the movable part of the window and the other on the fixed part of the window or door. In case of change of position (closing or opening), the sensor is sending a signal to the Z-Wave network controller (Gateway). It is a part of HRW Smart Home prototype which was only used in 3rd winter pilot.

27.2.3.2 Odroid Key Features

Odroid is the powerful Linux computer with 1.7 GHz Quad Core processor and 2GByte RAM 10/100 Mbps XUbuntu 13.10 or Android 4× Operating System. With its size 83 × 48 mm, weight 48 g including heat sink it is very handy. Odroid is used as a gateway for the HRW home automation prototype.

27.2.3.3 Gas Meter Impulse Counter

A gas meter impulse counter was used in 25 households (including the households with RWE Smart Home, HRW smart home prototype and LUQA Monitor) for recording the heat energy consumption in the course of the day and to find out more information about the use of the home automation system.

The IN-Z61 (Fig. 27.7) is a retrofittable low frequency pulse generator for all Elster diaphragm gas meters BK-G2.5 to BK-G100. It is a pulse sensor for all Z6 indexes in Elster-Instromet diaphragm gas meters. A generator magnet in the first

Fig. 27.7 IN-Z61 gas meter impulse counter used for 0.01 m^3/impulse measurements (http://www.elster-instromet.com/de)

or second powered roller of the Z6 index switches a reed contact in the IN-Z61. It is fastened to the index with a separate seal.

27.2.3.4 Comfort Dail by TU Delft, Netherlands

In four households sensor nests were installed (in two rooms each). We also gave the possibility to the participants to give us feedback about their personal perception of the thermal comfort in the room. Furthermore, it was possible for the users to give us a feedback about what they changed in order to improve this comfort level if it was not satisfactory. For this purpose participants used Comfort Dail and the app.

27.2.3.5 Wireless Weather Station at Campus Bottrop

Davis Instruments Vintage Pro2 Plus was installed on the roof of the new university building in Bottrop (Fig. 27.8), Germany and provides the local weather data (temperature, relative humidity, wind speed, wind direction and solar radiation) of the Bottrop area.

Fig. 27.8 Davis instruments Vintage Pro2 Plus

27.3 Results

Besides the CO2 concentration and temperature the energy consumption was measured. To compare measurements in different months, the energy consumption was corrected by using heating degree days and estimating the yearly energy consumption.

27.3.1 Limitations of the Design

The use of smart home system was not facile for all participants. Some daily profiles were adjusted by the participant at a much higher temperature than before measurements. By manual intervention, the automation system has been prevented in the correct function. Some participants, who received the system by post, were not able to set up the system. Some participants did not manage to assemble the radiator thermostats. This experience shows us that such systems are not appropriate for 100 % of the population and that much more is necessary to support individual users.

The programming of the HRW smart home system with the Z-Wave wireless technology was delayed. It could be used only in 5 households. These 5 households had used the heating almost continuously over the day until late in the spring due to the cool weather period.

Both Smart Home systems (RWE and HRW) had, in some cases, problems with the accessibility of the gateway through the window sensors.

In many cases mistakes were made in installing or in the settings of the systems by the participant. Therefore it is recommendable to check if the used system really does what it should do and if the settings are correct. In many cases participants could install the devices themselves, but in 60 % of the cases they needed help. It cannot be concluded based only on a field test if the use of such devices as the LUQA IAQ monitor is of permanent benefit. The feedback from the participants regarding the use of the devices was in any case very positive and many participants aim to purchase such devices in the future. The saving potential we identified was significant enough to justify further research.

27.3.2 Winter Pilot 1—Monitoring the Base Line

Measurements of indoor air quality were carried out in 80 households between November 2012 and March 2013. Motivation of participants was not only based on prospect of energy and costs savings but also on increasing comfort. We asked the participants not to change their normal ventilation and heating behaviour during our measurements. The apartments were equipped with mobile data loggers in different rooms for 1–2 weeks. Indoor air temperature, relative humidity and CO_2concentration were recorded with a time resolution of 3 min. After collecting the loggers the data was transmitted to a SQL data base.

In Fig. 27.9 is to see the average Indoor air temperature in the dwellings referring to calculated thermal energy consumption. The y axis refers to the measured indoor air temperature and the x axis refers to the calculated energy consumption for heating. Calculated energy consumption represents the forecasting of measured energy in 2 weeks of the pilot without using any assisting device by correcting for heating degree days.

On average the households with automatic temperature control have a lower indoor air temperature than households without automation (see black rectangle marker in Fig. 27.10).

Our results show that ventilation habits influence energy consumption more than the temperature level of the room. Good ventilation habits can be termed as often but short ventilation periods and bad ventilation habits can be termed as long periods of ventilation through opening the windows. Automatic temperature control is used to reduce room temperature level during the absence of the user.

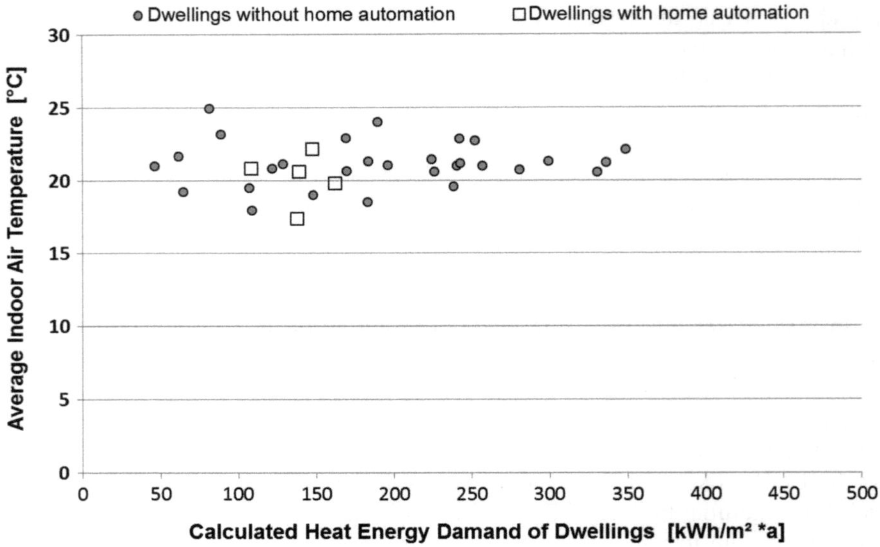

Fig. 27.9 Average indoor air temperature referring to calculated thermal energy consumption (forecast based on heating degree days) (Grinewitschus et al. 2013)

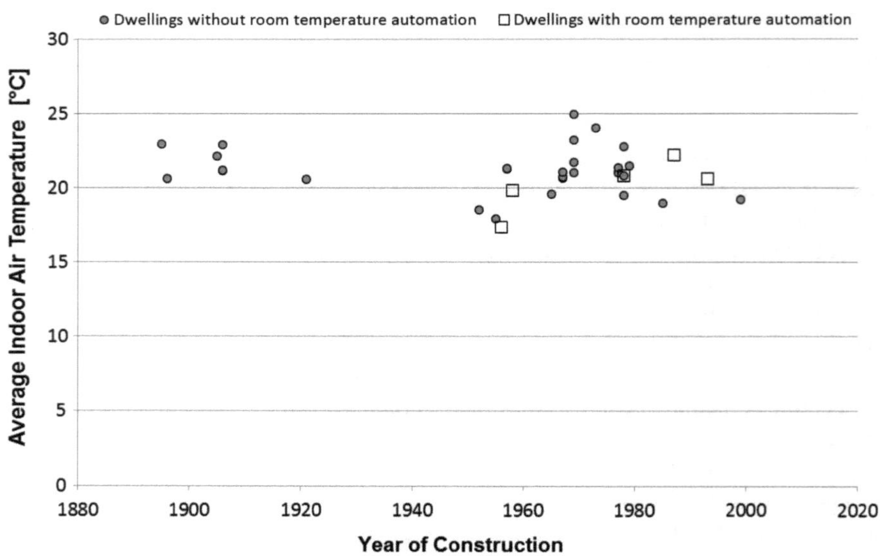

Fig. 27.10 Average indoor temperature referring to year of construction (Grinewitschus et al. 2013)

This improves energy efficiency even further. We have found automatic temperature control systems only in 5 % of the households investigated. These households have energy consumption lower than average and can be characterized by very efficient ventilation behaviour.

27.3.3 Winter Pilot 2

Measurements were repeated in a new field test (winter pilot 2) for a period of two weeks in 40 households. The first week was used to discover the baseline indoor air quality and heat energy consumption. In the second week two systems were deployed: a LUQA IAQ monitor and RWE smart home system. Using a LUQA IAQ Monitor the indoor air quality of 50 % of households rose significantly.

To compare measurements in different months, the energy consumption was corrected by using heating degree days and estimating the yearly energy consumption.

27.3.3.1 Results Using RWE Smart Home

Many participants set the room temperature to a higher level than before and kept it at a high temperature for much longer than they normally did before. This explains why in many cases additional energy consumption (negative bars in Fig. 27.11) resulted in week 2. We compared the calculated heat energy

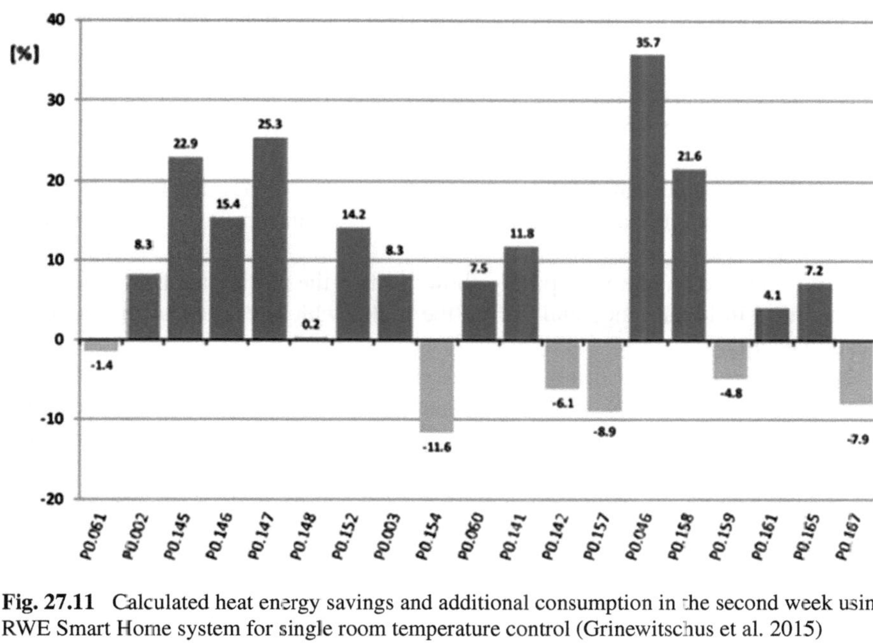

Fig. 27.11 Calculated heat energy savings and additional consumption in the second week using RWE Smart Home system for single room temperature control (Grinewitschus et al. 2015)

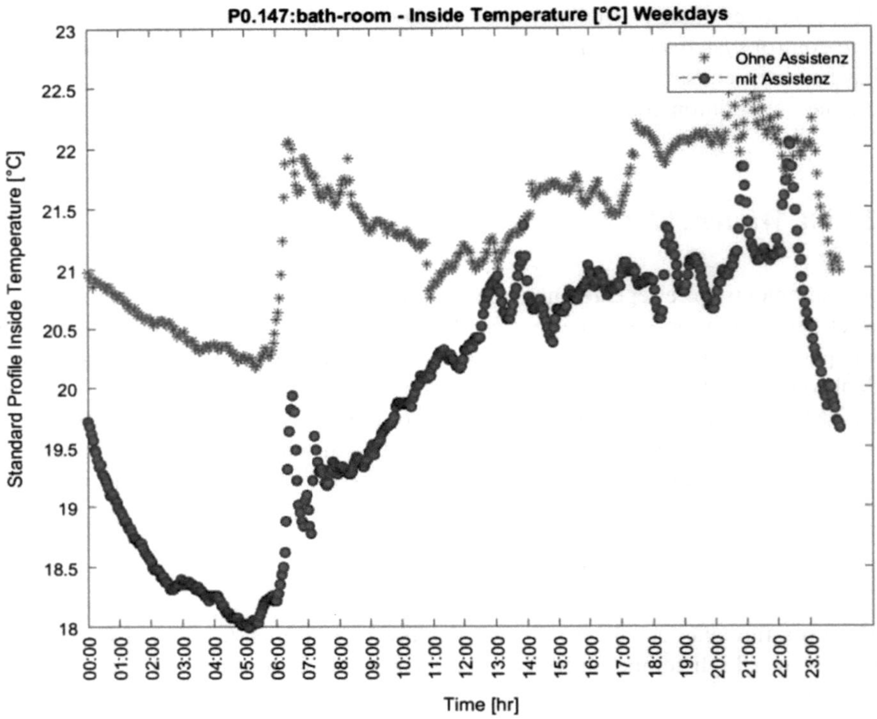

Fig. 27.12 Average indoor air temperature in the bathroom of the participant P0.147 from Monday to Friday, with assisting device (*round marker*) (RWE Smart Home) and without the assisting device (*asterisk marker*) (Grinewitschus et al. 2015)

consumption in the first and second week to find the savings and additional consumption.

Finally, we found that dwellings where savings were recorded through the use of home automation had a pronounced day-temperature profile as shown in Figs. 27.12 and 27.13.

A pronounced temperature profile shows us that the participant had lowered the temperature in his absence and during the night, which resulted in the saving of heating energy.

Savings on energy for heating resulted from using both systems. A hundred per cent was not achievable nevertheless about 77 % of the households achieved from 1.4 to 31 % energy savings using a LUQA monitor, even if the LUQA monitor was designed only for assisting in indoor air quality. About 67 % of the households that used the smart home system achieved energy savings of between 5 and 25 %.

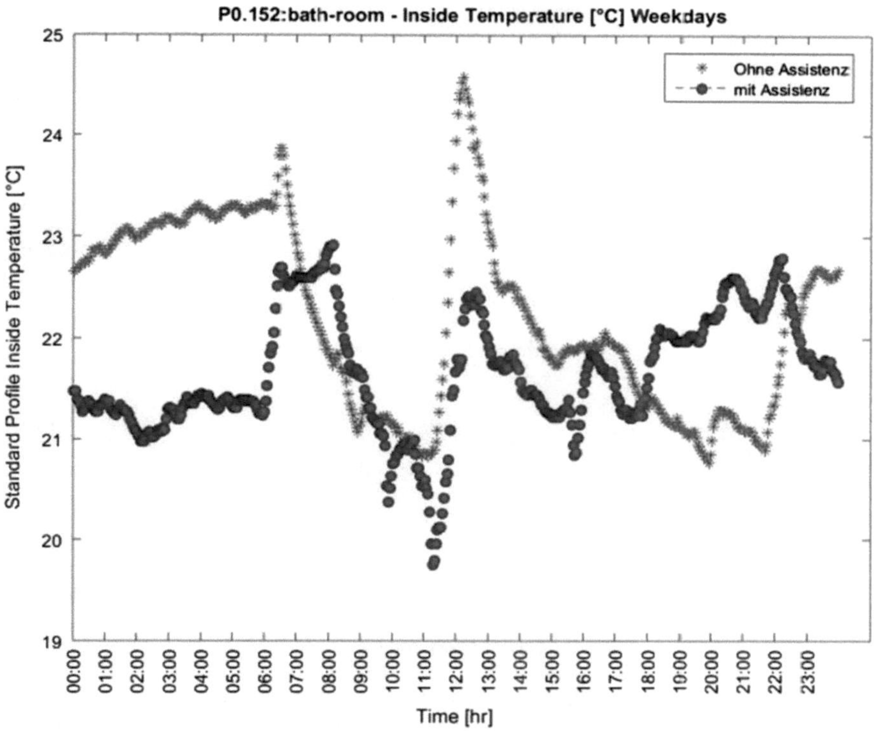

Fig. 27.13 Mean indoor air temperature in the bathroom of the participant P0.152 for Monday to Friday, with assisting device (*round marker*) (RWE Smart Home) and without the assisting device (*asterisk marker*) (Grinewitschus et al. 2015)

The influence or the use of the smart home system is shown in Fig. 27.14 for the household P0.033. The distinct reduction of indoor air temperature during the night and also during the day (absence time) is evident in the temperature heat map. Overheating periods are also not as distinctive as in week one.

27.3.3.2 Results Using LUQA Monitor

Via feedback to the user, the indoor air quality improved in houses that previously had a poor air quality. Using the CO2 and relative humidity sensors as an indication of indoor air quality, it is possible to increase the energy efficiency with a more purposeful airing behaviour. Users recognized, through the indoor air quality assistance device, that airing was needed and in doing so they could prevent moisture damage to the building. Households with very good indoor air quality and relatively high heating energy consumption now were able to allow shorter airing periods and so achieved higher energy efficiency.

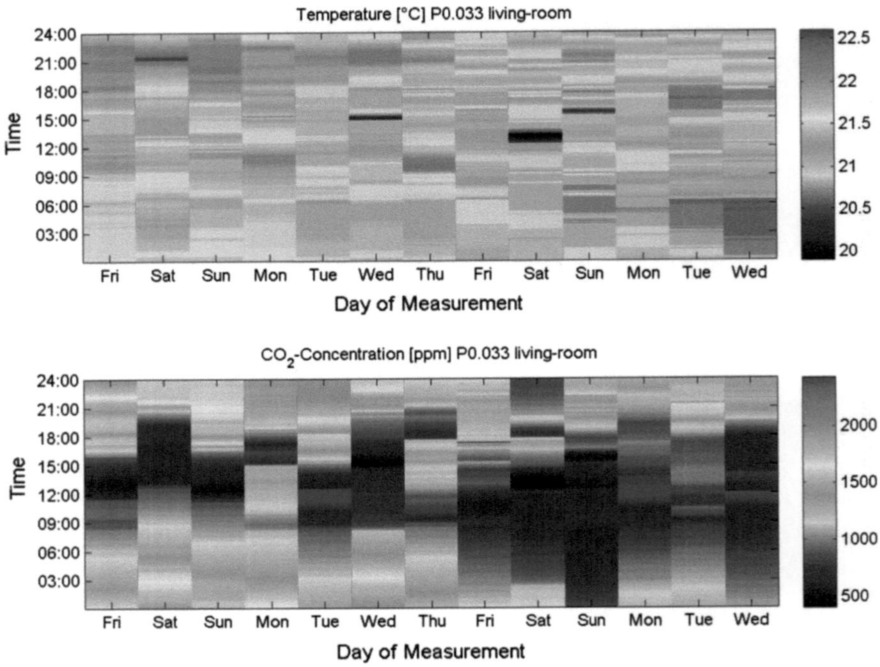

Fig. 27.14 Temperature and CO2-Concentration heat map of the indoor air in the living room of the participant P033—two assisting devices: RWE Smart Home and LUQA Monitor (Grinewitschus et al. 2015)

The following conclusion was reached regarding the average CO2 concentration of the indoor air evaluated. The measurements were divided in 2 weeks: the first week of measurement (status quo week) and the second week using home automation or LUQA IAQ monitor. The improvement was detectable in most households which used a LUQA monitor.

As shown by the average CO2 concentration Due to the mean CO2 concentration (Fig. 27.15), the households that used a LUQA IAQ monitor showed more frequently a higher indoor air quality (lower CO2 concentration) in the second week. Therefore we can identify two groups of indoor air quality: the first group of participants with, on average, a rather low CO2-concentration (<800 ppm) and the other group with, on average, a rather high CO2-concentration (>900 ppm). While the households group with low CO2-concentration of indoor air did not change very much during the second week, the group of households with higher CO2-concentration experienced remarkable changes in the second week. Four out of five households with high CO2-concentration improved the indoor air quality reducing on average from 120 to 330 parts per million CO2.

The shifting of the CO2 concentration density from higher level in the week 1 to lower level in the week 2 is to see in the Fig. 27.16.

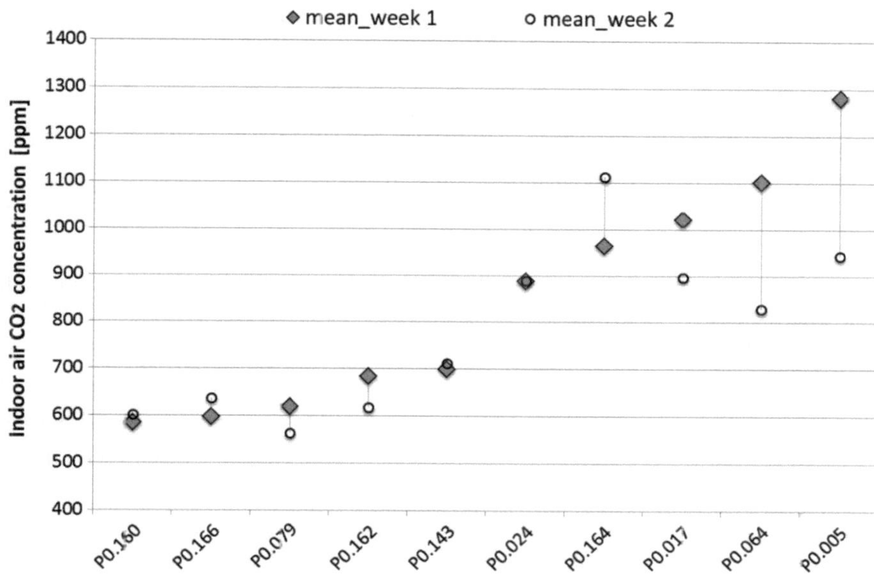

Fig. 27.15 Mean CO2 concentration in the week 1 (without LUQA monitor) and week 2 (using LUQA monitor) in the living rooms of the households (Grinewitschus et al. 2015)

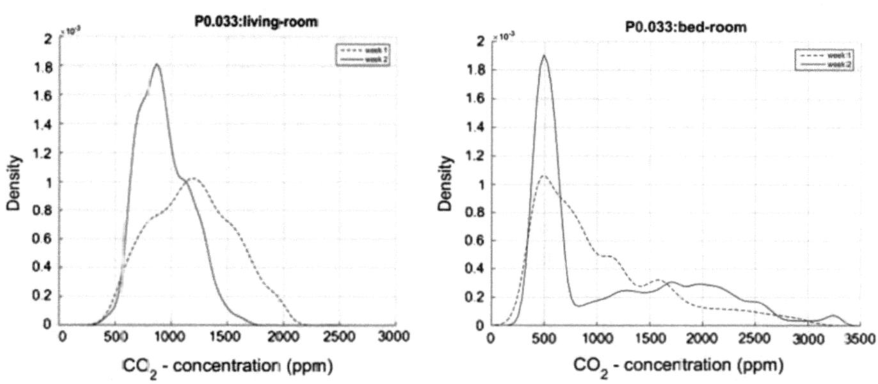

Fig. 27.16 Density of CO2-Concentration in two weeks with (*continued line*) and without assisting device (*dashed line*) in living room and the bed room of the participant P0033 (Grinewitschus et al. 2015)

27.3.4 Winter Pilot 3—Prototyping of Assisting Function

For the prototyping we tested different options for user assistance:

1. Visualization of CO2 level (supporting airing behaviour)
2. Visualization of energy consumption (gas metering)
3. Indoor air temperature control by home automation
4. User feedback about comfort level (comfort dial)

In the first part, HRW developed the indoor air sensor with logging function and graphical user interface (CO2, indoor air temperature, relative humidity). Requirements for this were:

- Temporal scalable measurement and persistent storage of the parameters CO2 content of the ambient air (ppm), air temperature (C) and relative humidity (%).
- Possibility of local storage of the data, data analysis and transfer to a central.

An approach for a graphical user interface was developed based on the analysis and storage of data. The participants could achieve on this way visualization of the current indoor air quality, which can be provided by desktop computers as well as tablets and smart phones. The integration in the wireless network to network the sensors could take place, so that a central visualization of data was made possible.

In the winter 2015 RWE Smart Home System, home automation system and the HRW prototype home automation system with GUI feedback of the indoor air quality and heat consumption will be used in each 10 households. For HRW prototype Raspberry PI combined with z-wave-USB-stick (Aeon) was used as main controller. The Raspberry Pi is an ultra-low-cost, deck-of-cards sized Linux computer, developed by the Raspberry Pi Foundation in the UK. Gas meter impulse counter is used in 30 households for recording the heat energy consumption in the course of the day and to find out the use of the home automation systems.

The software MatLab has been created to investigate the correlations measured. These visualizations include carpet plots and scatter plots. The environment includes thereby the automated reading of values (connecting to the database, query of the values) as well as the processing and storage of files.

At the start of the measurements in the winter pilot 2014/15 we hypothesized following:

Hypothesis 1. Participant airing behavior significantly improves if the user receives information from an assistant device.

Hypothesis 2. Not only indoor air quality improves but also participants are able to conserve heating energy.

Hypothesis 3. Using a home automation system means significant savings can be made on heating energy.

27.3.4.1 Gas Metering and the Meaning of the Heating Characteristic Curves

The heating characteristic curves are intended to represent the heating consumption in a household as the sum of the gas consumption of a day divided by the ruling average outdoor temperature. The analysis of consumption data is a statistical model for the reaction of the heating energy consumption on changing temperatures in the core. At the same time, the parameters of this model ("regression analysis") are parameters that describe the thermal state of the building with a high degree of precision (DIN EN 12831, bbl. 2 and DIN V 18599, bbl. 1). Through an extension of this linear slope, forecasts can be made for the consumption at expected falling or rising outside temperatures. The slope of this curve (in Germany "Energie Signatur") is an indication of the quality of the building fabric of a building in terms of transmission and ventilation heat losses. To analyse the daily consumption in the Winter 2014/15, we have looked at the heating energy consumption separately for two intervals, the interval with assistance device and the interval without assistance device, creating two curves. With the analysis we were able to determine the differences in the slope of the curves (with and without assistance systems).

27.3.4.2 Measurements with LUQA—Monitor

Figure 27.17 shows the characteristic curves for the two sections of a single-family terraced house from 1981 which used air quality monitor LUQA as an assistant device for energy-efficient ventilation.

In this house, underfloor heating and CHP by Innovation City is operating since 2014. The windows were also replaced by more efficient recently. In this case, there is usually someone at home throughout the day, which means that the heating has to be continuously on. The window is fully opened for ventilation for a short period.

The gradient of the curve is significantly lower in the assisting interval (dashed curve) than the slope in the interval without assisting device (solid curve). We could conclude that by observing of IAQ monitor light participants have reduced ventilation duration the and thus the heating energy consumption reduced.

Figure 27.18 shows that the indoor air quality has improved over the course of two weeks in the observed dwelling (P0753), with the exception of a peak. While CO2 concentration in the first week stayed under the 1500 ppm threshold (the orange light of the LUQA IAQ monitor) the participant kept CO2 concentration in the second week mostly under the 1000 ppm threshold (green light in the feedback device).

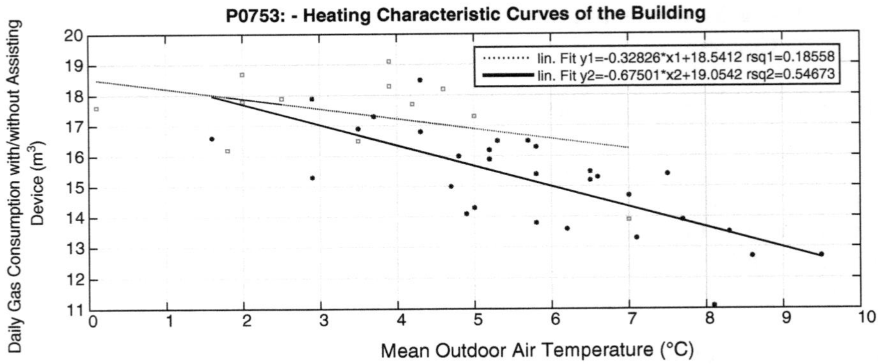

Fig. 27.17 Dwelling which used LUQA in first measurement interval (*dashed* gradient)

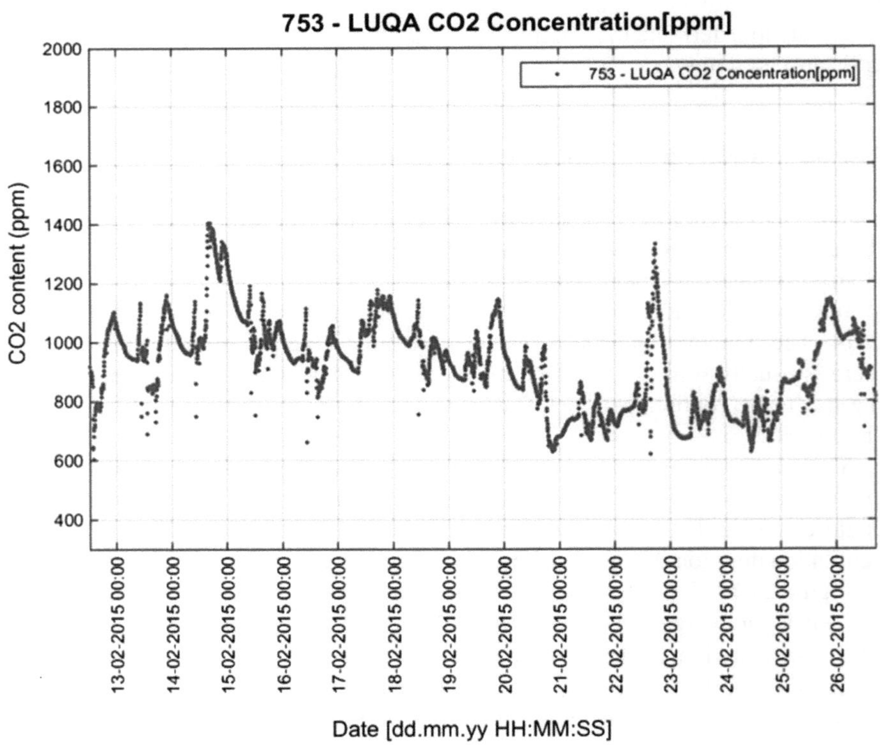

Fig. 27.18 CO2-Concentration in the living room of the dwelling P0753 during the 2 weeks using LUQA monitor

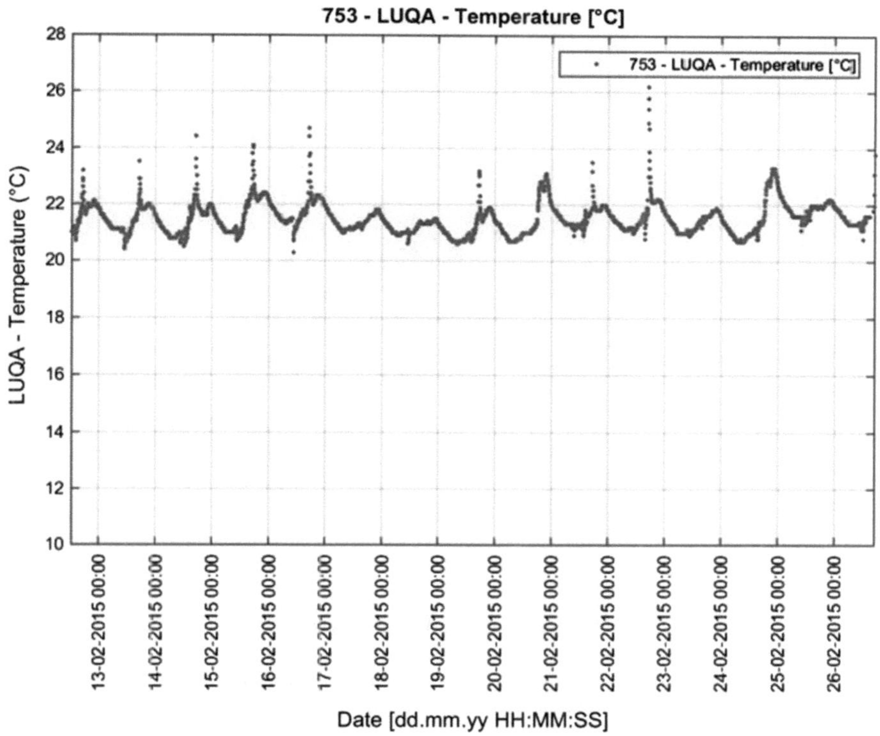

Fig. 27.19 Temperature in the living room of the dwelling P0753 during the 2 weeks using LUQA monitor

Temperature stayed at almost the same level during the two weeks as shown in the Fig. 27.19.

There were also dwellings in which the CO2 concentration had increased after the first week of using LUQA monitor. This effect of easing the good intentions has also to be considered when the use of such assistance systems in the general. For the improvement of indoor air quality through attention of the CO2 indicator is a matter of personal attitude, which shows us in turn the importance of user behavior and complexity of influencing behaviour.

Figure 27.20 shows heating characteristic curve of the dwelling refers to a 5 person's detached house from 1903. The cellar and attic are not heated. No insulation measures have been made lately but the windows were replaced. The heating energy consumption is the sum for space heating and hot water. There is always somebody at home (specified by the participant of the online survey).The window is widely opened by the participant for ventilation for about 5–10 min.

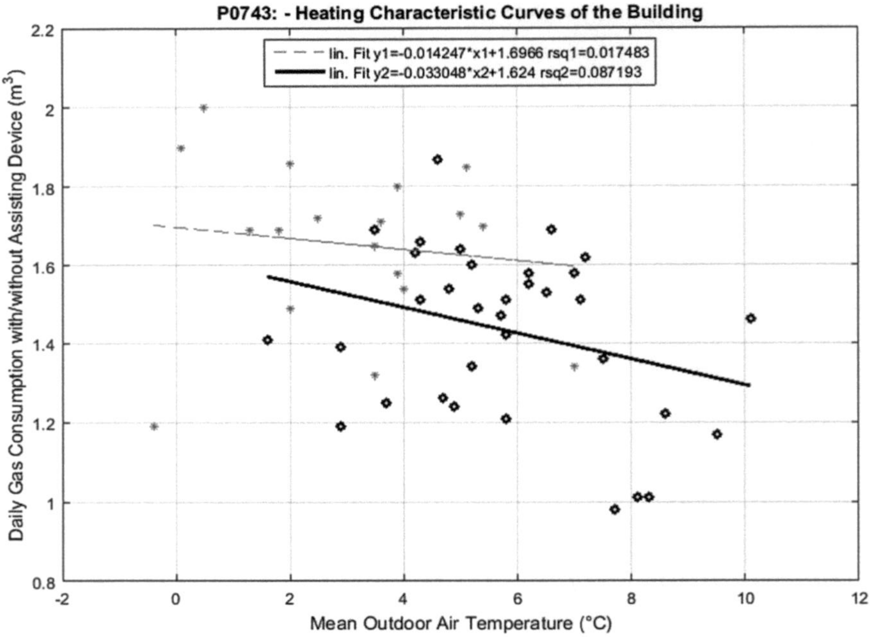

Fig. 27.20 Heating characteristic curves of the building for two intervals in household which used RWE Smart Home System in the first Interval (*dashed curve*)

27.3.4.3 HRW Smart Home Prototype

In five dwellings, the HRW Smart Home prototype was installed, which uses Z-Wave wireless radio wave. This installation could only be used towards the end of winter 2014/15 pilots. Figure 27.21 shows a set temperature time profile and the indoor air temperature in the course of about one and a half days in a living room of one participant.

In addition to the expected course of the indoor air temperature, the time points of the window opening are also clearly visible since a significant drop in the indoor air temperature can be recognized. In the periods of the lowered target temperature a gradient decline of indoor air temperature can be seen in the Fig. 27.21.

Heating characteristic curves in Fig. 27.22 refer to a 2 person's detached house from 1987. The age of the boiler is 17 years (1998), the cellar is heated and there is no attic in the house. Insulation measures have been made lately and the windows were replaced. The heating energy consumption is the sum for space heating and hot water. There are regular times when nobody is at home, at these times, the room temperature can be lowered (specified by the participant of the online survey). The window is tilted by the participant for ventilation for a period of 10 min to 30 min.

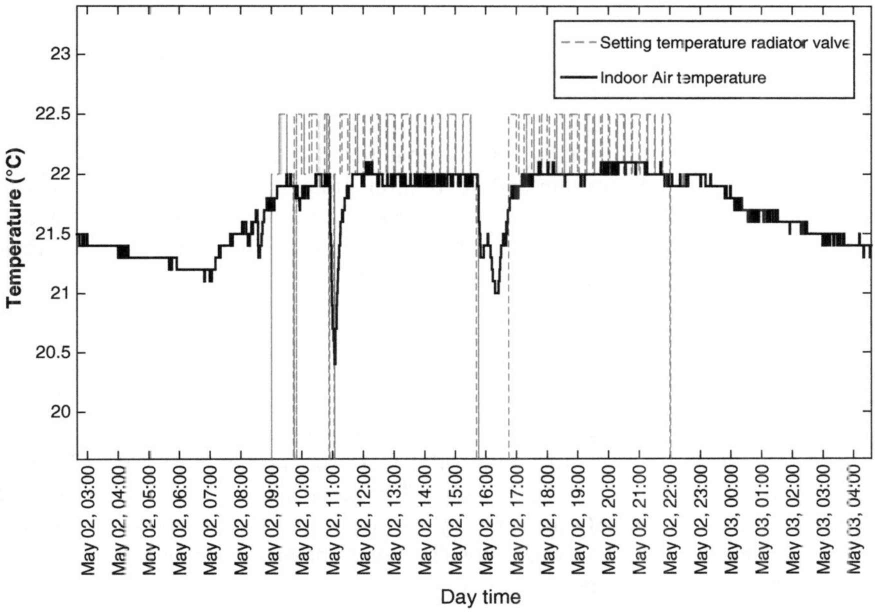

Fig. 27.21 Zoom in of radiator valve setting temperature (*dashed line*) and indoor air temperature (*solid line*) using HRW smart home system, P0734—living room

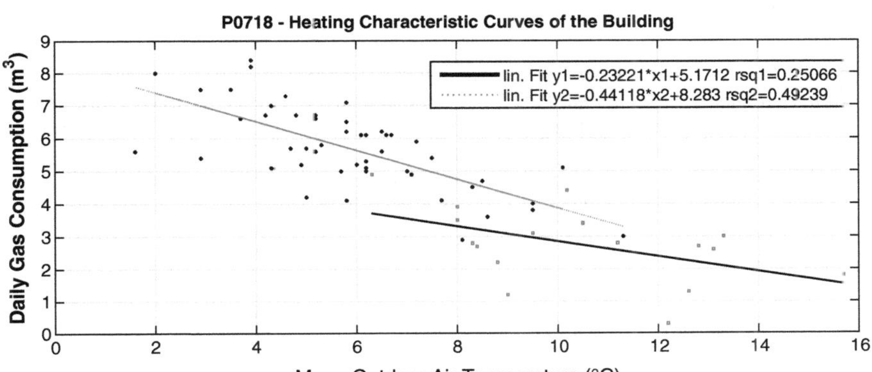

Fig. 27.22 Heating characteristic curves of the building for two intervals in household which used HRW-smart home system in the second interval (*solid curve*)

27.4 Conclusions

Over the past three winter pilots, we were able to gain a good overview of the situation in private households in the region Bottrop and test low-investment assistance systems. We initially noticed that the participants had a very strong assortment for different heating and ventilation strategies and accordingly widely differing ambient air qualities. The average daily temperature of the air seemed to have barely an impact on the demand for heating energy and the daily mean temperature does not depend on the age of the houses.

These results have confirmed our assumption that the user behavior plays a crucial role in the heating energy consumption. Furthermore, pointed households with home automation systems (single room heating control) had in the average lower heating requirements and lower indoor air temperatures. These households had already installed the home automation systems themselves.

The following 2nd and 3rd winter pilot series of measurements were carried out with the use of two assistance systems. The acceptance of both systems (home automation and LUQA IAQ monitor) by the users was very good. We compared the measurements of the first week (baseline) with the measurements in the second week (with assistance system). Results of the second winter pilot were on the one hand the improvement of indoor air quality in most homes through the use of airing traffic lights (indicating CO2 level) and on the other hand, heating energy savings through timed single room heating control in three rooms of households.

References

Grinewitschus, V., Lovrić, T., & Rumler, N. (2013).*Influence of user behavior and home automation on energy consumption*. Paper, 7th International Conference on Energy Efficiency in Domestic Appliances and Lighting (EEDAL'13), 11–13 September 2013, Coimbra, Portugal.

Grinewitschus, V., Lovrić, T., Lacombe, J., & Beblek, A. (2015).*Assisting user airing behavior for saving energy*. Paper, 8th International Conference on Energy Efficiency in Domestic Appliances and Lighting (EEDAL'15), 26–28 August 2015, Horw-Lucerne, Switzerland.

Part VIII
The Way Forwards: Business Models for Living Labs

Chapter 28
Concept House Village; A Next Step in the Development of Sustainable Housing in the Netherlands

Sacha Silvester, Bert Hooijer, Ria van Oosterhout and Floor van der Kemp

Abstract To be able to develop and implement high-impact sustainable innovations in the built environment, researchers, product/-service developers and policymakers in the region of Rotterdam felt the necessity to work with users to prototype, test and validate potential solutions in real life situations. In collaboration with European partners initiatives, the Concept House Village in Heijplaat Rotterdam was developed. The inner-city docklands of Rotterdam is an area in transition and forms an ideal setting for a real life test bed for future sustainable urban living and working. In this chapter the history, ambitions, context and the partnership is described. Furthermore the accompanying research and development program is enlightened. Additionally the business modeling of Concept House Village as a R&D facility is considered.

Keywords Living lab · Sustainable building (New building and retrofit) · Sustainable living · Business modeling

---

S. Silvester (✉)
Delft University of Technology, Delft, The Netherlands
e-mail: s.silvester@tudelft.nl

B. Hooijer · R. van Oosterhout
Rotterdam University of Applied Sciences, Rotterdam, The Netherlands
e-mail: h.hooijer@hr.nl

R. van Oosterhout
e-mail: r.van.oosterhout@hr.nl

F. van der Kemp
Building Changes, Rotterdam, The Netherlands
e-mail: fvdk@buildingchanges.nl

D.V. Keyson et al. (eds.), *Living Labs*, DOI 10.1007/978-3-319-33527-8_28

28.1 Introduction

In most of the European countries sustainable building is on the policy agenda since the early nineties of the last century. Radical improvements needed to meet the formulated policy ambitions repeatedly failed in the past because of the very complex structure of the building sector. The complexity is caused by:

- *The structure of the branch*; many different actors, in a variety of roles, mutual dependent to accomplish the projects,
- *The nature of the product*; the possibilities of mass fabrication are limited and the objects are composed of different products or subsystems with divergent life cycles.
- *The measures from other policy-fields*; besides the direct building regulations, sustainable building is influenced by environmental policy, urban planning, social housing, energy etc. as well.

The Dutch government promotes sustainable developments in the built environment systematically by applying a wide array of policy instruments. Besides the more traditional instruments like regulation and financial incentives, a second generation of instruments became popular like covenants, agreements, network management and specific use of communication measures. One specific type of 2nd generation instrument is the *demonstration project.* A demonstration project can be referred to as a project in which innovative technologies are being used in more or less normal situations to foster the development and diffusion in the regular market of these technologies (Buijs and Silvester 1996).

A series of demonstration projects on energy efficient and sustainable housing was initiated by different Dutch national governmental agencies till the mid nineties. For example in the Netherlands the national REGO—program was initiated in 1982 to stimulate the efficient energy use in the built environment (Silvester 1996). In the international arena the International Energy Agency fostered among others energy efficiency research and innovation programs in the built environment since its establishment in 1974 (IEA 2014). According to Buijs and Silvester (1996) a demonstration project can fulfill different functions in the development and diffusion of innovations:

- Concrete formulation and visualization of government policy.
- Collection and transfer of developed knowledge and experience.
- Development of strategic alliances within the concerning networks.
- Providing knowledge for regulation.
- Assessment of the technical feasibility of innovations.
- Signaling of the possible side effects of innovations at an early stage.
- Examining acceptance by different ‘users’ of the innovations.
- Examining the financial feasibility of the innovations.
- Market development through enlargement of the scale of the demonstration projects.

One of the important lessons learned from those series of projects is heterogeneous state of development of the applied innovations and that it is therefore crucial to make a clear methodological distinction between prototypes, pilot plants, experiments and demonstration projects etc. In fact essential products within the sustainable housing concepts were often not market-ready yet. Some of them were still not technically functioning the way they should and some of the innovations were not used or appreciated by the occupants according the way the designers thought. Balanced ventilation technology for example, was rejected by a part of the Dutch population as a solution for energy-efficient ventilation because of malfunctioning, interaction problems and perceived health impacts (de Jong 2014, Hasselaar 2009).

It was concluded that, to be able to develop more high-impact innovations, product developers and policymakers need to work with users to test and develop products, services and regulation (EU 2009). As a result of the EU funded Design Study, the Living Lab approach was developed as next generation of policy, research and design instrument (Eijk 2010).

This European the LIVING LAB Design Study (FP7) initiated the start of the Rotterdam based Concept House Village.

28.2 Location

The Concept House Village is urban area in Rotterdam dedicated for prototype houses that will be used, tested and evaluated by actual users from surrounding neighborhoods. The Village is a real life test bed for the latest technological applications in sustainable energy, water, sanitation and home automation on both building and district levels. Prototyping is taking place both for the retrofit as well as new building markets.

Concept House Village is situated in the middle of the Rotterdam Port Area (Fig. 28.1). It is part of the neighborhood Heijplaat. This rather isolated living area on the south side of the river Nieuwe Maas (part of the Rhine) was developed in the beginning of the 20th century as a living estate for the employees of the *Rotterdamse Droogdok Maatschappij* (RDM) shipyards. Heijplaat became a rather 'green village' within a port landscape. After the Second World War the living area was extended with multifamily, social housing estates. After the decline of the shipbuilding and connected maintenance activities and finally the closure of the RDM Shipyard in 1996, the living conditions at Heijplaat deteriorated.

In 2011 a new urban plan for the greater Rotterdam Docklands was approved by the Rotterdam City Council (Rotterdam 2011). According to this plan the Docklands will become the connection between the city and port. The Docklands will be transformed into an attractive, sustainable living and working environment. Within that plan the RDM/Heijplaat will gradually transform into a district in which still impressive port activities take place, but where the RDM premises will become an innovative campus site for new industries, education and research. The

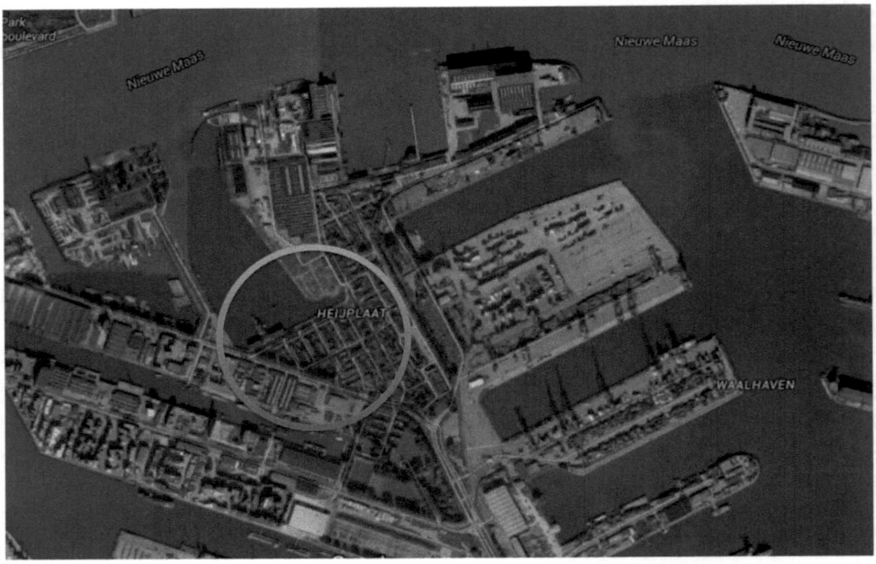

Fig. 28.1 The Heijplaat district in the Docklands of Rotterdam (Google Maps)

Rotterdam University of Applied Sciences is one of the leading partners in this transformation of the former shipyard. The abbreviation RDM is being re-used and stands since 2010 for Research, Design and Manufacturing (http://www.rdmcoe.nl/english).

At the same time the Urban Plan foresees a revitalization of Heijplaat as a village-like, sustainable living environment. As a consequence the post-war low quality multifamily complexes are demolished in recent years. Of the total of 465 houses in Heijplaat, 296 will be replaced. Ambitions for the new dwellings are:

- High energy efficiency, contributing to an energy neutral area,
- Local renewable energy production
- Climate adaptive building
- Expand local nature in the area
- The use of recycled and or bio-based materials

Due to the economic crisis—reflected in the postponement of investments in new building activities—the execution of the Heijplaat revitalization has been retarded.

This situation of an area ready for new developments on the border of the old Heijplaat village, together with the high ambitions in terms of sustainability offered an interesting starting point for the establishment of the Concept House Village Living Lab in 2011.

28.3 Themes

At the start of the Concept House Village Living Lab development in November 2011 a workshop was organized with the research communities from both involved knowledge institutes, Delft University of Technology (TUDelft) and University of Applied Sciences Rotterdam (HR) with additional input from other stakeholder in the built environment. As a result of this workshop an ambitious multi-annual research portfolio was formulated and is still valid. The major research issues are presented in the matrix in Fig. 28.2. In this matrix distinction is made between the system levels (City/District—Building—Installation/Appliance) on the vertical axis and the aspects like interaction, processes and technical functioning on the horizontal axis.

The knowledge development within the SusLabNWE project was mainly focused on the cells of matrix on the left bottom part; the levels 'building' and 'installations/products' and the 'human-product' interaction aspects. SusLabNWE provided the tools and methods for executing further longitudinal research in the Concept Houses that are completed and those to be built in the near future. The same households will inhabit all the prototypes as much as possible during the monitoring period of at least 2 years.

The development of the three first Concept Houses delivered already a lot of insights into the building process and technical functioning of installations and

Research Portfolio CHV	human-product	building & end-of-life	technical functioning
city-/district	• Commitment of inhabitants with the sustainable development of Heijplaat Rotterdam and choices of measures taken • What is the potential in urban areas for sustainable public products and services? • How can sustainable measures contribute to the creation of social interaction? • What role can a social community oriented energy feedback system play? • Comfort and experience of the public space and social interaction	• Define crucial elements for local metabolism • Define optimal ecological infrastructure and socio-technical organization of it. • The logistics of building materials • (Re-)Use of local materials.	• Possibilities of energy-exchange and –buffering at city and district level
building	• Commitment of (future) inhabitants with the development of dwellings (co-creation) • Comfort and experience of dwellings (monitoring & design interventions). • Influence of behavior on the energy efficiency. • Long-term monitoring on behavioral and technical/material adaptations of sustainable buildings (level of needed flexibility)	• Evaluation of permit-, building- and end-of-life processes of sustainable buildings. • Quality assessment of dwellings • Possibilities of energy-exchange and –buffering on building level • Business modeling for sustainable and zero-energy building concepts	• Optimal level of sustainable measures (individual or collective) • Energy efficiency performance during the seasons. • Maintenance aspects of innovative sustainable buildings • New business modeling for maintenance services
installation-/product	• Comparison of routines and practices related to management of energy and sustainable sources with the former living/working situation. • Comfort and experience of installations and interfaces. • Habituation to installations, interfaces and products. • Role of interfaces and related services on the acceptance of sustainable measures.	• Quality assessment of installations, products and measures • Flexibility and adaptability of installations, products and measures concerning practices and lifestyles of inhabitants.	• Detailed analysis of energy-use and comfort related to the use of the installation and products • Explaining differences in energy-use (socio-, demographic, cultural, behavior, building type, measures, appliances etc.) • Maintenance of installations and products

Fig. 28.2 Research portfolio Concept House Village

specific sustainable (recycled and bio-based) products applied. For the Prototype 1 and CHIBB concept house extensive monitoring of the development phase took place and is documented (Eekhout and van Timmeren 2013) (http://chibb.nl/).

Furthermore the SusLabNWE tools and methods are being applied within the 2nd Skin project, an innovation development project aimed at the large-scale refurbishment of post-war apartment complexes in Europe. This project is part of the BTA Flagship of the Climate KIC (BTA).

Besides the longitudinal monitoring mentioned, the research within CHV will foster more research on district and city level. Together with the Municipality of Rotterdam several community projects started like the Groenkleed, IKS and ICOR projects that are focusing on sustainable innovations in the public space (http://concepthousevillage.nl/locatie.html).

On city-/district level research will be continued by the TUDelft, HR within CHV on the challenge of sustainable materials and circular building. This is a follow-up of the REAP + -program of CHV in which an approach is developed to link streams (materials, energy, water) on different system levels to support the development of a regional circular economy (Geldermans and Rosen-Jacobsen 2015).

28.4 Prototypes

Since the start in 2011, three temporary prototypes of energy-neutral dwellings are operational, and four prototypes for the refurbishment of porch apartment are under development (see Figs. 28.3, 28.4, 28.5 and 28.6). Other networks of consultants, companies and knowledge institutes are developing permanent prototypes for a new plot in the Village coming available for re-development see Fig. 28.7. All networks focus on different sustainable innovations. The surrounding urban area—Stadshavens Rotterdam—offers the potential for up-scaling the results from the test beds through the involvement of the local community and to monitor the longitudinal effects of the innovations. Recently Concept House Village is appointed to play an important role in the acceleration of the energy-neutral refurbishment of the Rotterdam Building Stock the coming decades.

28.5 Partnership

Concept House Village (CHV) is initiated by the City Ports Academy Rotterdam and is executed by Rotterdam University of Applied Sciences, Delft University of Technology and the Social Housing Corporation Woonbron. The partners have a long-lasting agreement on the exploitation of the CHV-facility signed by the boards of executing partners.

Fig. 28.3 Four concepts for Energy Neutral Refurbishment of existing post-war apartments at CHV. (realization planned in 2016)

28.6 Financing the CHV-Facility and Prototypes

During the first period of its existence, in the period 2010-2015, the facility of Concept House Village is financed both from public and private sources. With the public financial support from the Interreg SusLabNWE project, the Dutch Ministry of Economic Affairs (Peaks in the Delta) and the Municipality of Rotterdam it was possible to establish the CHV-facility and to manage the acquisition of new prototypes, to guide the building processes of the prototyping, to coordinate the research program and foster the knowledge transfer.

The basic approach is that the involved consortia finance the prototypes themselves. The two involved knowledge institutes guided the development of the first two prototypes (*CH Prototype 1* and *CHIBB*). Although the partners of the two consortia financed the main part of the research, development and building of prototypes, the realization of those prototypes would have been impossible without the crucial, additional subsidies from SusLabNWE, Peaks in the Delta and the Municipality of Rotterdam.

The two prototypes serve as a laboratory for the research programs of both knowledge institutes. The on-going research is mainly financed by the institutes'

Fig. 28.4 Maskerade prototype by a consortium led by Maarten van der Breggen since 2015

research funds. The prototypes also serve as a learning environment in the curriculum of the bachelor and master students of the institutes. Besides the whole design and development of CHIBB by students of the *Rotterdam University of Applied Sciences*, students of the *Albeda Polytechnic* did the actual building of the CHIBB prototype. The CHV-facility has agreements with the consortia behind the prototypes about the involvements of students and research staff of both institutes in the development, monitoring, testing and validation of the dwellings.

28.7 Business Development

In 2015 The Concept House Village Facility is revising its business model, to become less reliant on structural public funding, to accelerate the real market application of the results of the prototyping and to become a regional focal point of sustainable building to support the realization of the political ambitions.

Whereas the first two prototypes were initiated by the involved knowledge institutes and forcefully supported by public funding, the role of the industry had

Fig. 28.5 Concept House Prototype 1 by TU Delft at CHV since 2011

to become more prominent for a sound continuation of Concept House Village as a facility. The public regional financers indicated that structural financial support of the facility was not feasible.

After a series of interviews with important stakeholders from the whole building value chain and the demand side by a consultancy firm and follow-up negotiations with potential partners for an extended Concept House Village Facility, a model was decided as a backbone for the business modeling (Ong et al. 2014). The model is illustrated in Fig. 28.8.

The CHV facility—in its renewed arrangement—is a close-fitting collaboration between the academia (TUDelft & HR), the building industry (represented by branch organizations), the local government and representatives from local neighborhood organizations. This *collaboration* is formalized by a steering group with representatives of the organizations mentioned. During the monthly meetings the general direction is being established and the decisions are taken about new experiments in the facility. Furthermore the steering group will play an active role in the influencing of the policy- and R&D-agenda on sustainable urban development in the region and beyond.

Fig. 28.6 CHIBB Prototype by Rotterdam University of Applied Sciences at CHV since 2014

The acquisition of new consortia willing to use the facility for developing, testing, validating and demonstrating their innovations is managed by delegates from a consultancy of the SME in the Dutch Building Industry, called *Building Changes*. The facility is offering supporting management capacity—if needed—for the consortia. Lessons learned from the start of the facility showed the need for this additional project management capacity for the—often—ad hoc innovation networks.

The actual research and knowledge development in the CHV Living Lab is closely linked to the research agendas of the two universities involved (see Fig. 28.2). The execution of the research is partly in parallel with the education curriculum. Besides the regular research and design courses of the curricula, an other noteworthy and recognized option for the execution of the research in the CHV Living Lab is the cooperation of bachelor and master students in *innovation teams* under supervision of the Community of Practice (CoP) Circular and Sustainable Building, in which not only students but also alumni, researchers, lecturers and industry are working more permanently together on the challenge of building, living and working in a circular economy.

CHV is part of several European Networks of Sustainable Living Labs, like ENoLL (European Network of Living Labs), SusLabNWE and BTA/Climate KIC. These connections are of importance for the smooth diffusion of knowledge,

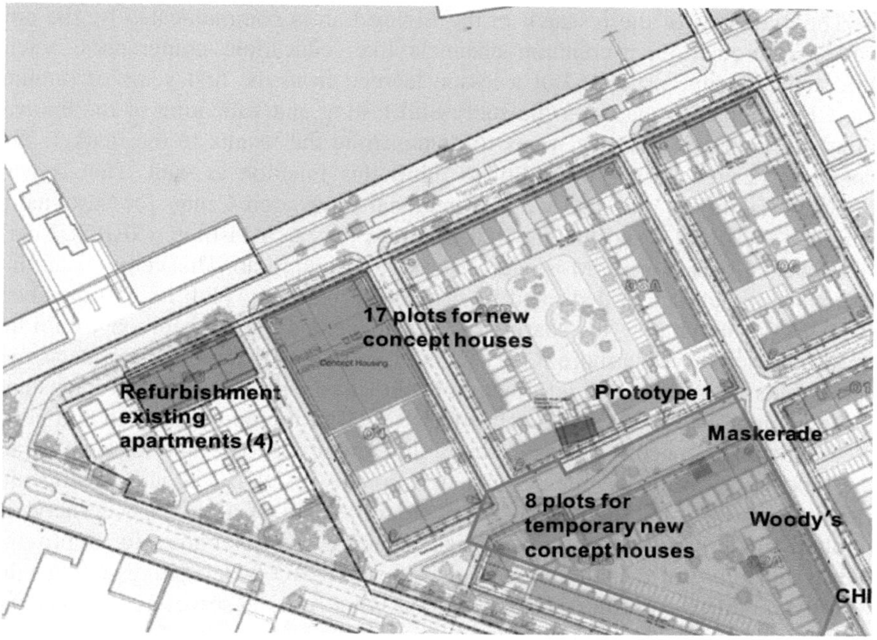

Fig. 28.7 Urban plan for redevelopment Heijplaat with marked areas for refurbishments, permanent and temporary Concept Houses as part of the CHV-Living Lab

Fig. 28.8 Business functions of Concept House Village facility

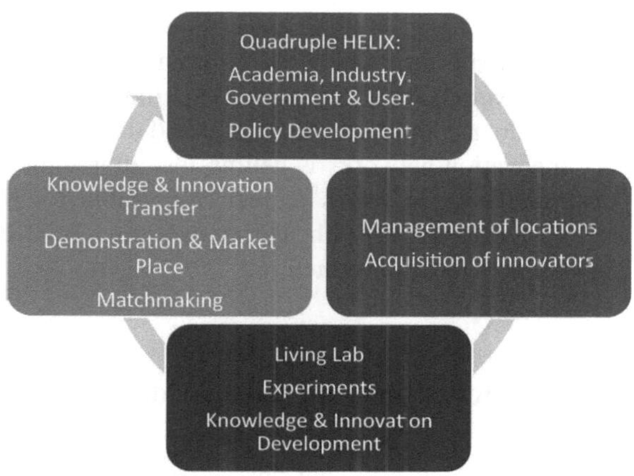

methods and tools needed for the acceleration of sustainable innovations. Moreover these European networks also offer the market- and matchmaking place to connect a broad array of actors for the speeding up the large scale implementation of the developed innovations and insights.

The outcomes of the research in the Living Lab is communicated by the universities through their common channels like, education, conferences, scientific and academic articles. But a lesson learned from the first years of running the CHV facility is that after the successful testing and validating of the innovations, the involved industry wants to demonstrate the results to the market. The consortia desires the CHV facility to fulfill this function as well. That is why the CHV facility is affiliated with the regional Innovation Center for Sustainable Building (ICDUBO) and the national Building Research Institute (SBRCURnet). *ICDUBO* is located at RDM is nearby the CHV Living Lab. This center is aiming at connecting all actors dedicated to sustainable building. With its 250 members ICDUBO is stimulating the sharing of knowledge and the cooperation within the quadruple helix. Furthermore ICDUBO is offering a showroom for architects and consultants with fine-tuned information about sustainable innovative products and installations. The CHV Living Lab offers ICDUBO the opportunity to add complete integrated concept houses to their 'showroom portfolio' and to demonstrate the innovations in a real life setting. Additionally ICDUBO is also hosting the *Woonwijzerwinkel* for the region of Rotterdam as part of a nation wide network of shops for private consumers looking for advise and inspiration for improving the sustainability of their living environment. ICDUBO is delegating staff for the widespread diffusion of the results of CHV Living Lab and serves as a market and matchmaking place to accelerate the adoption of the innovations of CHV.

SBRCURnet is an independent knowledge institute for the building sector in the Netherlands. This institute takes care of securing the knowledge developed in the consortia and makes it available for the different stakeholders in the sustainable building sector. A portal called *DuboLab* is developed to support the consortia, operating in CHV, in their knowledge management and -transfer and to provide an internet site thru which the information will be available in a structured way. Lessons learned from decades of experimenting and demonstrating with sustainable building in the Netherlands showed that the information of the mostly temporary innovation networks is getting lost after some time. The networks are rather loose and nobody seems to feel responsible for the developed knowledge and its transfer anymore. With the involvement of SBRCURnet this loss of valuable knowledge will be prevented.

28.8 Conclusions and Discussion

The Concept House Village can be recognized as one of the first Living Labs aiming at sustainable building in the world. Started in 2011 the development of the facility delivered a lot of insights into the feasibility, the incentives and barriers of these new generation of R&D infrastructures. Lessons learned:

- The knowledge institutes (Rotterdam University of Applied Sciences and TU Delft) played a crucial role in the initiation phase of the Living Lab facility.

- European and regional public funding was making the initiation of the Living Lab possible. This funding was especially aimed at the acceleration of **innovation**. The role Living Labs can play in generating more **fundamental knowledge** on the complex interaction of users and the sustainable solutions is still neglected and hard to get financed. Attempts to get the Living Lab recognized as a national research facility for more fundamental research on interactive R&D for sustainable transitions in the home environment and open for the whole Dutch research community unfortunately did not succeed up to now.
- It is very hard to create a sound business model for a 'physical and long-lasting' Living Lab facility. The investments in the concept houses are vast and for the temporary prototypes the depreciation is as well. In the Concept House Village business model it is therefore that a mix of temporary and permanent houses has been chosen as an answer. The temporary concept houses offer the opportunity to keep experimenting with new challenging innovations. Those houses of parts of them can be replaced and guarantee a dynamic orientation on more fundamental knowledge development and still experimental innovative trajectories. The permanent concept houses are designed and developed in such a way that they intent to have a high value on the real estate market. For these concept houses no depreciation costs have to be reserved in the financial program of the consortia that develop these houses. Moreover these permanent houses offer the unique possibility to gather in-depth insight into the longitudinal interaction between user and the house, its appliances and the urban environment
- The expectation of being a living lab that could play a role in the development of *individual products* has been very modest so far. The Prototype 1 played a role in the ideation and first prototyping of water saving showers and reflective energy feedback systems. Based on the results of the first years of Concept House Village Facility it can be concluded that CHV is especially fostering the business development of *integrated sustainable housing concepts*.
- It showed very important for the commitment of the numerous partners in the consortia involved in the concept houses that in the region a lot of challenges in the built environment are at stake. For the region of Rotterdam for example the refurbishment of the port area is such a challenge. Not only should be thought about the new buildings but also on the buildings to be renovate in the near future.

References

BTA. http://www.climate-kic.org/programmes/building-technologies-accelerator/.

Buijs, A., & Silvester, S. (1996). Demonstration projects and sustainable housing. *Building Research & Information, 24*(4), 195–202. doi:10.1080/09613219608727529.

de Jong, E. (2014). *User-experiences with balanced ventilation systems (in Dutch)*. Rotterdam: WONO Housing Research.

Eekhout, M., & van Timmeren, A. (2013). *The development and realisation of Concept House Delft Prototype* (p. 155). Delft: TU Delft, Faculty of Architecture and the built environment.

Eijk, D. J. v. (2010). *Living Lab, Design Study for the LIVING LAB Research infrastructure, to research human interaction with, and stimulate teh adoption of, sustainable, smart and healthy innovations around the home*. Brussels: European Commission.

EU. (2009). *Design as a driver of user-centred innovation*. Brussels: European Commission.

Geldermans, R. J., & Rosen-Jacobsen, L. (2015). *Circular material & product flows in buildings* (A. E. T. Architecture and The Built Environment,Trans.). Delft: Delft University of Technology.

Hasselaar, E. (2009). Health risk associated with passive houses; an exploration. In *Indoor Air 2008: Proceedings of the 11th International Conference on Indoor Air Quality and Climate, 1–8. (2008)*: Technical University of Denmark.

IEA. (2014). *Energy Building Community*. http://www.iea-ebc.org2014.

Ong, K., Groenewoud, R., & Van Rijn, M. (2014). *Concept house village, chances for diffusion of innovations*. Barcelona: Bax & Willems.

Rotterdam, S. (2011). *Structuurvisie Stadshavens Rotterdam*. Rotterdam: Stadshavens Rotterdam.

Silvester, S. (1996). *Demonstration projects and energy efficient housing*. PhD, Erasmus University Rotterdam, Rotterdam

Chapter 29
Commercial Consortia

Maria Ådahl

Abstract Science parks constitute an important intermediary in the innovation chain since they provide a location where government, universities and industry can cooperate and collaborate. In this chapter, Johanneberg Science Park which is located in Gothenburg city, Sweden is presented as well as its collaboration model through which development, transfer and commercialization of technology is taking place. More specific, the concept of science park's "open arenas" is described as well as their connection to living labs. This chapter is also addressed to a unique collaboration housing project within the open arena of urban development at Johanneberg Science Park. The project is called HSB Living Lab and encloses the link between open arenas of science parks and the creation of living labs. The uniqueness of this project lies on the fact that a number of partners with different backgrounds are gathered within a single organisation, enter into a 10-year partnership agreement and get involved to a joint research and development project. In the last part of this chapter, the value proposition of HSB Living Lab for the academics, the different companies and the inhabitants being involved is discussed.

Keywords Science parks · Collaborative projects · Living lab · Innovative housing · Urban development · Open innovation

29.1 Johanneberg Science Park—Collaboration Model

Science parks are characterised as important intermediaries in the innovation chain since they provide connectivity between science, business and social society. Johanneberg Science Park brings industry closer to the cutting edge of science

M. Ådahl (✉)
Johanneberg Science Park, Göteborg, Sweden
e-mail: Maria.adahl@johannebergsciencepark.com

D.V. Keyson et al. (eds.), *Living Labs*, DOI 10.1007/978-3-319-33527-8_29

to make excellent explorative research more accessible. The mutual proximity between university students, teachers, researchers and industry will give all parties involved a better insight into their individual needs for competence. In this way the science park promote interactions and technology transfer and thus create the right circumstances for inducing regional growth. The science parks structure has been used for developing triple helix projects within the SusLab NWE project.

Johanneberg Science Park creates open and attractive collaboration platforms for interdisciplinary exchange of ideas and knowledge between academia, industry, research institutes and public bodies, with the aim to:

- Develop and make available knowledge and expertise relating to new attractive products and services with a high technology and knowledge content, to achieve sustainable economic growth.
- Create environments for research and production that together lead to increased competitiveness by preserving, developing and generating new employment opportunities.
- Ensure that competitive education is provided, as well as good quality fundamental and applied research.
- Promote start-up and development of knowledge and technology intense companies and contribute to the creation and marketing of an environment that is attractive to business.
- Solve the increasingly complex challenges of society, in collaboration with different stakeholders.

Johanneberg Science Park is developing Open Arenas within its profile areas; Urban Development, Energy and Materials & Nanotechnology.

Open Arenas are neutral places for collaboration between industry, academia and society. They are meeting places where researchers, business people, students, designers, project managers and others can get together to brainstorm, exchange experiences, generate new knowledge, network, and make the most of the opportunities provided by the arena. Johanneberg Science Park's organization acts as a project broker and rigs research and development projects within the three Open Arenas mentioned above. An Open Arena offers participation in activities and processes as well as physical infrastructures, such as test beds, laboratories and premises. The premises are designed to stimulate both spontaneous meetings and innovation-creating activities. All the employees of companies that are established in Johanneberg Science Park have access to Open Arenas and can use them for meetings or as temporary workplaces. For example, employees can have informal business meetings or a meeting at any creative space of the building.

Small and medium-sized enterprises (SME) will be the future growth engines for both the region and the nation, which is why we want to be a resource for these companies, helping them to benefit from new knowledge and expertise. We work with companies in all sectors, with the drive, potential and will to develop.

Within the framework of the Arena, external stakeholders, in particular SMEs, are invited to interact with projects in different ways, for example through workshops, project brokering, innovation contests, etc. The Arena will enable a two-way flow of ideas and innovations, where businesses can both make use of ideas born within the projects and spread their own ideas to other practitioners. The Arena will also offer support and inspiration when new projects are initiated. (Website Johanneberg Science Park)

Living labs are an emerging Public Private Partnership (PPP) concept in which firms, public authorities and citizens work together to create, prototype, validate and test new services, businesses, markets and technologies in real-life contexts,[1] in our specific case the HSB Living Lab. The real-life and everyday life contexts will both stimulate and challenge research and development as owners, partners and inhabitants will not only participate in, but also contribute to the whole innovation process.

The connection between the Open Arena and Living Labs lies in the labs' full value chain and encompasses the creation of ideas in workshops in which the different stakeholders (and end users) have input and create ideas based on common interests or purposes and different viewpoints and knowledge. These ideas are then prototyped in the living lab, where they are used and tested, on which the academics report. Improved prototypes and iterative processes with the involved stakeholders will then lead to tested and validated products

HSB Living Lab is a collaboration project within the Open Arena of Urban Development, which currently has nine partner members. These are active within different sectors and have both shared and individual aims. To gather all of them within a single organisation is a challenge, but this is also one of the main objectives of the project. The project organisation must not focus solely on current needs, but must also be flexible enough to be able to meet the challenges of tomorrow and to include future partners. This is organised in a hierarchal three-layer steering organisation that sees a group for project development and operational issues, a decisions making group and a owners' group that looks after issues. Due to its proximity to the university (Chalmers University of Technology) and its owners (HSB) the focus of HSB Living Lab (HSB LL) has been on building materials (BTA) and other areas around living in a house (i.e. the social environment around washing machines). The market segment is this area with its companies or academics that are active in these fields. In a way the partners in the HSB Living Lab represent the market-segmentation well. The partner companies involved in HSB Living Lab will be able to use all relevant findings in their respective operations, to make it possible for their customers and partners to also benefit from the results. Organisations choose to take part because the project offers a unique opportunity to be involved in the creation of products and solutions for the future. Each partner has a contract with the HSB that sets its relation to the Living Lab, IP issues and revenue streams. The set up of a consortium agreement between the partners is under consideration. Except for the collaboration project of HSB Living Lab, Johanneberg Science Park's activities have resulted in a number

[1]State-of-the-Art and Good Practice in the Field of Living Labs, Veli-Pekka Niitamo, Seija Kulkki, Mats Eriksson, Karl A. Hribernik.

of interesting projects that bring added value to the area and directly contribute to progress. A few high-profile examples are Riksbyggen's Positive Footprint Housing and ElectriCity project with its new electric bus.

29.2 HSB Living Lab—Consortium and Collaboration

HSB is a cooperative association for housing in Sweden for which, collaboration is key, not only within its own organization and with its members, but also with external companies and organizations outside. HSB has existed for almost 100 years, with the mission to build and manage housing for our 550,000 members in all parts of Sweden. When developing its housing stock, HSB has always been driven by challenges. When HSB was first launched the challenges were focused around providing good quality homes for everyone or helping women to enter the labour market. Today company faces both traditional and new challenges. The challenge to provide good homes for everyone remains, but they now also have to deal with new types of issues, including climate change.

HSB decided to become a partner of Johanneberg Science Park in 2011, as this would give them the opportunity to work together with other companies and universities. During an EU collaboration project on living labs, in which both Johanneberg Science Park and Chalmers University of Technology were partners, the idea of starting a collaboration project together with HSB where a demonstration environment in the campus area of Chalmers would be developed, grew up. This idea was the base for Johanneberg Science Park, Chalmers University and HSB to together set up the HSB Living Lab. From HSB's side, the aim is to create an arena where they can improve homes in real time, working with the residents. The property management company hopes that this project will help a relatively slow-moving sector, the building industry, to move forward faster. By testing innovations in real life and together with the residents, HSB will learn lessons that can then be implemented in all the properties the company builds and manages in different parts of Sweden.

HSB, Chalmers University of Technology and Johanneberg Science Park have together set up the HSB Living Lab. The main project partners are currently working to recruit more partners to the project. As they believe in collaboration, their view is that by working together with other businesses, the partners together will be able to make much faster progress in the development of future homes. This is why the three main project partners are working actively to recruit more partners into the project, businesses with the ambition and ability to see the opportunities on offer. The aim is to recruit 10 partner companies from different sectors: architecture, white goods, construction, technical expertise, energy, interior decorating, etc.

So far, the three main partners have managed to recruit 10 partners to the project: Bengt Dahlgren technical consultants, white goods manufacturers Electrolux, the energy company Göteborg Energi, the building contractors Peab, Tengbom firm of architects, one of Sweden's largest property companies Akademiska Hus, the kitchen supplier Vedum and the EU's largest public-private innovation partnership focused on climate innovation, Climate-KIC. Further partners are being sought and found in other sectors, such as IT, with the company Tieto, Scandinavia's largest IT supplier and Elfa that makes entrée solutions. The companies are selected based on their areas of operation and their potential impact on the homes of the future. Focus is also on finding businesses with which partners and project want a long-term partnership. These could be companies with which partners already have a relationship, or companies with whom partners would like to strengthen their collaboration further.

HSB enters into 10-year partnership agreements with the participating companies. The reason is that they are looking for long-term collaboration, in which the partners depend on each other and are involved in both research and development within the HSB Living Lab. The project has a structure where all partners make an annual payment into a joint research fund. This fund will be used for joint research projects, agreed between the partners. The structure offers good potential for attracting external funding for joint research projects, due to the large number of partners and the available co-funding in the form of both time and money from the joint research fund. (Website HSB Living Lab)

The value proposition for the HSB Living Lab for the academics is that they create data that they can use for scientific papers and to attract research funds, for companies that they get ideas from the open innovation processes and a testing facility for products, and for the inhabitants that they have a place to sleep and contribute to an experiment.

As the whole concept is built around the integration of different organizations, there is no simple costs-income table that shows a clear profit. It is in fact easier to see how each stakeholder makes a revenue generation mechanism that has a profit.

For the academics the value of the HSB Living Lab is that it is a great research infrastructure, which costs can come in as co-financing to specific projects, as an opportunity to meet stakeholders that need research, and as an object of research itself. One could see that the profit comes back to Chalmers in the research that will take place. Parts will come from the HSB LL Research fund, brought together by the partners.

For the commercial organizations the access to the research facility and research may have lower costs then in-house development. Furthermore the open-innovation may bring out stuff that they simply cannot do themselves. Finally there is the access to top-students, access to stakeholders and marketing value that may have higher profits than costs. In general, the revenue generation for these partners would be more in the field of low costs, than actual cash flow.

More concrete is that each partner will donate at least SEK 25.000 a year on research in the HSB LL until 2025. The current idea is that this is a cash contribution and that projects are being proposed and paid out the fund. It is—currently-expected that participating partners are very modest around their own costs in these proposed projects. This gives a lot of room for growing the projects with contributions in kind and then asking for subsidies.

HSB, Chalmers University of Technology and Johanneberg Science Park have great ambitions for HSB Living Lab. HSB expects to see multiple results in a number of different areas. These could be anything from experiences of using different types of materials, to the residents' views on indoor climatic conditions and comfort. This unique opportunity to improve people's homes in collaboration with researchers and residents is something HSB will make the most of. Experiences and lessons from the project will be shared with HSB's operations nationwide—the project will not have succeeded until the knowledge gained from the HSB Living Lab is utilised in everything built, managed and renovated by HSB. As there are so many partners involved, the partners expect the results to benefit the entire industry, as they will be spread outside HSB, Chalmers University of Technology and Johanneberg Science Park. This is particularly important now, as we are facing the challenge of climate change; there is not enough time for each of us to develop separate solutions. The challenges are too big and the time too short. (Video HSB Living Lab—Homes of the Future)

The experiences from HSB Living Lab has been chaired to the SusLab NWE consortium and vice versa. This has been an important input in developing the labs. Study visit tours have been arranged to other labs in the SusLab NWE consortium and HSB employee has been participated to conferences in the SusLab NWE project.

References

Video HSB Living Lab—Homes of the Future. https://www.youtube.com/watch?v=AqRyYP1xxuc.
Website HSB Living Lab. https://www.hsb.se/kampanjer/hsblivinglab.
Website Johanneberg Science Park. http://www.johannebergsciencepark.com/en/what-we-do/arenas-collaboration/project-arena/hsb-living-lab.

Chapter 30
Business Models for Sustainability in Living Labs

Mike Burbridge, Gregory M. Morrison, Menno van Rijn, Sasha Silvester, David V. Keyson, Lali Virdee, Carolin Baedeker and Christa Liedtke

Abstract There are an estimated 170 active living labs across the globe. All have common elements but not all of them contribute to the delivery of sustainable living. Here we consider the business models of sustainability in living labs (SusLabs). Specifically we review four active living laboratories that are part of the SusLab North West Europe network. We show that the business cases are different for at least two reasons. One is that each SusLab project has a specific focus even though all are seeking to develop energy efficient innovative products, services or systems. Examples of focus include demonstration projects, knowledge generation through research and business to business development. The other is

M. Burbridge · G.M. Morrison (✉)
Curtin University Sustainability Policy (CUSP) Institute, Curtin University, Perth, Australia
e-mail: greg.morrison@curtin.edu.au

M. Burbridge
e-mail: mike.burbridge@postgrad.curtin.edu.au

M. van Rijn
Bax and Willems, Barcelona, Spain
e-mail: m.vanrijn@baxwillems.eu

S. Silvester · D.V. Keyson
Industrial Design, TU Delft, Delft, The Netherlands
e-mail: S.Silvester@tudelft.nl

D.V. Keyson
e-mail: d.v.keyson@tudelft.nl

L. Virdee
Institute for Sustainable Futures, London, UK
e-mail: lali.virdee@instituteforsustainability.org.uk

C. Baedeker · C. Liedtke
Wuppertal Institute for Climate, Environment and Energy, Wuppertal, Germany
e-mail: carolin.baedeker@wuppertal.org

C. Liedtke
e-mail: christa.liedtke@wuppertal.org

D.V. Keyson et al. (eds.), *Living Labs*, DOI 10.1007/978-3-319-33527-8_30

that each came about for different reasons which might include significant public or private sponsorship, or through academia-business co-creation, and this too is reflected in the business case. We also show that the business cases are not static, but may evolve over time as opportunities are created and as partners develop a clearer understanding of the potential of each SusLab. We propose that, based on a common definition of a SusLab, theoretical considerations and societal needs, as well as insights from the cases, it should be possible to build a business case for a SusLab which draws on knowledge rather than learning-by-doing.

Keywords Living lab · SusLab · Business case · Triple helix · Sustainable innovation · Path dependency · Transition · Social practice theory

30.1 Introduction

There are at least 170 active Living Labs across the globe (ENoLL 2015). All living labs have specific attributes in common (Salter and White 2013). One is that they aspire towards the innovation of new products or services; they involve close relationships between at least two of the following three sectors: academia, business and society or consumer; they involve co-creating the innovation bringing to bear the expertise of stakeholders on the innovation. However, not all Living Labs contribute to the delivery of sustainable development (Vanaker 2014), nor necessarily progress towards the delivery of some of the recently adopted sustainable development goals (UN 2015).

This chapter only deals with business models of sustainability in Living Labs (SusLabs). Specifically we have reviewed the experience and evidence from four active Living Labs which are part of the SusLab North West Europe (SusLab NWE) network.

30.2 SusLab Case Studies

The SusLab network covers facilities in North West Europe and provides acceleration of innovation for the building industry and society. SusLabs may demonstrate energy efficiency (some are even energy producers) and focus on energy use in and around the home, but they also offer innovation in terms of user and practice, as well as water and resources other than energy. They offer the opportunity for partners to undertake insight research, product prototyping and field testing in a variety of building typologies that represent society at large.

At present the network includes four linked hubs:

1. HSB Living Lab, which is a unique international facility on the Chalmers University of Technology campus in Gothenburg, where researchers and

societal actors can co-create ideas and initiatives for products and services which will enable sustainable living.
2. SusLabNWE Living Lab which is on the London Sustainable Industries Park (London SIP), East London and has been designed, supplied and built by Climate Energy Homes. SusLabNWE commissioned the Institute for Sustainability, working in partnership with the Royal College of Art (RCA) and Imperial College London to procure and run the SusLab.
3. The Concept House Village which is an urban area in Rotterdam dedicated for prototype houses that has been and will be used, tested and evaluated by actual users from surrounding neighborhoods.
4. The SusLab North-Rhine Westfalia (SusLab NRW) Infrastructure which consists of a Smart Home Lab, real home environments and showcase apartments.

The SusLab network is transnational and has dealt with all aspects of valorization through partner and stakeholder dialogue. According to Carlsson (2006) there are few studies of the degree of internationalization of innovation systems so the SusLab NWE is an interesting case to study.

In the following sections we provide a brief analysis of the theoretical underpinning of the SusLab model (Sect. 28.4) before offering an analysis of the different business models that have been developed by the SusLab members (Sect. 28.5). We discuss some of the issues that are common amongst the SusLabs and some that are distinct. Although all SusLabs concern themselves with delivering evidence-based sustainable innovation they all have quite different business models. These different business models are historical and have partly arisen as a result of different motivators for the creation of the SusLab. Interestingly too, it appears that the business model can change over time as relationships and understandings of the potential benefits that can be co-created also develop (Franzen 2015).

30.3 Theory

The ideas behind the development and running of living laboratories vary depending on purpose and this has been reviewed by Dell'Era and Landoni (2014). Their definition of a Living lab concerns the methodology, which does not underpin a business case or model well. We have modified their definition to see the Living Lab as a place rather than a methodology.

> A Living Lab is a real-life place for user[1] co-creation of innovations in knowledge, products, services and infrastructures.

Based on this definition we have identified three theories in particular that are worth further consideration in developing a business case and model for a SusLab.

[1]User is used in general terms and may refer to those living in the lab, if there are any, but equally well to stakeholders from business, society and academia.

The first is the triple helix. This theory owes its existence largely to the work of Henry Etzkowitz and Loet Leyesdorff (2000) from Stanford University who, along with others, developed the notion that business, academia and government (or society in general) all share common interests (Etkowitz et al. 2000; Etzkovitz 2013; Ranga 2015; CEC 2006). Put simply academics produce research which can help business develop innovative products which, if successful, enhances the products competitive advantage, which in turn can deliver societal benefit, such as regional economic development, employment (Nyman 2015; Hessels 2008; Erosa 2012; Gebhardt 2015) and perhaps even the transition to a low carbon economy (Pohl 2008; Hadorn 2006). Although Etzkowitz and his colleagues did not initially dwell on the sustainability issues they argued that if business, society and academia can work closely together then the potential for innovation could be enhanced, delivering benefits to all partners (Etzkowitz 2000). For the partnerships to be successful it is necessary to recognise the different cultures of each organisation (Max-Neef 2005; Nicolescu 2015; Erosa 2012) and different ways of working within different organisations (Jantsch 1970; Pohl 2011). This is why Living Labs are so interesting (Sunitiyoso et al. 2012).

The second is path-dependence theory which is important to understand because it influences the creation of the different business models (Nee and Cao 1999; Bednar et al. 2015; Malm et al. 2012). Path dependency theory sets out how decisions made today are influenced by how previous decisions have been made (Christensen 1997; Senge 1995; Dolan 2015). This does not mean that people cannot change the way they do things, but rather that how they respond will be conditioned by how they have responded in the past (Dolan 2015; Senge 1995). This turns out to be an important issue when we are looking at the different business models adopted by SusLab participants (Vanaker et al. 2014).

Finally, the issue of socio-technical change is important (Geels and Kemp 2007) and in this respect, social practice theory (see Sect. 1.5 for fuller discussion) helps participants to conceptualise environmental behaviour and awareness and design sustainable product-service-systems around the home (Baedeker et al. 2014; Liedtke et al. 2015). This is important because studies in failed innovations have shown that the benefits derived from an eco-product are not fully realised if they have been designed without input from users (Spaargaren 2011). It is for this reason that co-creation is central to the SusLab approach. Involving end users in the design of the product helps to reduce negative rebound effects (i.e. making both the product and the innovation process more efficient) and it is argued that the design of product service systems (PSS) will help with the transition to a low carbon economy (Liedtke et al. 2015 and see Sect. 1.3 for further discussion of these issues). The purpose of the PSS is to focus on the service that is being delivered (for example, the outcome desired might be a warm home) rather than what is producing it (for example, the process of producing and dispersing the output, heat, is the central heating system) and so design focus is on the outcomes service users want produced by a low carbon system. (Liedtke et al. 2015).

30.4 The Business Case

The business case for the SusLabNWE comprises two components. One is the value of being part of SusLabNWE; the other component is the development and maintenance of the SusLab infrastructure itself.

The four SusLab partners have utilised different models, partly as a result of how the individual SusLab came into existence, and partly as the result of what the SusLab was intended to deliver.

As the labs are at the early to medium stage of development (all are active but some of the infrastructure has not yet been built) it is not possible to give details of the benefit outcomes that have been accrued. However, it is possible to set out the value that is expected to be delivered and that forms the basis of the investment that partners have committed. The business cases deal with investment (to create the SusLab), income (from the delivery of services) and value (financial and non-financial benefits derived from the SusLab).

Although each SusLab, and therefore each business case, is different each share some common elements (Etzkowitz and Leydesdorf 2000; Etzkowitz et al. 2000; Etzkowitz 2013). The common elements come largely from being part of the transnational SusLab North West Europe (NWE) network.

30.5 SusLab NWE—The Value of the Network

The SusLab network provides real value to the labs themselves but importantly also to business (both active and future) as well as society. The network was created as the result of a grant from the European Union's Regional Development Fund through INTERREG which is designed to help member states with the development and sharing of information across borders.

Some of the key shared benefits to the member SusLabs of the network is set out in Table 30.1. Business and society also benefit from the value that the network and partnership generate (Ranga 2015). Business benefits from a rigorous, shared process for development of market ready innovation and it follows that business will then be confident that the methodologies are academically robust representing the latest developments in research. Business also gains value from working with potential end-users who are not tied to any organisation in the development of their product or service at various stages (in-sight research, product prototyping and field testing) in the product lifecycle, all of which takes place in a real living place (or near-living in the case of the UK SusLab)—cf. definition of a Living Lab under 8.1.

Society gets value from the network by the development of products, services and product-service-systems that contribute to a sustainable lifestyle (the studied SusLabs are largely focussed on reducing energy usage and carbon emissions) as well as economic benefits associated with successful development of the SusLab

Table 30.1 SusLabNWE—value to partners in the SusLabs of the Triple Helix

Communications	Document ongoing work on SusLab website, including findings from HSB Living Lab, and the Living Lab at the Institute of Sustainability Finalise SusLab book and promote methods at scientific and professional forums as well as at network events
Academic	Work with network to share and co-develop new SusLab methodologies Publication of results and methodology in academic journals Continue to develop SusLab sensor tool kit to support new work (for example Building Technologies Accelorator—building occupancy certification system (BTA-BOCS) pilot projects in office buildings) Continue to develop the SusLab tools such that the toolkit can be easily deployed in the field by sustainable building researchers and practitioners Each SusLab has a measuring and monitoring element and all four can be connected to a single data store and analysis tool to enhance rigour and ensure learnings are shared
Outreach	Maintain network of living labs linked to SusLab website Link SusLab work to regional networks (for example, TU Delft and AMS, Chalmers and HSB housing association) Leverage SusLab network combined with new partners to develop new joint projects Value/weight to partners from increased potential to attract further funding (public and private) for SusLab and product development Enhanced business and academic brand value

and successful product-service-systems (Maassen and Stensaker 2010). This latter value is likely to include new jobs, economic diversification and regional development (Etzkowitz et al. 2000; Leydesdorff and Deakin 2011).

30.5.1 HSB Living Lab, Chalmers University of Technology

The HSB Living Lab model builds on the assumption that the costs for building the facility can be more or less returned to HSB (HSB is a Swedish national housing association) over a longer time horizon. It also builds on the facility attracting companies who are prepared to commit to knowledge generation and development for their own and societal interest. Finally, the model also builds on the facility being attractive to a whole range of researchers at Chalmers, as well as across the SusLab facility (Table 30.2).

The relationship between the core partners (Chalmers, HSB and Johanneberg Science Park) is maintained through a binding ten year agreement signed between the President of Chalmers and the CEO of HSB in Gothenburg. Further, a business to business partnership has been established with 10 partners along the value chain (including architect, building company, IT specialist, bathroom supplier and white goods specialist) who have made a financial and resource commitment to the facility.

The total cost for the building was expected to be fairly typical for a pre-fabricated modular building with a land footprint of 420 m^2 and four floors. However,

Table 30.2 Summary—HSB Living Lab, Chalmers University of Technology

Business model	Knowledge to business and business to business
Partners	HSB Housing Association, Chalmers University of Technology, Johanneberg Science Park. 10 other business and societal partners
Sectoral interest	Local government, academia
Duration	10 years minimum
Purpose	Attracting companies who will commit to research and development for their own and societal interest
Motivation to create the lab	First building demonstration for Johanneberg Science Park

the building has been designed as a flexible living lab and this has meant extra costs which are unlikely to be covered in the lifetime of the building. HSB as an organization was prepared to accept an annual loss for the building over the 10 year period, after which the value of the modules will be roughly 50 % of the new construction cost.

Chalmers researchers have been provided with initial co-funding which has thereby initiated the idea generation between the SusLab partners during the formative process of the HSB Living Lab. Co-funding for the sensor networks based on the SusLab winter pilot has also been an important contribution. Chalmers researchers have obtained and are seeking further national and international research funding to bring the facility up to some 20 active researchers, technical staff and PhD students working in the HSB Living Lab.

The financial sustainability of the HSB SusLab will rely in part on securing the research grants set out above but also by generating an income from working with businesses to develop products and services. This will, in all likelihood, be based on a positive feedback loop whereby successful product development will lead to wider relationships with business which will in turn lead to more grants and which will then lead to more capacity to develop more products. The accommodation itself will be rented out to students (who are happy to be part of the SusLab) at market rates which will cover the costs of running and maintaining the accommodation.

30.5.2 SusLab NRW

The SusLab in North-Rhine Westfalia—led by the Wuppertal Institute—builds on real home environments (in the City of Bottrop). It consists of 4 main pillars: real home environments, the Smart Home Lab, showcase apartments and a new concept for energy efficiency consulting (product-service-system) (Table 30.3).

Insight research and prototyping in real home environments is substantial in a SusLab Infrastructure. One key component for the exploitation has been the development of a methodology for testing products and prototypes in real home

Table 30.3 Summary—SusLabNRW

Business model	Business to business and business to consumers. Embedded in existing, lived-in homes
Partners	Wuppertal Institute, Hochschule Ruhr West, Innovation City Management GmbH
Sectoral interest	Local government; academia; technical college
Duration	On-going
Purpose	Testing products and services in real home environment
Motivation	Developed as part of Ruhr region's commitment to reduce CO_2 emissions by 50 % as well as in response to local economic downturn

environments. The handling and the accuracy of fit of products and services has to be tested within the daily practices of households. The sensor infrastructure consists of data loggers for measuring room climate (temperature, humidity, CO_2 concentration) which are connected to the internet, an internet connected database at the Hochschule Ruhr West (HRW) and software for evaluation of the measured values. This infrastructure was tested in cooperation with the companies RWE and Deutsche Telekom by evaluating the usability and efficiency of their smart home products and systems. The HRW will offer this infrastructure to other companies for further testing of assistive technologies. The aim is to offer investigations in living labs for product improvement as a service to different companies, allowing full exploitation of the facility. The smart home is a 30 m^2 sized lab that is used for testing of the prototypes (proof-of-concept) and for tests of user-interaction before installing them in real homes.

In order to be able to fund the infrastructure, in addition to contracts for co-creating products and services, the German partners HRW, Wuppertal Institute and Innovation City Management GmbH developed the concept of energy efficiency services for communities. The product-service-system is a new concept for energy efficiency consulting for tenants and homeowners. The concept is based on a pre-analysis of the energy efficiency potential by using the evaluation methods developed in the SusLab project. The analysis of room climate delivers a first impression on user behavior, heating system functionality and building insulation characteristics. Based on the results a suitable, economically optimized proposal for increasing energy efficiency can be given to tenants and homeowners.

30.5.3 SusLab Living Lab in London

The SusLab Living Lab in London is a procured element of the project and the Institute for Sustainable Futures is responsible for delivering and managing the facility. Through a competitive process, Climate Energy Homes were chosen as the preferred supplier to design, supply and build the facility. The Institute has worked with others to deliver this facility and improve the exploitation (Table 30.4).

Table 30.4 Summary—SusLab Living Lab in London

Business model	Demonstration project
Partners	Institute for Sustainability (NGO)—responsible for build and management of the living lab, Imperial College, Royal College of Art
Sectoral interest	NGO, Academia, commercial
Duration	Through to 2024
Purpose	Procured project to demonstrate potential of collaboration
Motivation	To provide a living laboratory that is closer to a home environment

The UK has very few facilities which can act as a SusLab outside academic focused facilities, which are usually science laboratory based and not near to a living environment. Hence this facility is welcomed by academics, housing developers and manufacturing companies based in the south east of England. Some wider interest from health authorities has been shown to conduct care in the community type simulations. This facility will be able to emulate a high proportion of the building typology in the UK and therefore participants can relate more easily to experiments or assessments which can be run in the SusLab. This choice is also important as many highly elaborate building systems and cutting edge technologies usually have been tested or have their own research and development back office.

The SusLab aims to attract stakeholders, locally, nationally and EU-wide who are interested to commit to research and development of innovative energy control and awareness technology for their own and the wider benefits associated with being more effective with energy use. The SusLab also aims at attracting research and design practices from local and international academic institutions and institutions that are interested in co-creation and experimentation of solutions in relation to daily life practices.

The original concept was to enable free and full use to encourage the core aspirations of the SusLab project to be delivered over the life of the facility. Therefore, encouragement to use is the key feature and interested parties will be required to reinstate the facility to its current condition on completion of their activities. Encouragement will be on extending use and maintaining the SusLab in at least its current condition, rather than generating an income from activity, other than to cover the operational costs including utilities, cleaning and routine maintenance.

30.5.4 Concept House Village

The Concept House Village (CHV) is an initiative from City Ports Academy Rotterdam and is executed by Rotterdam University of Applied Sciences, Delft University of Technology and the Woonbron Social Housing Corporation. The facility is financially and in-kind supported by the Municipality of Rotterdam.

Table 30.5 Summary Concept House Village

Business model	Sponsorship
Partners	Delft University of Technology, Woonbron Social Housing corporation
Sectoral interest	Local Government, Academia
Duration	Up to 10 years
Purpose	Procured project to demonstrate potential of collaboration. Students involved in design and build
Motivation	To provide a living laboratory that is closer to a home environment

The partners have a long-lasting agreement on the exploitation of the CHV-facility signed by the boards of executive partners which provides a sustainable cost model for exploitation (Table 30.5).

During the first period of its existence in the Netherlands, CHV was financed from different sources. With the financial support from the INTERREG SusLab NWE project, the Dutch Ministry of Economic Affairs (Peaks in the Delta) and the Municipality of Rotterdam it was possible to establish the CHV-facility and to manage the acquisition of new prototypes, to guide the building processes of the prototyping, to coordinate the research program and foster the knowledge transfer.

The basic approach is that the involved consortia finance the prototypes themselves. The two involved knowledge institutes lead the development of the first two prototypes (CH Prototype 1 and CHIBB). The partners of the two consortia financed these prototypes with important additional subsidies from SusLabNWE and Peaks in the Delta.

Those two prototypes serve as a laboratory for the research programs of both knowledge institutes. The on-going research is mainly financed by the institutes' research funds. The prototypes also serve as a learning environment for the Bachelor and Master's students of the institutes. Besides the whole design and development of CHIBB by students of the Rotterdam University of Applied Sciences, students of the Albeda Polytechnic built the CHIBB prototype.

The CHV-facility has agreements with the consortia behind the prototypes about the involvement of students and research staff of both institutes in the development, monitoring, testing and validation of the dwellings.

In 2015 the CHV facility started revising its business model, to become independent from structural public funding, to accelerate the real market application of the results of the prototyping and to become a regional focal point of sustainable building to support the realization of the political ambitions.

30.6 Conclusion

SusLabNWE is in the early stages of development. Some of the infrastructures are still being built but to an extent that will always be the case as the SusLabs will be forever changing. It is therefore not possible to include all values in each of the SusLab's business models. In part they are benefits that have been valued and accepted as part of the development of the network, but they have not, yet, necessarily been realised. Clearly the business models are evolving. For example London SusLab and CHV are actively refreshing theirs and the SusLab NWR in Germany has identified a new stream of potential income which they are currently developing.

One interesting insight to come from this study of the business models is how the genesis of the SusLab has influenced the business model. For example, part of the motivation for the development of the SusLab NWR was to encourage economic diversification in the Ruhr. As such it initially attracted interest and assistance from Government to develop itself as a SusLab. Similarly the CHV was initiated on the backdrop of public funding, but now that it is established it is seeking to become independent of public funds. Both of these demonstrate, in real terms, the impact of path dependence theory (Bednar et al. 2015).

The other interesting, and unexpected, result of the work has been how relationships with business varies over time, and between businesses (Franzen 2015). By this we mean that different models suit different businesses.

Some are seeking involvement in research that will help to deliver sustainable lifestyles and to change social practices into more sustainable ones. All SusLabs share this opportunity, but the Concept House Village was developed with this market in mind (although it is revising its business model). Effectively CHV set the research program and then invited industry to fund it if they are interested. This is a widely adopted model to fund research in universities.

On the other hand, HSB Living Lab is looking to work with business to develop innovative products and services. The HSB model involves a closer, symbiotic, working relationship between business and academia with both sharing the same project goals but with each having slightly differentiated project deliverables. For example Chalmers researchers will also be drawn to a greater extent to developing the knowledge base within their fields of expertise; business will be drawn by the opportunity to develop and test products using the latest research. All parties seek to deliver the best co-created product, service or system and will gain individual benefits from working together. These outcomes reinforce the benefits predicted by the theory (Pohl 2008; Nicolescu 2015).

Recently completed research has also demonstrated how relationships, values and benefits can change over time (Franzen 2015). A relationship that might start on the basis of developing a new market for a business can with time enable a business the opportunity of developing a product line contributing to a sustainable lifestyle as the relationship between partners develops and matures.

Despite there being close to 170 active Living Labs there is a need to undertake further analysis of the various business models to identify common themes in the business cases, what is included and excluded and if Living Labs with the same aims share the same style of business case. There is the need to share an understanding of the business case development in order that the next generation of Living Labs do not have to re-learn the same lessons that the current generation have learnt. Further research as outlined above will also help to start to address the deficiency in studies of transnational innovation systems as identified by Carlsson (2006).

References

Baedeker, C., Greiff, K., Grinewitschus, V., Hasselkuß, D., Keyson, D., Knutsson, J., et al. (2014). *Transition through sustainable Product and Service Innovations in Sustainable Living Labs: application of user-centred research methodology within four Living Labs in Northern Europe*. Paper for presentation at the 5th International Sustainable Transitions (IST) Conference, August 27–29, 2014 Utrecht, The Netherlands.

Bednar J., Jones-Rooy, A., & Page, S. (2015). Choosing a future based on the past: Institutions, behaviour, and path dependency. *European Journal of Political Economy* (in print).

Carlsson, B. (2006). Internationalisation of innovation systems: a survey of the literature. *Research Policy, 35*, 56–67.

Christensen, C. (1997). The innovators dilemma. *Harvard Business Review*, USA.

Commission of the European Communities. (2006). *Delivering on the modernisation agenda for universities, education, research and innovation*. COM(2006) 208 final, Brussels.

Dell'Era, C., & Landoni, P. (2014). Living Lab: A methodology between user-centred design and participatory design. *Creativity and Innovation Management, 23*, 137–155.

Dolan, P. (2015). *Happiness by design*. London: Plume.

ENoLL (European Network of Living Labs. Retrieved November 17, 2015, from http://www.openlivinglabs.eu/.

Etzkowitz, H., & Leydesdorf, L. (2000). The dynamics of innovation: from national systems and mode 2 to a triple helix of university-industry-government relations. *Research Policy, 29*, 109–123.

Etzkowitz, H., Webster, A., Gebhardt, C., & Terra, B. (2000). The future of the university and the university of the future: Evolution of ivory tower to entrepreneurial paradigm. *Research Policy, 29*, 313–320.

Etzkowitz, H. (2013). Anatomy of the entrepreneurial university. *Social Sciences Information, 52*(3), 486–511.

Erosa, V. (2012). Dealing with cultural issues in the Triple Helix Model implementation: A comparison among Government, University and Business Culture. *Social and Behavioural Sciences, 52*, 25–34.

Franzen, S. (2015). Sustainable transitions at home: HSB living lab project member organisations' view on sustainability and the living lab itself. Dissertation, Chalmers University of Technology.

Gebhardt, C. (2015). The spatial dimension of the triple helix: the city revisited—towards a mode 3 of innovation systems. *Triple Helix,* 2(11).

Geels, F., & Kemp, R. (2007). Dynamics in socio-technical systems: Typology of change processes and contrasting case studies. *Technology in Society, 29*, 441–455.

Hadorn, G., Bradley, D., Pohl, C., Rist, S., & Wiesmann, U. (2006). Implications of transdisciplinarity for sustainability research. *Ecological Economics, 60*, 119–128.

Hessels, L., & van Lente, H. (2008). Re-thinking knowledge production: A literature review and research agenda. *Research Policy, 37*, 740–760.

Jantsch, E. (1970). Inter- and transdisciplinary University: A systems approach to education and innovation. *Policy Sciences, 1*, 403–428.

Leydesdorff, L., & Deakin, M. (2011). The triple helix model of smart cities. A neo-evolutionary perspective. *Journal of Urban Technology, 18*(2), 53–63.

Liedtke, C., Baedeker, C., Hasselkuß, M., Rohn, H., & Grinewitschus, V. (2015). User-integrated innovation in Sustainable LivingLabs: An experimental infrastructure for researching and developing sustainable product service systems. *Journal of Cleaner Production, 97*, 106–116.

Malm, A., Ljunggren, O., Bergstedt, O., Pettersson, T., & Morrison, M. (2012). Replacement predictions for drinking water networks through historical data. *Water Research, 46*, 2149–2158.

Maassen, P., & Stensaker, B. (2010). The knowledge triangle, European higher education policy logics and policy implications. *Higher Education, 6*, 757–769.

Max-Neef, M. (2005). Foundations of trandisciplinarity. *Ecological Economics, 53*, 5–16.

Nee, V., & Cao, Y. (1999). Path dependent societal transformation: Stratification in hybrid mixed economies. *Theory & Society, 28*, 799–834.

Nicolescu, B. (2015). How can we enter in dialogue? Transdisciplinary methodology of the dialogue between people, cultures and spiritualities. *Dialogue and Universalism, 1*, 9–19.

Nyman, G. (2015). University-business-government collaboration: From institutes to platforms and ecosystems. *Triple Helix, 2*(2).

Pohl, C. (2008). From science to policy through transdisciplinary research. *Environmental Science & Policy, 11*, 46–53.

Pohl, C. (2011). What is progress in transdisciplinary research? *Futures, 43*, 618–626.

Ranga, M., & Etzkowitz, H. (2015). *Triple Helix Systems: an analytical framework for innovation policy and practice in the knowledge society*. H-STAR, Stanford University

Salter, R. & White, S. (2013). *Collaborative research in the real world: review of living labs*. Living Lab Framework Project, CRC for Low Carbon Living

Senge, P. (1995). *The Fifth Discipline: The art and practice of the learning organisation*. New York: Doubleday.

Spaargaren, G. (2011). Theories of practice: Agency, technology, and culture. Exploring the relevance of practice theories for the governance of sustainable consumption practices in the new world-order. In *Global Environmental Change, 21*, 813–822.

Sunitiyoso, Y., Wicaksono, A., Utomo, A., Putro, U., & Mangkusubroto, K. (2012). Developing strategic initiatives through triple helix interactions: Systems modelling for policy development. *Social and Behavioural Sciences, 52*, 140–149.

United Nations. (2015). *Transforming our world—the 2030 agenda for sustainable development*. New York: United Nations.

Vanaker, T., Manigart, S., & Meuleman, M. (2014). Path dependent evolution versus intentional management of investment ties in science based entrepreneurial firms. *Baylor University*,. doi:10.111/etap.12007.

Chapter 31
Reflecting on LivingLabs and Future Trends

David V. Keyson and Carolin Baedeker

Abstract As technology has advanced so has the future role of living labs evolved. With the increased availability of wireless monitoring technologies including climate and activity sensors as well as self-reporting tools, the capability to convert virtually every house or building into a living lab has become a reality. ICT enables now the connectivity and merging of data sets across multiple living labs and monitored homes, providing a unique infrastructure for accelerating the adoption and marketing of innovations focused on sustainable living. The LivingLab approach is gaining more and more importance as a mechanism to study and shape sustainable behavior from the public and private perspective.

Keywords Living labs · Communities

31.1 Reflecting on LivingLabs and Future Trends

The convergence of globalization, shifting demographics, new services and urbanization is transforming many aspects of our lives. There are many new challenges relating to where and how we work, live, travel, communicate, and ensure health and safety. The advancement of services and products impacting our everyday living and work requires real-world research methods that can deal with the inherent system complexities.

D.V. Keyson (✉)
Delft University of Technology, Delft, Netherlands
e-mail: d.v.keyson@tudelft.nl

C. Baedeker
Wuppertal Institute for Climate, Environment, Energy, Wuppertal, Germany
e-mail: carolin.baedeker@wupperinst.org

D.V. Keyson et al. (eds.), *Living Labs*, DOI 10.1007/978-3-319-33527-8_31

As technology has advanced so has the future role of living labs evolved. With the increased availability of wireless monitoring technologies including climate and activity sensors as well as self-reporting tools such as those described in this book, the capability to convert virtually every house or building into a living lab has become a reality. The means to rapidly prototype design interventions in the real-world context combined with non-obtrusive measurement devices has blurred the delineation between experimental prototypes versus an actual product as experienced by the user. In many ways the physical world of living labs resembles the real world deployment of fully functional beta software by ICT companies for user feedback.

The use of methods such as experience sampling, which actually embeds measurement in the product itself not only inform the user, but act as a means of self reflection thus shaping the user experience itself. Through experience sampling, products can respond and adapt to user preferences. For example, a query on a building occupant's comfort level may actually result in changes in the heating, ventilation or cooling settings. Machine learning based upon user inputs can lead to environments that are capable of predicting to an extent user needs.

Dedicated living labs facilities require an infrastructure, which can be maintained and financed. To this extent, models such as the HSB Living Lab at Chalmers University provide an economically viable model for future living labs. The lab serves as a demonstration, test environment, and student housing.

One of the key challenges facing the Living Labs, whether for dedicated labs or existing homes, is the recruitment of subjects and the involvement of actual households in the research and development of new products and services for sustainable living. Similar to beta test groups of software solutions, it is conceivable that in the future a community of living lab households could be developed. For example, households could subscribe to a living lab network and in return gain access to the newest technologies to be tested in their living environment. The community of Living Lab participants in this manner would act as lead users and could help promote and accelerate the adoption of new innovations. Similarly a network of living labs for other sectors in the building industry could be developed. Currently consortiums such as Amsterdam Metropolitan Solutions have linked academia, the Amsterdam municipality, and industry partners in such as way that the city of Amsterdam can serve as a large living lab. Two key aspects of AMS are the notion of open data at the city level relating to transportation, energy and resource use combined with the spirit of open innovation.

Large communities of users and professionals can be accessed through a network of living labs targeting on corporate and smaller to medium size companies. In particular, there is a new generation of young entrepreneurs and startup companies with limited resources who are in need of mechanisms to facilitate the testing adoption and acceleration of innovations for sustainable living and work. Living Labs can serve as access to otherwise unattainable market access.

Several product groups, technologies and services have been identified as areas in which LivingLabs and methods could be applied. Table 31.1 below lists the identified product groups, technologies and services with high resource efficiency

Table 31.1 Overview of relevant fields of application identified for Living Labs for Sustainable Development (*Source* Rohn and Leismann, in Geibler et al., 2014: 584–585)

Field of application	Research perspective	
	Product/service innovations	User behavior
Life and Work	**Building and infrastructure,** e.g., security, heating and energy supply, insulation, e-energy/energy assistance **Food,** e.g., chilling, storage, preparation, assistants **Health and hygiene,** e.g., medical care, fitness, medical technology **Furnishings of living and working spaces,** e.g., design of electric and electronic equipment, furniture, textiles **Information management,** e.g., communication in the home/out of home, ICT products and their use **Substitution of physical mobility by** "ICT mobility", connection to logistics systems, Smart Grids	**Behavior at home and workplace,** e.g., health and exercise, energy consumption **Nutrition,** e.g., food wastage, shopping, health **Phase of life appropriate design of home/workplace,** e.g., autonomous life at old age, user acceptance of innovations **Integrated design,** e.g., in the area of fields of demand or service design **Furnishings of living and working spaces,** e.g., new workplace concepts, ways of utilization, cascading systems, ICT **Service and time management,** e.g., being mobile, eating healthy, exercise
Town, Region and Mobility	**Out of home catering,** e.g., delivery services, drive-in restaurants, etc. **Mobility,** e.g., efficient mobility options (logistics), freight, public transport linkages, design of mobility options **Regional networks/"location promotion",** e.g., health support systems, urban planning, communication systems, regional energy supply, tourism, sharing and renting options	**Mobility,** e.g., use and user acceptance of resource efficient mobility options **Communities/networks,** e.g., urban agriculture, barter systems, neighborhood networks, service concepts and suburb development **Leisure/holiday behavior,** e.g., regional tourism **ICT services,** e.g., integrated ICT, mobility and logistics management
Retail and Gastronomy	**Furnishings,** e.g., electric and electronic equipment, lighting, media, online shopping, design **Mobility,** e.g., efficient mobility options **Nutrition,** e.g., food labelling and declaration **Support at old age,** e.g., intelligent appliances	**Intelligent appliances,** e.g., digital product memory **Choice of products,** e.g., influence of advertisement and information campaigns

potential and high potential for development in Living Labs. Three particularly relevant areas of application are: ((1) Living and working; (2) Town, region and mobility; (3) Retail and gastronomy), as well as the two research perspectives "user behavior" and "product innovation" (Geibler et al. 2014).

Given the high and still rising resource and energy consumption, an indispensable challenge in the 21st century is to achieve sustainable development, meaning that society's welfare generation has to happen within the natural system's boundaries. To enhance these needed changes in lifestyles the involvement of end-users to better respond to their needs can enable the development of value-added solutions and act as a driver for Europe's innovation system.

The Living Lab approach has been defined as "a user-centric innovation milieu built on every-day practice and research, that facilitates user and actor influence in open and distributed innovation processes engaging all relevant partners in real-life contexts, aiming to create sustainable values". The approach is gaining more and more importance on different scales, including recent European and international projects focused on smart cities combined with open innovation, open data, and direct citizen participation.

31.2 Outlook

The Sustainable infrastructure will continuously be further developed and enlarged (all over Europe, including the ENOLL – Network = European Network Of Living Labs) and also used for different fields of action (food/nutrition, mobility, housing, ICT) and for different foci like e.g. user-integrated product-services innovation in official buildings. The European SusLab infrastructure provides a platform for large scale demonstration projects supporting co-creation by experimenting and demonstrating new business models addressing all forms of innovation, combining technological, organisational, societal, cultural and behavioural innovation, and strengthening the participation of civil society. The SusLab living lab approach enables new forms of collaboration between end-users, producers, researchers and regional/city actors to accelerate the introduction and adoption of meaningful technological innovations and models of sustainable living.

Reference

Geibler, J. v., et al. (2014). Exploring the potential of a German living lab research infrastructure for the development of low resource products and services. *Journal Resources,* 3, 575–598.

MIX
Papier aus verantwortungsvollen Quellen
Paper from responsible sources
FSC® C105338

If you have any concerns about our products,
you can contact us on
ProductSafety@springernature.com

In case Publisher is established outside the EU,
the EU authorized representative is:
Springer Nature Customer Service Center GmbH
Europaplatz 3, 69115 Heidelberg, Germany

Printed by Libri Plureos GmbH
in Hamburg, Germany